FAREWELL THE TRUMPETS

An Imperial Retreat

Viceroy's Palace, New Delhi: Edwin Lutyens' vision (frontispiece)

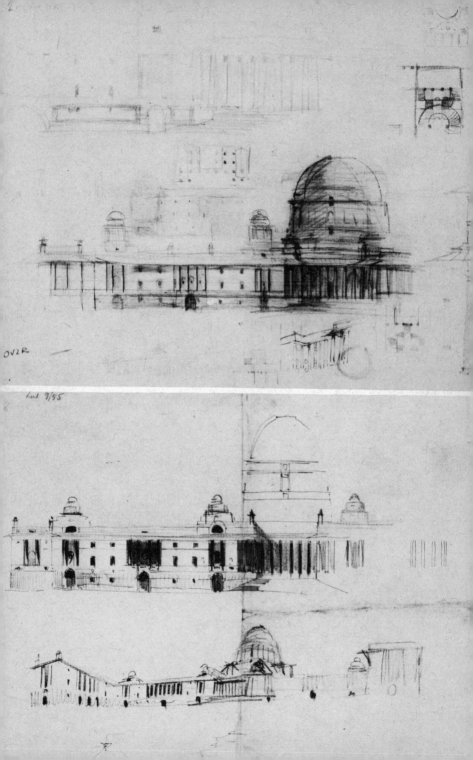

James Morris

FAREWELL
THE TRUMPETS

An Imperial Retreat

HBJ

A Harvest/HBJ Book
A Helen and Kurt Wolff Book
Harcourt Brace Jovanovich
New York and London

Printed in the United States of America

The lines from "The Circus Animals' Desertion" are quoted from
Collected Poems by William Butler Yeats, copyright 1940 by
Georgie Yeats, renewed 1968 by Bertha Georgie Yeats, Michael
Butler Yeats and Anne Yeats, and are reprinted with permission of
Macmillan Publishing Co., Inc.; those from "Hong Kong" are
from Collected Poems by W. H. Auden, edited by Edward
Mendelson, copyright © 1976 by Edward Mendelson, William
Meredith and Monroe K. Spears, executors of the Estate of W.H.
Auden, and are reprinted by permission of Random House, Inc.

Library of Congress Cataloging in Publication Data

Morris, James, 1926–
Farewell the trumpets.

(A Harvest/HBJ book)
"A Helen and Kurt Wolff book"
The 3d volume of the author's trilogy; the 1st of which
is Heaven's command; 2d, Pax Britannica.
Bibliography: p.
Includes index.
1. Great Britain—Colonies—History.
2. Great Britain—History—George VI, 1936–1952.
3. Great Britain—History—Elizabeth II, 1952–
I. Title.
[DA16.M595   1980]   909'.09'71241082   79-24253
ISBN 0-15-630286-1

First Harvest/HBJ edition 1980

A B C D E F G H I J

For

MARK MORRIS

Pax Tibi, Marce

Set in this stormy Northern sea,
  Queen of these restless fields of tide,
England! what shall men say of thee,
  Before whose feet the worlds divide?

OSCAR WILDE

INTRODUCTION

This book is the right-hand panel, so to speak, of a triptych depicting the rise and decline of Queen Victoria's Empire. The centre panel, *Pax Britannica*, portrayed the Empire at its moment of climax, which I have taken to be the Diamond Jubilee of 1897. The left-hand panel, *Heaven's Command*, traced its nineteenth-century development to that apogee. *Farewell the Trumpets* completes the ensemble with a narrative picture of the imperial retreat from glory. Taken together, the three are intended to be an impressionistic evocation, subjective and often emotional, of a great historical movement. I have been concerned not so much with what the British Empire meant, as what it felt like—or more pertinently, perhaps, what it felt like to me, in the imagination or in the life.

For towards the end of this volume I become an eye-witness, and immediately less reliable. I do not come from an imperial family, and could write about the nineteenth-century Empire with absolute detachment, but in the twentieth century few of us have been immune to the imperial effects. Even my poor father was gassed for his Empire. Even my poor Uncle Geraint, fresh from his cello at Monmouth School, was whisked away to the Indian Expeditionary Force in France, and never came home again. Even I found myself, for a decade of my life, embroiled in the imperial mesh, as I followed the retreating armies of Empire from one after another of their far-flung strongholds. Most of us were imperialists in the end, however gentle our instincts, and hardly a reader of this book will feel altogether aloof to its narrative, or impartial to its judgements.

Mine is an aesthetic view of Empire, and there is no denying that as the flare of the imperial idea faded, so its beauty faded too. It had not always been a pleasant kind of beauty, but it had been full of

9

splendour and vitality, and when the Empire lost its overweening confidence and command, its forms became less striking and its outlines less distinct. The world was swamping it, and an incident like the invasion of Tibet, which gives me an entire chapter in 1903, would hardly provide a paragraph in 1943. My book is therefore sad without being regretful. It was time the Empire went, but it was sad to see it go: and so these pages too, while I hope they are not blind to the imperial faults and weaknesses, are tinged nevertheless with an affectionate melancholy—

> *Men are we, and must grieve when even the shade*
> *Of that which once was great is passed away.*

The volume begins with Victoria's Jubilee and concludes with Churchill's funeral. I hope my readers will discover in themselves, between these ceremonial book-ends, at least some of the mingled sensations of admiration, dislike, amusement, pity, pride, envy and astonishment with which I have watched and pictured the passing of the British Empire.

Say farewell to the trumpets!
You will hear them no more.
But their sweet sad silvery echoes
Will call to you still
Through the half-closed door.

Part One
THE GRAND ILLUSION: 1897–1918

Part Two

THE PURPOSE FALTERS: 1918–1939

Part Three
FAREWELL THE TRUMPETS: 1939–1965

PICTURES

Sources: 1 *Royal Institute of British Architects*; 2 *Gwynedd Record Office, Cymru*; 3, 4 *Royal Photographic Society*; 5, 10 *Imperial War Museum*; 6 *Thomson Organisation Picture Service*; 7 *India Office Library*; 8 *Illustrated London News Picture Library*; 9 *Royal Geographical Society*; 11 *Popperfoto*; 12 *Keystone Press Agency Ltd*; 13 *David Holden, Esq*; 14, 15 *Associated Press*

PART ONE
The Grand Illusion
1897–1918

CHAPTER ONE

The Dial to the Sun

QUEEN Victoria of England went home happy on her Diamond Jubilee day, June 22, 1897. History had humoured her, as she deserved. The sun had shone all day—'Queen's weather', the English called it—and there was nothing artificial to the affection her people had shown during her hours of celebration. She had passed in procession through London intermittently weeping for pleasure, and studded her diary that evening with joyous adjectives: indescribable, truly marvellous, deeply touching.

It was more than a personal happiness, more even than a national rejoicing, for the British had chosen to commemorate the Diamond Jubilee as a festival of Empire. They were in possession that day of the largest Empire ever known to history, and since a large part of it had been acquired during the sixty years of Victoria's reign, it seemed proper to honour the one with the other. It would mark this moment of British history as an Imperial moment, a Roman moment. It would proclaim to the world, flamboyantly, that England was far more than England: that beneath the Queen's dominion lay a quarter of the earth's land surface, and nearly a quarter of its people—literally, as Christopher North the poet had long before declared it, an Empire on which the sun never set.

So the day had been a proud, gaudy, sentimental, glorious day. This was *fin de siècle*. The public taste was for things theatrical. Statesmen and generals were actors themselves, and here was the brassiest show on earth. Through the grey and venerable streets of the capital—'the greatest city since the ruin of Thebes'—there had passed in parade a spectacle of Empire. There were Rajput princes and Dyak headhunters, there were strapping troopers from Australia. Cypriots wore fezzes, Chinese wore conical straw hats. English gentlemen rode by, with virile moustaches and steel-blue eyes, and Indian lancers jangled past in resplendent crimson jerkins.

Here was Lord Roberts of Kandahar, on the grey Arab that had taken him from Kabul to Kandahar in his epic march of 1880.[1] Here was Lord Worseley of Cairo and Wolseley, hero of Red River, Ashanti and Tel-el-Kebir. Loyal slogans fluttered through the streets—'One Race, One Queen'—'The Queen of Earthly Queens'— 'God Bless Her Gracious Majesty!' Patriotic songs resounded. Outside St Paul's Cathedral, where the Prince of Wales received the Queen in her barouche, a service of thanksgiving was held, with archbishops officiating and an Empire in attendance.

That morning the Queen had telegraphed a Jubilee message to all her subjects—to Africa and to Asia, to the cities of the Canadian West and the townships of New Zealand, to Gibraltar and Jamaica, to Lucknow and Rangoon, to sweltering primitives of the rain-forests as to svelte merchant princes of the milder tropics. The occasion was grand. The audience was colossal. The symbolism was deliberate. The Queen's message, however, was simple. 'From my heart I thank my beloved people', she said. 'May God bless them.'

2

'My people'. If to the Queen herself all the myriad peoples of the Empire really did seem one, to the outsider their unity seemed less than apparent. Part of the purpose of the Jubilee jamboree was to give the Empire a new sense of cohesion: but it was like wishing reason upon the ocean, so enormous was the span of that association, and so unimaginable its contrasts and contradictions. Some of its constituents were complete modern nations, the self-governing white colonies in Australia, Canada, New Zealand and South Africa. Some were Crown Colonies governed, in one degree or another, direct from London. Some were protectorates so isolated and naive that the very idea of Empire was inconceivable to most of their

---

[1] And who wore on his bridle that day the Afghan Medal, awarded him by the Queen's express command. Roberts bought Vonolel, who was named after a dissident Assamese chieftain, from an Arab horse-dealer in Bombay in 1877, and buried him in 1899, aged twenty-seven, in London.

inhabitants. At one extreme was India, a civilization in itself: at the other was Ascension, a mere speck in the South Atlantic, uninhabited by any kind of vertebrate until the British arrived. Every faith was represented in the British Empire, every colour of skin, every philosophy, almost every branch of human history. Disraeli had called it the most peculiar of all Empires, and so it was, for it was a gigantic jumble of origins, influences, attitudes and intentions.

The inhabitants of Tristan da Cunha (for example) had no government at all, and no written laws either. In India 1,000 British civil servants, protected by 70,000 British soldiers, ruled 300 million people in a sub-continent the size of Europe. In Cairo the residence of the British Agent and Consul-General was known to Egyptians as *Beit-al-Lurd*—House of the Lord. On Norfolk Island in the South Pacific citizens saluted each other with their traditional greeting 'Whataway you!' On Pitcairn the descendants of the Bounty mutineers were governed by their own President of Council.

In Mauritius that year crops were threatened by the plant pest *Cordia macrostachya*, brought there in 1890 from British Guiana. In Zanzibar the entire economy depended upon the cultivation of cloves, taken there in 1770 from Mauritius. Scottish gorse thrived on St Helena, Irish donkeys in South Africa, English stoats, hedgehogs, rooks and mice in New Zealand. The descendants of Canadian convicts, transported to Australia, still lived in Sydney. Mr Dadabhai Naoroji was Member of Parliament for Finsbury. In Aden the Parsees had their Tower of Silence, in Cape Town the Malays had their mosque, in Calcutta race-horses were habitually called Walers because they came, with their jockeys, from New South Wales. The bubonic plague had recently been introduced to India, by rats on board a ship from Hong Kong.

When Major Allan Wilson and thirty-two of his men were trapped on a river-bank by Matabele tribesmen in Rhodesia in 1893, they sang 'God Save the Queen' as they mustered back to back to defend themselves.[1] When the gunboat HMS *Wasp* approached Tory Island off western Ireland to collect rates in 1897, the islanders revolved maledictive stones, and pronounced curses upon the vessel.[2] When the Bishop of Gibraltar was received in audience by

[1] They all died.   [2] It sank.

the Pope, the Pontiff remarked: 'I gather I am within your Lordship's diocese.'[1]

If there was one characteristic diffused throughout this bewildering gallimaufry, it was an almost feverish enthusiasm. The mood of Empire in 1897 was *bravura*—'an attempt', as the painter Constable once defined it, 'at something beyond the truth'. The British Empire was a heady outlet for the imagination of a people still in its prime. Its subjects were of all races: its activists were nearly all British. Through the gate of Empire Britons could escape from their cramped and rainy islands into places of grander scale and more vivid excitement, and since the Queen's accession at least 3 million had gone. By 1897 they were everywhere. There were Britons that year commanding the private armies of the Sultan of Sarawak, organizing the schedules of the mountain railway to Darjeeling, accepting the pleas of runaway slaves in Muscat, charting the China Sea, commanding the Mounties' post on the Chinook Pass in the Yukon, governing the Zulus and the Wa, invading the Sudan, laying telegraph wires across the Australian outback, editing the *Times of India,* prospecting for gold in the valley of the Limpopo, patrolling the Caribbean and investigating the legal system of the Sikhs—all within the framework of Empire, and under the aegis of the Crown.

All this the Diamond Jubilee reflected. It was truly the Empire in little, as its organizers intended: a grand and somewhat vulgar spectacle, reflecting a tremendous and not always delicate adventure, and perfectly expressing the conviction of Cecil Rhodes, the imperial financier, that to be born British was to win first prize in the lottery of life.[2]

3

The origins of the British Empire, like the form of it, were random. There had been British possessions overseas since the days of the

[1] He was: it extended from Portugal to the Caspian. The Archdeacon of Bloemfontein told me this story.

[2] Recorded, Mr Peter Ustinov tells me, by Colonel Weston Jarvis in his *Jottings from an Active Life,* together with a parallel aphorism from Lord

Normans, who brought with them title to the Channel Islands and parts of France, and who presently seized Ireland too. Since then the imperial estate had fitfully grown. Sometimes possessions had been lost—the thirteen colonies of America, for instance, or the ancient possessions of the French mainland. Often they had been swapped, or voluntarily surrendered, or declined. Tangier, Sicily, Heligoland, Java, the Ionians, Minorca had all been British at one time or another. Costa Rica had applied unsuccessfully for a British protectorate, and Hawaii was British for five months in the 1840s.[1] During Victoria's reign the expansion of the Empire had been more consistent. 'Acquired in a fit of absence of mind', the historian J. R. Seeley had said of it in a famous phrase, but in fact its piecemeal development had been conscious enough. Each step had its own logic: it was the whole resultant edifice that had an absent air.

Essentially most possessions were acquired for profit—for raw materials, for promising markets, for investment, or to deny commercial rivals undue advantages. As free traders the British had half-convinced themselves of a duty to keep protectionists out of undeveloped markets, and they were proud of the fact that when they acquired a new territory, its trade was open to all comers. Economics, though, must be sustained by strategy, and so the Empire generated its own extension. To protect ports, hinterlands must be acquired. To protect trade routes, bases were needed. One valley led to the next, each river to its headwaters, every sea to the other shore.

To these material, if often misty impulses were added urges of a higher kind. At least since the start of the nineteenth century the British Empire had regarded itself as an improvement society, dedicated to the elevation of mankind. Raised to the summit of the

---

Milner: *'Everyone can Help'*. 'If only we carry these two declarations of two great men in the forefront of our minds,' Colonel Jarvis commented, 'there is very little doubt that democracy can still be educated along the right lines.'

[1] To this day its State flag contains the Union Jack in its upper quarter, while until her deposition in 1893 the last of the Hawaiian monarchs, Queen Lydia Liliuokalani, modelled her court upon that of Queen Victoria (except perhaps for its ceremonial robes, which were made from the tufted feathers of the o–o bird).

world by their own systems, the British believed in progress as an absolute, and thought they held its keys. AUSPICIUM MELIORIS AEVI was the motto of the imperial order of chivalry, The Most Distinguished Order of St Michael and St George—'A Pledge of Better Times'. The British way was the true way, free trade to monarchy, and it was the privilege of Britons to propagate it across the world. Through the agency of Empire the slave trade had been abolished, and on the vehicle of Empire many a Christian mission had journeyed to its labours.

The desire to do good was a true energy of Empire, and with it went a genuine sense of duty—Christian duty, for though this was an Empire of multitudinous beliefs, its masters were overwhelmingly Church of England. Sometimes, especially in the middle of the nineteenth century, their duty was powerfully Old Testament in style, soldiers stormed about with Bibles in their hands, administrators sat like bearded prophets at their desks. By the 1890s it was more subdued, but still devoted to the principle that the British were some sort of Chosen People, touched on the shoulder by the Great Being, and commissioned to do His will in the world.

And of course, as in all great historical movements, the fundamental purpose was not a purpose at all, but simply an instinct. The British had reached an apogee. Rich, vigorous, inventive, more than 40 million strong, they had simply spilled out of their islands, impelled by forces beyond their own analysis. In this sense at least they were truly chosen. Destiny, an abstraction the imperialist poets loved to invoke, really had made of them a special kind of nation, and had distributed their ideas, their language, their ships and their persons uniquely across the world.

4

They were uniquely selected: were they uniquely qualified? Certainly by the end of the nineteenth century the British had fallen into an imperial posture, an imperial habit perhaps. Technically they were as well fitted as any to govern a quarter of the world. Their own country had escaped the social convulsions that shook the rest of Europe, to provide a model of liberal but traditional stability. Their original mastery of steam, and all that came from it, had given them a technological start over all other nations, an advantage they put to imperial uses. The flexibility of their unwritten constitution was handy for an expansionist State. The semi-divine nature of their monarchy gave them a mystic instrument that was often useful. Being islanders, they knew more about the world than most of their neighbours: they possessed more ships than all other nations put together, and there were few British families who had not sent a man abroad, if not to settle, at least to sail a vessel or fight a foreign war. They were an immensely experienced people. Compact, patriotic, paradoxically bound together by an ancient class system, theirs had been an independent State for nearly a thousand years, and this gave them punch and phalanx.

Over the years they had, too, created an imperial elite to whom Empire was a true vocation. Everybody knew its members. They were products of those curious institutions, the English public schools. Within the last century the traditional schools of the landed gentry, Eton, Harrow, Winchester, had been widely copied, until all over England were the cricket pitches, the tall chapels, the cloisters and the dormitories of the Old School, whose friendships, slangs and values often lasted a man through life. These were the nurseries of Empire—as Sir Henry Newbolt wrote of his own school:

> *The victories of our youth we count for gain*
> *Only because they steeled our hearts to pain,*
> *And hold no longer even Clifton great*
> *Save as she schooled our wills to serve the State.*[1]

They taught a man to be disciplined, tough, uncomplaining, reserved, good in a team and acclimatized to order. The prefect system, in which boys exerted much of the school's authority, gave a man an early experience of command. The cult of the all-rounder taught him to put his hand to anything. The carefully evolved code of schoolboy conduct told him when to hold his tongue, when a rule was made to be broken, and even something about the nature of love—for love between men, generally platonic but often profound,

[1] Not a very ancient purpose. Clifton was founded in 1862, the year of Newbolt's birth.

was an essential strain of the imperial ethic. The stiff upper lip, the maintenance of appearances, the sense of inner brotherhood, the simple code of fair play—all these provided a potent ju-ju for the few thousand Englishmen who, in the 1890s, ruled so much of the known world.

This was the imperial class. Its members stood to gain directly from the existence of Empire, in jobs, in dividends, or at least in adventurous opportunity. The mass of the British people were far more remote from the imperial enterprise, and until the last decades of the century had in fact taken little notice of their Empire, except when they wished to emigrate or join the Army. But the grand sweep of Victorian history had by 1897 turned the whole nation briefly into enthusiasts. The new penny press, preaching to a newly literate and newly enfranchised audience, was stridently propagandist, and the events of the past twenty-five years had swept the people into a highly enjoyable craze of Empire.

What events they had been! Anybody over thirty, say, at the time of the Diamond Jubilee had experienced a period of British history unexampled for excitement. What theatre! The tragedy of Isandhlwana, the thrilling defence of Rorke's Drift! Gordon martyred at Khartoum! 'Dr Livingstone I presume'! The redcoats helter-skelter from the summit of Majuba, Sir Garnet Wolseley burning the charnel-houses of Kumasi! Never a year passed without some marvellous set-piece, of triumph or of tragedy. Champions rose to glory, the flag forever flew, the Empire grew mightier yet.

And across the world the graveyards spread, as generation after generation contributed its quota to the imperial sacrifice. Young men died in battle, young women died in tropical childbirth, children died of smallpox or cholera, heatstroke or food poisoning, a hundred thousand expatriates died of the climate, or of homesickness, or of plain exhaustion.[1] The Empire was a pageant, but it was reality too. Its pretences were all on the surface. The knowledge of its power and of its responsibilities gave a corporate pride to the British people, buttressing their sense of family, so that a sigh passed across the nation when a hero died or a regiment was

[1] Three Mourning Warehouses advertised themselves in Hart's Army List, 1887.

The Dial to the Sun: Queen Victoria goes to sea

humiliated, and on Jubilee night the bonfires burnt brightly on hilltop and beacon from Cornwall to Cromarty.

5

There were few in 1897 to question the morality of the British Empire. It was grand, and it was honourable. What it did for the nation materially, nobody really knew: its profits were great but so were its expenses, and the burdens of it matched the assets. But there was no denying its stimulation to the national spirit. In the 1890s Imperialism had reached an ebullient and aggressive climax. The politicians, habitually aloof to the Empire and its causes, had taken it up, and a hazy movement called the New Imperialism was busily publicizing the glory of it all. In 1895 the Conservatives and their Liberal Unionist allies, the party of Empire, had won an overwhelming victory over the Liberals: the nation talked Empire, thought Empire, dreamed Empire. Two geniuses, Rudyard Kipling and Edward Elgar, were translating the emotions into art, and a thousand lesser practitioners were putting it into jingle, march or tableau.

There was calculation to this climax, of course, the cunning of financiers, the opportunism of politicians, the ambitions of soldiers, merchants and pro-consuls. By their own best standards the British of the 1890s were beneath themselves, their patriotism coarsened and their taste debased. This was hardly the England that Burke had idealized, 'sympathetic with the adversity or with the happiness of mankind, [feeling] that nothing in human affairs was foreign to her'. The England of the Diamond Jubilee was essentially insular, for its people saw the whole wide Empire, even the world itself, only as a response to themselves.

Yet it was not a conscious arrogance, and the New Imperialism was seldom malicious. The British Government of the time was a fastidiously aristocratic regime, one of the last in Europe: Lord Salisbury the Prime Minister, the last to sit in the House of Lords, saw the Empire more as an instrument of diplomatic policy than a source of glory—to calculate its worth, he once mordantly observed, 'you must divide victories by taxation.' The unprecedented expan-

Following the Flags: military parade in London

sions of the last half-century, especially in Africa, were not part of any concerted policy of aggrandisement, but occurred haphazardly, often in reflex, generally for *ad hoc* reasons of economics or strategy. The British as a whole would have been shocked at any notion of wickedness to their imperialism, for theirs was a truly innocent bravado. They really thought their Empire good, like their Queen, and they were proud of it for honest reasons: they meant no harm, except to evil enemies, and in principle they wished the poor benighted natives nothing but well.

These were brittle times—times of change and sensationalism, of high stakes and quick fortunes, outrageous fashions and revolutionary ideas. Socialism was an intellectual fad, the New Woman smoked her cigarettes ostentatiously in the Café Royal, and only a month before the Jubilee Oscar Wilde had ended his sentence in Reading Gaol. The grand Victorian synthesis of art, morals and invention was already fading, and with it would presently fade the certainty and the optimism. Only a sexually restrained society, warned the psychologist J. D. Unwin, would continue to expand: and there were many Britons in 1897 who, looking around them at the feverish high jinks of the capital, saw omens of disillusionment to come. The times were too gaudy to be safe. The mood could crack, or be shattered by a stray note.

Part of the triumph was bluff anyway. The people might think themselves citizens of the happiest, richest, strongest and kindest Power: their leaders knew that Great Britain was no longer beyond challenge. The Germans and the Americans were fast overtaking her in technique, brute power and public education. She had few friends in the world, and no allies. Her creed of Free Trade, which had served her well in the days of absolute supremacy, was not so infallible in a world of competitive tariffs. The basis of her immense prestige was fragile really. Bismarck said the German police force could easily arrest the British Army, and there was nothing sacrosanct to the British command of the seas—any Power could defy it, if prepared to put enough money into a fleet. The very state of the world was increasingly precarious to the Empire: Germany, France and Russia were all potential enemies, the moribund Ottoman Empire was a perpetual problem, an unstable

Austria-Hungary threatened instability to everyone else, a derelict China seemed an incitement to colonial rivalries.

To seers, then, there was a detectable element of disquiet to the celebrations of 1897, an unease not often declared, nor even perhaps realized, but intuitive. It was a thunderstorm feeling—a heaviness in the air, an unnatural brightness to the light. Queen's Weather it might have been on Jubilee Day, but the outlook was changeable.

6

Still the public at large felt no premonitions, for the Empire was grand above all in the idea of it, in the grand illusion of permanence and paramountcy. Its strongest loyalties were loyalties to a Crown, or a Throne, or a Way, or a Duty, or a Heritage, and all over the world people responded to its call emotionally, out of their hearts. In India that Jubilee year people sacrificed goats before images of Her Majesty. In Canada Red Indians swore oaths by the Great White Queen. In Kansas City, Missouri, the children of the Brown family, recent emigrants from England, 'assumed a lofty and haughty air', while in Milton, Massachusetts, another English exile must surely have infuriated some of her neighbours, even at Jubilee time, with the song she so often loved to sing around the place:

> *Long may that brave banner flutter on high,*
> *O'er mountain, o'er desert, o'er sea,*
> *A beacon to friends but a terror to foes,*
> *The most glorious banner there be.*
>
> *And if there's a despot who dares to defy*
> *The most glorious banner that ever did fly,*
> *We'll show him an Englishman knows how to die*
> *For the Union Jack of Old England.*[1]

The Diamond Jubilee might be contrived, as a boost to the imperial confidence—the British were past-masters at the suggestive

[1] I heard about the Browns, whose family name I have changed, from a neighbour of theirs at the time ('the boy was Arthur, the girl, poor plain thing, was Muriel') while the song of the Milton patriot was kindly sent me by her granddaughter.

display—but its emotions were deeper than its intentions, and were to survive, in a clutch at the throat, a chill down the spine, a cross on a distant grave, when the physical structure of Empire was dismantled and discredited. As the inscription said upon the sundial at Government House in Mauritius, one of the Queen-Empress's least necessary dependencies:

> *God Save the Queen!*
> *For loyalty is still the same*
> *Whether it win or lose the game,*
> *True as the Dial to the Sun*
> *Although it be not shone upon.*[1]

Sensations profounder still, too, were aroused by the Diamond Jubilee, for the Queen's embodiment of the imperial power reached far back into the people's folk-memory, conjuring atavistic spirits out of the past. Hardly less than peasants of India or Australian aboriginals, simple Britons, especially country people, regarded the power of the Throne with an almost superstitious veneration. Old gods were honoured by the majesty of the Jubilee, by the welling-up of the corporate enthusiasm, and by the spectacle of the aged Queen, her black moiré dress embroidered with silver symbols, attended by her marshals, clerics and statesmen through the streets of London. The bonfires that blazed that night were like rituals of this instinct. A watcher in Worcestershire counted more than forty, flickering far into the distance on beacon hills across the breadth of England: and their scattered lights in the darkness, their glow in the night sky, were reminders of older urges behind the pride of Empire, beliefs and battles long ago, mysteriously linking the very soil of the imperial island with reef and tundra, desert and distant veld.[2]

[1] It comes from Samuel Butler's *Hudibras*, 1678, and is still there. So are the Virgilian quotations inscribed on the garden seats by a former governor of classical tastes, Sir George Bowen, who had been president of the University of Corfu, and who as author of Murray's *Handbook for Greece*, 1854, gave his countrymen the immortal assurance: 'Any Englishman having the usual knowledge of ancient Greek will be able to read the Athenian papers with ease.'

[2] If any resilient reader would like more in this vein, the entire central volume of this triology, *Pax Britannica* (London and New York, 1968) is devoted to an evocation of the Empire, what it was and how it worked, at the time of the Diamond Jubilee.

CHAPTER TWO

'An Explorer in Difficulties'

THE tumult and the shouting slightly died, as Jubilee year
came to an end, but on the frontiers the British Empire
tremendously proceeded—especially in Africa, the last undeveloped
continent, where the imperial dynamic was providing a whole new
pantheon of heroes, saints and martyrs. Two of them in particular
were in the public mind, for far away on the Upper Nile General Sir
Herbert Kitchener, the rising star of the British Army, was avenging the death of 'Charlie' Gordon, 'the noblest man who ever lived'.

Since 1882 the British had been effectual rulers of Egypt, and had
thus become concerned in the affairs of the Sudan, an Egyptian
dependency of a million square miles immediately to the south. For
years the Sudan had been in a state of rebellion under a fiery Sufi
mystic who called himself the Mahdi, 'The Leader', and who
formally announced the End of Time, a conception particularly
unwelcome to the British just then. In 1884 it had been decided to
abandon the country, and to organize the withdrawal the British
Government sent to Khartoum, the capital, General Charles
Gordon, Royal Engineers, everyone's archetype of the Christian
soldier, 'not a man but a God'. Trapped in Khartoum by his own
death-wish, in January 1885 Gordon was killed by the Mahdists, and
so capped his already legendary career with an imperial apotheosis.

It had been one of the great romantic tragedies of the Victorian
age. Ever since the British had dreamed of recovering the Sudan,
and avenging the memory of the martyr. The Mahdi died in 1885,
but his successor, the Khalifa, held similarly apocalyptic views, and
by the 1890s the Reconquest was at hand. The obvious man to
conduct it was Kitchener, whose hooded eye, huge figure and commanding bearing were imperial factors in themselves. Kitchener was

made Sirdar, Commander-in-Chief, of the Egyptian Army, which was in effect an imperial force, and for years he grimly planned the operation. A complicated man, sometimes hesitant, a bachelor of somewhat dilettantish tastes, he was made for the retributive role. His forte was organization, and with infinite care and thoroughness he prepared the campaign, designing his own gunboats for the passage up the Nile, and commissioning his own railway to take his armies out of Egypt towards Khartoum.

It was slow, but it was inexorable. By the end of 1896 Kitchener had an army of 25,000 men, 8,000 of them British, the rest Egyptian and Sudanese, deep in the Sudan. His method of campaign was barbarically deliberate and symbolic. The soldiers went into action crying 'Remember Gordon!'. Gordon's nephew directed the shelling of the Mahdi's tomb at Omdurman, and Kitchener seriously thought of keeping The Leader's skull as a souvenir. It all went like very slow clockwork. By Jubilee Day Kitchener was preparing his advance upon Khartoum, and by the autumn of 1898 he had annihilated the Mahdist army in the battle of Omdurman, killing at least 10,000 Sudanese for the loss of 28 Britons.[1] On the morning of Sunday, September 4, 1898, he crossed the Nile into the ruined capital, where the shattered remains of Gordon's Residency lay as a wreck of rubble and undergrowth beside the river; and there, in a famous Victorian moment, we shall join the conqueror ourselves.

Beside that sacred ruin, on the Nile, the British sealed their victory with a requiem. Its altar was the Residency itself, upon whose surviving walls, their windows still barricaded with bricks and sandbags, the Union Jack was triumphantly hoisted, together with a very much smaller Egyptian flag. Moored at the bank were two of Kitchener's gunboats, swirled in steam, and beside them were assembled men from every regiment and corps in the campaign—British guardsmen, Egyptians in white tarbooshes, pipers in sun-helmets and sporrans, dismounted cavalrymen holding their pennanted lances.[2]

---

[1] But grumbling, for he was very economical, at the ammunition expended in shooting the wounded as they lay on the ground.

[2] They were from the 21st Lancers, a regiment which had its baptism of fire at Omdurman, after so long a history of peaceful soldiering that its regimental motto was said to be 'Thou Shalt Not Kill'.

Many celebrities of Empire were in the congregation, some already famous, some tipped for fame to come—Colonel Reginald Wingate, Kitchener's brilliant intelligence chief, Colonel John Maxwell, the most promising younger officer of the Egyptian Army, young Douglas Haig, its most dashing cavalryman. Brigadier 'Andy' Wauchope of the Black Watch was there, 'the pride of Scotland'. So was Lord Edward Cecil, the Prime Minister's son and one of the wittiest men in the Empire. That faintly oriental figure in the front rank, with slit eyes and long moustaches, is the disturbing young firebrand Charles Townshend, hero of the Chitral siege on the Indian frontier; the young naval lieutenant who hoisted the Union Jack so reverently is C. M. Staveley, who is certain to go far; elbowing his way to a better view of the ceremony, we may be sure, is the most bumptious subaltern of the whole army, Lieutenant Winston Churchill.

And at the head of his men, ramrod stiff, one hand on the hilt of his curved scimitar, one booted foot raised upon a convenient boulder, Kitchener himself stood impassive and immaculate. A salute was fired by a gunboat at the quay.[1] Three cheers for the Queen were called. As the solemn men's voices sang the old words of 'Abide with Me', Gordon's favourite hymn, to the uncertain harmonies of a Sudanese band, a tear was seen to roll down the Sirdar's brown and flinchless cheek. 'The sternness and harshness had dropped from him for the moment', wrote one of the war correspondents, all of whom he despised, 'and he was gentle as a woman.' The parade had to be dismissed by the Chief of Staff, so incapacitated was the victor by his emotions.

When he returned to his camp at Omdurman across the river, though, General Kitchener was recalled at once to harsher realities. He knew that sacramental revenge was not the true purpose of the Army of the Nile. On the previous day he had opened sealed orders from London, to be read immediately after the capture of Khartoum. They required him to proceed at once still further up-river, to forestall any French annexation of the Upper Nile. Gordon had been given his memorial service, but a more truly imperial monument

[1] With live shot, there being no blank: the gunners aimed high over the river into the desert beyond, where nobody who mattered was likely to be.

would be British control, once and for all, of the entire White Nile and its headwaters.

2

Britain's was not the only European empire. Portugal, the Netherlands, Belgium, Germany, Italy, all had overseas possessions of their own. The chief arena of their ambitions was Africa, in which there was then proceeding the unlovely process of grab and self-justification known as the Scramble. Most of the imperial Powers were concerned with this free-for-all, and though it had been to some degree regulated by international agreement in 1885, still it was *au fond* an exercise in which few holds were barred. Black Africans, in those days, hardly counted as real people, and the idea of Europeans simply seizing African territories, to rule, improve or exploit them by their own methods, was generally considered quite justifiable. It was also perhaps inevitable, as the technical power of the West sought out, almost despite itself, vacuums and victims. Most of the Empire-building was peaceful anyway. Africans were persuaded into submission with promises or treaties, or awed into it with demonstrations. Sometimes they asked to be taken under imperial protection: only occasionally did they have to be bludgeoned.

The British had got the lion's share. They had possessed footholds in West and South Africa for generations. By the 1890s they were also established in Egypt, Kenya and Uganda, and in the vaguely defined territories between the Limpopo and the Zambezi. They had gained their ends by a variety of means—diplomacy, economic pressure, deceit, gasconade. Sometimes they acted openly, sometimes conspiratorially, sometimes as servants of the Crown, sometimes as agents of commercial companies, sometimes in the name of the Khedive of Egypt, or even of his hypothetical overlord, the distant Sultan of Turkey.[1]

[1] Asked what were the 'Sultan's rights' mentioned in one particularly obscure African treaty, the Under-Secretary of State for the Colonies replied that he wasn't sure what they were, but whatever they were, they were reserved.

At the end of the century the general direction of their expansion was north-south. Their most remarkable activist, the South African financier Cecil Rhodes, foresaw a British axis running from Cairo to the Cape, fed by access lines to the coast east and west, and giving the Empire effective domination of the whole continent. Though this scheme was blocked for the moment by the presence of the Germans in Tanganyika, still the proposed railway line was already north of the Limpopo River at one end, south of the Egyptian frontier at the other: the first of its feeder lines, from Mombasa to Lake Victoria, was nearly finished, and Kitchener had presciently built his Sudan railway to the South African gauge. Essential to the vision was British control of the whole Nile Valley, and to secure this without war was Lord Salisbury's principal imperial purpose. 'If you want to understand my policy in any part of the world,' he said himself, 'in Europe, Asia, Africa or the South Seas, you will have constantly to remember that.'

The French, who were the principal contenders for African mastery, thought transversely, east to west. Besides their large possessions in North Africa, they were strongly established on the Niger, in the west, and had an east-coast port at Djibouti, in Somaliland. They looked always across the continent, and they dreamed of uniting their eastern and their western footholds to establish their supremacy throughout Central Africa. This ambition clashed with the British, and took the two Empires on a collision course. By their occupation of Egypt the British had staked a claim to the whole Nile valley, and the further their forces advanced up the river, whether their purposes were sentimental, intuitive or purely practical, the less the French chances of a corridor across Africa.

The French overseas empire possessed a brilliance all its own, chiefly because of the extraordinary individuals who administered it in the field, but it had been weakly supported from France, where Governments succeeded each other in febrile succession, and it lacked the strong economic and technical base that gave the British Empire its power. Empire was a sideline for the French: between 1880 and 1889 Britain had seven different Ministers responsible for the Colonies, but France had twenty-one. In 1894, though, the formidable Gabriel Hanotaux became Foreign Minister, and for the first time the imperial urge in France was given a forceful and daring direction. Fortified by a new alliance with Russia, the French turned their eyes upon the Upper Nile. In Hanotaux' opinion it belonged to nobody. The southern Sudan was in a state of rebellion, and any Power, or at least any *civilized* Power, had a right to step in. From their base at Brazzaville on the Congo, the outpost of their central African activities, the French set to work on a 'drive to the Nile', and the French Press openly discussed the chances of reaching its headwaters from the west or east before the British could get there from the north or south.

In 1893 a well-known French hydrologist, Victor Prompt, had suggested that the key to the control of the Nile valley might lie in the area, some 300 miles south of Khartoum, where the River Sobat joined the greater river. There was nothing much there except an isolated riverain fort called Fashoda, used by the Mahdists as a penal colony, and an attendant hamlet of the Shilluk tribespeople: but Prompt suggested that a dam there might effectively control the flow of water into Egypt. Since Egypt depended entirely upon the flow of the White Nile, control of Fashoda could mean command of Egypt: a French presence there, it was argued, could paralyse British activities down-river, and give the Quai d'Orsay an almost unanswerable bargaining power in Africa.

So as the British strengthened their hold on Egypt, and majestically advanced southwards through the Sudan, the French resolved to make a race of it, and prepared an expedition to travel from Brazzaville clean across Africa to the Upper Nile. Reports of the plan greatly disturbed the British. 'The advance of a French expedition . . . from the other side of Africa', Sir Edward Grey of the Foreign Office had told the House of Commons, 'into a territory over which our claims have been known for so long, would not merely be an inconsistent and unexpected act, but it must be perfectly clear to the French Government that it would be an unfriendly act, and would be so viewed by England.' The French responded merely by hastening their preparations, and in the summer of 1897, Jubilee summer in England, Captain Jean-Baptiste Marchand of the French Marines set out to cross the continent and 'establish French claims in the region of the Upper Nile'. He took with him 12 Frenchmen and 150 Senegalese riflemen, and his destination was Fashoda.

3

The day after the Khartoum memorial service, while Kitchener was still considering his secret orders, British outposts on the river south of the city intercepted a small steamer flying the crescent flag of the Mahdists. Its crew, who were unaware that Khartoum had fallen to the British, were taken ashore for questioning, and said they had been far up the Nile foraging for the Khalifa's armies. Near the mouth of the Sobat, they said, they found a strange flag flying over the old fort at Fashoda, and had been fired upon by white men. Several of their crew had been killed.

The British interrogators were startled by this tale. Were the strange Europeans British, working down the Nile from Uganda, or were they foreigners? Wingate asked the Mahdist captain to draw in the sand the flag he had seen at Fashoda, and to describe its colours, and thus learnt that it was the French tricolour. So the French were there already! Five days later Kitchener sailed southward from Omdurman with five gunboats and a dozen barges, with 100 Cameron Highlanders and 2,500 Sudanese askari, with field guns and Maxims and Lord Edward Cecil, and orders to proceed at his own discretion, but to dislodge the French.

So one of history's famous meetings came about. Purposeful up the Nile went the imperial flotilla, its trim little steamers in line ahead—*Dal, Nasir, Sultan, Abu Klea, Fateh*.[1] Lashed alongside or towed behind were the barges that carried the troops, the Camerons lounging in the sunshine with sun-helmets over their faces, the askaris jostled, cheerful and sometimes breaking into song—'Oh, them golden slippers' was a particular favourite of the Sudanese. Kitchener wore civilian clothes, and spent much of his time sprawled

[1] The captain of the *Fateh*, 128 tons, was Commander David Beatty, aged twenty-seven. Twenty years later he was to have under his direct command 41 battleships, 9 battle-cruisers, 54 cruisers, 4 aircraft carriers and nearly 200 destroyers.

on a deck-chair beneath an awning on the *Dal*, contemplating the baked brown landscape streaming by, or talking to Wingate and Edward Cecil. It was a week's voyage from Omdurman to Fashoda, and the flotilla made good speed, the stern-wheels of its steamers frothing the muddy waters, the smoke from their funnels billowing far away downstream, to disperse hours later as a murky black cloud across the desert.

Sometimes great storks and cranes flapped away from their passage. Sometimes hippopotami emerged muddy from the swamp. At villages along the way notables flocked to the water's edge to offer their submission, and intelligence officers went ashore to scribble them notes of pardon. It rained a lot as they sailed further south, the mosquitoes were terrible, and soon the steamers were labouring through the floating mass of decayed and pestilent vegetable matter called the Sudd. At night, though, when the cool descended over Africa, and the helmsmen looked along the banks for somewhere to tie up, marvellous sounds of beast, bird and whirring insect reached the men on board, and made them feel they were penetrating great mysteries (for not even the ship's officers had been told why they were making this equatorial voyage).

Kitchener was a Francophile. He spoke fluent French, he liked the company of Frenchified women, he delighted in the French style of things. He would be reluctant to dislodge any French outpost by sheer force—there would be no primitive triumph at Fashoda, and Colonel Marchand's skull ran no risk of immortality as a table ornament. But he did not know the strength of the French force, he had no idea how truculent it might be, and he decided to move cautiously. When they were about twelve miles from Fashoda two Sudanese orderlies were put ashore with a message addressed to the 'Chef de l'expédition Européenne à Fashoda'. It announced the news of Omdurman, thus implicitly declaring the British to be suzerains of the Upper Nile, and said that General Kitchener hoped to be in Fashoda the following day. Kitchener signed it not as a British general, but as Sirdar of the Egyptian Army, and at Wingate's suggestion he ordered that only the Egyptian flag would be flown by the flotilla as they approached Fashoda, and that he and his officers would wear their Egyptian uniforms. The impact would be less pointed, and the suggestion of a clash between two great Empires less direct.

Next morning as the ships steamed slowly on, the lookouts saw approaching them a small rowing boat, flying at its stern an enormous tricolour, and carrying a black sergeant in French uniform. He brought a reply from Fashoda:

Mon général, I have the honour to acknowledge the receipt of your letter dated 18 September 1898. I hear with the greatest pleasure of the occupation of Omdurman by the Anglo-Egyptian army, the destruction of the Khalifa's hordes and the final defeat of Mahdism in the Nile Valley. I shall be the very first to present the sincere good wishes of France to General Kitchener, whose name for so many years has epitomized the struggles of civilization against the fanatical savagery of the Mahdists—struggles which are today successful. These compliments therefore I send with all respect both to you and to your valiant army.

This agreeable task completed, I must inform you that, under the orders of my government, I have occupied the Bahr-el-Ghazal as far as Mechra-er-Req and up to its confluence with the Bahr-el-Jebel, also all the Shilluk territory on the left bank of the Nile as far as Fashoda. . . . I signed a treaty on 3 September with Abd-ed-Fadil, their Reth, placing all the Shilluk country on the left bank of the White Nile under French protection. . . . I have forwarded this treaty to Europe, via the Sobat-Ethiopian route, also . . . by Mechra-Er-Req, where my steamer the Faidherbe *is at the moment with orders to bring me such reinforcements as I judge necessary to defend Fashoda. . . .*

Again, I give you my good wishes for a happy visit to the Upper Nile. I also note your intention to visit Fashoda, where I shall be happy to welcome you in the name of France.

Signed MARCHAND

4

This engaging persiflage, full as it was of meaningless treaties, non-existent reinforcements and unenforceable claims, paved the way for a meeting between the two commanders, and later in the day, proceeding southwards through the desolation of the Sudd, the

British sighted Fashoda. Forlornly above the rotted swamp, stretching away as far as the eye could see, the little fort stood half-derelict upon a peninsula, with a few conical huts of the Shilluk outside its walls, a group of palms, and a soggy garden of vegetables. It looked hot, wet and verminous, but the tricolour flew boldly above it, and at the water's edge, as the gunboats approached, an honour guard of Frenchmen and Senegalese stood bravely at the salute. The British were touched by this show of pride, and by the defiant isolation of the fort. 'It was a puny little thing', one officer wrote in reminiscence. 'Were we to be compelled to break it down?'

At this dismal spot the Empires met. Marchand and his men had reached it after a terrible overland journey, one of the most remarkable in the records of African travel, eight months on foot, three months in their leaky collapsible steamer, which they had dragged laboriously overland from the Niger to the Congo. Kitchener and his men had travelled there more magnificently, in the after-flush of a great victory. Marchand's force was pale and emaciated, after months among the toads, insects and fevers of the Sudd. Kitchener stood on the foredeck of the *Dal* bronzed and bulky, his soldiers well-nourished at his back, his gunboats spick-and-span.

At midday, September 19, 1898, the commanders met on board the *Dal*. Kitchener wore his Sirdar's regalia, with tarboosh. Marchand, a small bearded figure, wore no military insignia at all—wisely, perhaps, since he was only a captain. They sat with Wingate and Marchand's adjutant on the deck, watched intently from shore and ship by officers with binoculars. Peace and war hung in the balance, and their conversation was tense. Sometimes, the watchers thought, the talk seemed less than amicable, and Marchand was to be seen gesturing angrily at the Sirdar—'distinct signs of hostility', reported a British colonel to his colleagues. Presently, though, a steward climbed up the ladder to the deck carrying a tray of glasses, 'full of golden liquid', and a moment later Kitchener and Marchand, raising their glasses, were clinking them in agreement and good wishes—in relief too, no doubt, as they sat there, half in shade, half in sunlight, on the deck of the little ship.

It had been a close thing. Kitchener had declared flatly that the episode might lead to a European war—did Marchand, with such stakes at issue, really mean to prevent the representatives of Egypt from hoisting the Egyptian flag over an Egyptian possession? Marchand replied that obviously he was powerless to prevent it, since he was outnumbered ten to one, but that without contrary orders from France he could not retire from his position, and that all his men could do, if Kitchener insisted, was to die at their posts. Proud of his achievement, prickly in his patriotism, he was undoubtedly ready to defend his awful fort to the end—he had a 'terrible desire', he later said, to rebuff Kitchener altogether: but he was awed despite himself by the imponderables at stake, and perhaps even by the presence of Kitchener, and so as the servant with his drinks began his precarious ascent of the upper-deck ladder, an accord was reached.

Marchand would not be ejected from his outpost, and the French flag would continue to fly there, pending orders from Europe. The British would establish their own garrison at a discreet but practical distance—500 yards to the south, on Marchand's only line of retreat through the marshes. The British would formally take possession of the area, but in the name of the Khedive of Egypt, and only the Egyptian flag would fly above their own quarters. There the matter was left, in a compromise that seemed to protect everybody's face, and would allow the two imperial Governments, far away, to achieve a solution.

It was a soldier's formula, and the soldiers liked each other. They negotiated in French, but they talked the same professional language too, of flags and gunpowder, supplies and disciplines. After the talks the British mounted a parade, and the Egyptian flag was hoisted by a Sudanese detachment, to a twenty-one-gun salute from the artillery. Everybody saluted quiveringly, and did their best to be tactful. The warships flew Egyptian flags, most of the British officers wore Egyptian uniforms, and even the three cheers were in Arabic. The French, for their part, behaved with great courtesy: the officers wore fresh-laundered whites for the occasion, and their Senegalese soldiers, in red fezzes and jerseys, seemed even to the British to be commendably soldier-like.

Later Kitchener was entertained in Marchand's mess, bending

his gigantic frame almost double to enter it. The two sides toasted each other in sweet champagne, and exchanged fairly ornate pleasantries. A Senegalese guard was then inspected in the blazing heat, and as the Sirdar left the soil of Fashoda—Fort St Louis to the French—he was presented with a huge basket of vegetables and flowers from the garden: French flora from French soil, it was tacitly suggested as the last salutes were exchanged and the boats rowed out to the waiting gunboats.

Almost at once the flotilla sailed, leaving only a regiment of Sudanese and some guns in a bivouac on their mudflat, and soon the trail of black smoke was far away across the empty Sudd. Kitchener left behind him, nevertheless, a steely after-taste, for before he embarked he handed Marchand a formal letter of protest at the presence of the French in the Nile valley, and a list of stern restrictions on their movements. Not even private letters were to be sent down the river without British approval: in effect the French were to be imprisoned in their fort with their flowers, flags and vegetables. This was the iron within the glove. No pretence was made now that Kitchener was acting purely in Egyptian interests. The protest was made in the name of Great Britain, and Marchand and his Frenchmen were left in no doubt, as they watched the gunboats disappear, that the British Empire itself had passed that way.

5

The British also sent the French, as a parting gift, a package of newspapers. It was a Parthian courtesy, for the papers contained the news of the Dreyfus Affair, the cancer in French public life which festered around the imprisonment on Devil's Island, for alleged espionage, of the innocent Jewish Captain Alfred Dreyfus. 'You have achieved something remarkable, very remarkable,' Kitchener had told Marchand in his mess hut, but he had added enigmatically, 'but you know the French Government will not back you up.' This was why. The Dreyfus scandal, which divided French society from top to bottom, also hamstrung French foreign policy, while the Russians had chosen the moment to tell their allies that they wanted no part of the Nile dispute. Marchand was all on his own, far away in the Sudd: France had her mind on closer predicaments.[1]

Sure enough, war never came. The British were united behind Kitchener—they realized, Churchill wrote, 'that while they had been devoting themselves to great military operations, in broad daylight and in the eye of the world . . . other operations, covert and deceitful, had been in progress in the heart of the Dark Continent, designed solely for the mischievous and spiteful object of depriving them of the produce of their labours. And they firmly set their faces against such behaviour.' The negotiations between London and Paris were protracted, but the British held all the cards, if only because they really were ready to go to war over Fashoda. 'We've only got arguments,' said Theophile Delcassé, the new and more conciliatory French Foreign Minister, 'they've got troops.' Admiral Sir John Fisher, the British naval commander on the North American Station, was standing by to fall upon Devil's Island and snatch Dreyfus away to freedom, while at home the imperial propaganda machine worked at full blast. 'What has France to look for?' asked the *Illustrated London News*. 'Of our ultimate triumph, and of the utter disaster that would befall her, there can be no question whatever.' George Wyndham, one of the most promising of the younger Conservative imperialists, put the British attitude in a nutshell: 'We don't care whether the Nile is called English or Egyptian or what it is called, but we mean to have it and we don't mean the French to have it. . . . It is not worthwhile drawing distinctions of right and wrong in the matter, it is a matter entirely of interest.'

Salisbury mischievously described poor Marchand as 'an explorer in difficulties on the upper Nile', and in fact the French presence at Fashoda, invested north and south by British power, did

---

[1] Sir Ronald Wingate, whose father's advice to Kitchener was crucial to the Fashoda settlement, points out to me the striking coincidence that almost exactly a century before, in August 1799, Admiral Sir Sidney Smith sent Napoleon, after his victory at Aboukir, a packet of newspapers giving him his first intimation of crisis in France: he abandoned his army in Egypt at once, and, sailing secretly for home, presently became First Consul and Emperor.

become more and more ridiculous. By the time Paris recognized realities, in November 1898, the six Frenchmen and seventy Senegalese left at Fashoda felt themselves cruelly betrayed and humiliated. They withdrew that December, after five months in the fort, and were played away to Djibouti (they refused an easier passage down the Nile to Egypt) by the strains of the 'Marseillaise' from the Sudanese band of the British garrison—'sad yet proud', as one of them recorded, 'with moist eyes yet with our heads held high'.

'Now', wrote Churchill, 'the British people may . . . tell some stonemason to bring his hammer and chisel and cut on the pedestal of Gordon's statue in Trafalgar Square the significant, the sinister, yet the not unsatisfactory word, "Avenged".' In fact Gordon was doubly avenged, by the defeat of his murderers and by the final extension of British power to the headwaters of the Nile. In the end, too, the unhappy Fashoda episode had happier consequences, for it was the last dangerous clash between the British and the French Empires. Tempers cooled, as the two nations recoiled from the brink. The French recognized Britain's supremacy on the Upper Nile, the British accepted French paramountcy in Morocco. Within a decade the Fashoda incident, with all its acrimonies, had been succeeded by the Entente Cordiale.

Marchand, a French hero in the classic mould, was made a Commander of the Legion of Honour and promoted to Lieutenant-Colonel.[1] Kitchener was awarded the Grand Cross of the Bath, raised to the peerage as Lord Kitchener of Khartoum, voted a Parliamentary grant of £30,000 and appointed Governor-General of the Anglo-Egyptian Sudan. Fashoda itself disappeared from the maps. The dwindling remains of its fort remained there among the swamp, slowly disintegrating over the years, and for a while it was a relay station on the British telegraph line to Uganda: but in deference to French feelings the imperial surveyors changed the name of it, and it was known henceforth as Kodok, after the Shilluk settlement nearby—a fateful but forgotten spot, $9°51'N$, $32°07'E$, where a

---

[1] He died a general in 1934, after a life of mixed fame and obscurity. Hanotaux, the true instigator of Fashoda, became a somewhat disconcerting delegate to the League of Nations, and survived until 1944.

possible highway from Nigeria to Somaliland might cross a putative railway from Cairo to the Cape.[1]

[1] Later it became a local administrative headquarters, but when in 1955 an RAF aircraft flew me over it (the Sudan was still Anglo-Egyptian then) it looked as forlorn and isolated as ever, and when I told the pilot that it was Fashoda, where Kitchener and Marchand met, he had never heard of the place.

The Khalifa, having escaped the slaughter at Omdurman, was caught and killed a year later, but the Mahdist movement never lost its appeal for the Sudanese, and until the last years of British rule the Mahdi's own posthumous son, Abd ar-Rahman al-Mahdi, was a great man in the land, and also a Knight Commander of the British Empire.

CHAPTER THREE

Following the Flags

THE British at home were cock-a-hoop, glorying in Kitchener's successes and half hoping for war with France. ('It was a pity it could not have come off just now', wrote Admiral Fisher, regretfully putting away his charts of Devil's Island, 'I think we should have made rather a good job of it.') The bourgeoisie, in particular, revelled in their colonial wars, and a picture popular around the end of the century portrayed them doing it.

It was called *Following the Flags*. On the left sits Papa, wearing a frogged smoking-jacket and holding that morning's *Times*. On the right are grouped his family: mother in lace jabot and speckled muslin, daughter with ribboned hair over her striped blouse, son in Eton collar and kilt. On the table between them lies a map of the current campaign, wherever it happens to be, stuck about with Union Jacks, and as the paterfamilias reads out the latest despatch from the front, his children eagerly move the flags and assess the tactical situation—for all the world as though they are indulging in some favourite nursery game, whose conquered territories are only imaginary, and whose dead are make-believe. The British were not merely interested in imperial affairs. At this climactic moment of the Empire's history they were imperially brain-washed.

2

There was hardly a moment of the day then, hardly a facet of daily living, in which the fact of Empire was not emphasized. From exhortatory editorials to matchbox lids, from children's fashions to parlour games, from music-hall lyrics to parish-church sermons, the imperial theme was relentlessly drummed. Empire was the plot of novels, the dialogue of plays, the rhythm of ballads, the inspiration of oratorios. It was as though the whole nation was being deliberately disciplined into the imperial fervour.

That small kilted boy, for instance, when he returns to his preparatory school at the end of the holidays, will not often be allowed to forget that he is born to an imperial heritage. Every day's curriculum reminds him. School prayers, for a start, will doubtless include a prayer for the Queen-Empress and her subjects across the world, may well include a sermon about imperial responsibilities or missionary needs, and is very likely to end with some devotional hymn of Empire. Refulgent upon the classroom wall will hang the map of the world, on Mercator's projection preferably, for no other shows to such advantage the lavish slabs of red which mark the authority of the British—and perhaps a shipping chart, demonstrating by its distribution of boat-shaped blobs how overwhelming is the British maritime power—and possibly some inspiring steel engraving too, Caton Woodfield's famous picture of the Khartoum memorial service, or one of Lady Butler's celebrated reconstructions of British gallantry in the field.

Is it history this morning? No flabby internationalism in the 1890s. History was not only firmly Anglo-centric, it was also frankly designed to impress upon a pupil the superiority of all things British, and the privilege of being born to the Flag. Or is it Eng Lit? What better textbook than Arthur Stanley's recent anthology of patriotic verse, with its introduction by the Lord Bishop of Calcutta, late headmaster of Harrow? 'The song that nerves a nation's heart', as is very properly quoted from Tennyson on the title-page, 'is itself a deed', and within these pages art and action are certainly allied. Here is 'The Song of the English Bowmen', and 'Private of the Buffs'—here Drake's Drum sounds down the years, and here Wolfe dies once more on the Plains of Abraham—all to the glory of Britain and her Empire, for as the Bishop observes rather obscurely in his preface, an Empire lives not by bread alone.[1]

Luncheon nevertheless, in school hall, and perhaps there is some

---

[1] Things might have been worse. They could have been set C. R. Low's *Crécy to Tel-el-Kebir* (1892), which described in 500 pages of heroic triplets all the principal imperial battles for five centuries.

old boy at high table, home from the war, swaggering in red and gleaming brass, and reminiscing airily to the envious masters all about. To the cricket field next, nursery of England's style, where Newbolt's verses echo always, if not among the players, at least among the umpires—

> There's a breathless hush in the Close tonight—
> Ten to make and the match to win—
> A bumping pitch and a blinding light,
> An hour to play and the last man in.
> And it's not for the sake of a ribboned coat,
> Or the selfish hope of a season's fame,
> But his captain's hand on his shoulder smote—
> 'Play up! Play up! and play the game!'[1]

And when our little friend goes to bed after prep that night, to a good read under the blankets, there G. A. Henty awaits him no doubt, with his tales of healthy British adventure, or Rider Haggard's vision of Africa, or perhaps just the *Boy's Own Paper*, the *B.O.P.*, in which the pluck of British youth is for ever matched, and for ever victorious, against wickedness, savagery and foreignness.

3

There was no escaping it, breakfast to lights out, and to most people, in the full flush of Omdurman and Fashoda, it simply seemed inevitable. The Empire was as immemorial as the Palace of Westminster itself—which, though it had been built less than half a century before, was popularly assumed to be as old as anything.

This was partly because the long reign of Queen Victoria had given the British a sense of organic permanence. Foreign countries had coups or revolutions, invaded one another, replaced kings with Emperors. Britain progressed differently, quietly, steadily, as though

---

[1] The BBC have a recording of the poet reading these lines, and nothing could be more suggestive of lost values than the sound of his dry fastidious voice expressing, with a melancholy intensity, their now ridiculed convictions. Newbolt died in 1938, aged seventy-six, but Clifton College flourishes still, behind the Bristol Zoo.

the history of a nation was a direct paradigm of human life, and was simply part of the universal ageing process. So it surely was with the Empire. The idea that Kitchener had not really gone to the Sudan to avenge Gordon, but to forestall foreign competition in Africa, would strike most people as sophistry, if not actually sacrilege. Every Englishman was born, wrote George Bernard Shaw, with a miraculous power that made him master of the world. 'When he wants a thing, he never tells himself he wants it. He waits patiently until there comes into his mind, no one knows how, a burning conviction that it is his moral and religious duty to conquer those who possess the thing he wants. Then he becomes irresistible.'

Even the rank and file of the imperial activists, the soldiers, the administrators, the merchants in the field, did not generally think of imperialism as power politics. The Empire was there, it was patently beneficial, they would do their best for it. The old idea of an imperial trusteeship had been transmuted by Kipling into the more readily comprehensible image of the White Man's Burden—

> *Take up the White Man's Burden,*
> *In patience to abide*
> *By open speech and simple, an hundred times made plain*
> *To seek another's profit,*
> *And work another's gain.*

Most rank-and-file British imperialists, at the end of the nine-teenth century, would probably admit to these sentiments as their own, especially as they were tinged with complaint—'the blame of those ye better, the hate of those ye guard'. The Empire-builder often felt himself to be unappreciated. The ingratitude of subjects, the lack of material reward, the interference of politicians, the ignorance of intellectuals—all these figured often in letters home from Kodok or Kidderpore, and perhaps made the writers feel all the more noble. It was seldom, however, a grumble about the life itself. For many imperialists theirs was a true calling, often transmitted through generations in the imperial service.

For on this, the professional level of Empire, the idea of service really was paramount. The imperial classes, trained with such precision by the public schools and ancient universities, were bred to it, and Joseph Conrad the Pole thought love of service, even more than love of adventure, to be the first characteristic of the Englishman in his Empire.

4

And if to the public at home Empire was a craze, to the man in the field it was often a perennial fascination, for the imperial profession catered for every preference. The most esoteric speciality could be pursued somewhere under the Crown, and over the years the Empire had built up a vast corporate body of knowledge, scientific, anthropological, strategic, economic. Each extension of Empire widened this expertise. Hardly were the imperial soldiers on the Upper Nile than the imperial hydrologists were following them: after the gun came the butterfly net.

Only in the British Army, for instance, would there be supply officers ready to demonstrate that the number of camels needed to carry x loads y stages into Afghanistan might vary according to the formula $14x\left(\left(\frac{15}{32}\right)y-1\right)$. Only a British diplomat, perhaps, told by a Persian Prince that the ink-stain on a new treaty was 'a mole upon its face', would be able instantly to reply with a quotation from Hafiz—'I would give all Samarkand and Bokhara for the Indian-dark mole on the face of my lady-love.' The handful of Englishmen who supervised the affairs of Sikkim in the 1890s produced a huge folio volume listing everything there was to know about the State, its flora and fauna, its history, its legal code, its folk-customs, its geology, its religion, down to Scorpion Charms against Injury by Demons and Jungle Fruits, &c, Eaten by Lepchas.[1] The literary works of Elias Ney, British Agent at Meshed in 1896, included *An Apocryphal Inscription in Khorassan*, *A Journey Through Western Mongolia*, an introduction to the history of the Shans of Upper Burma, and a translation with learned notes of *The Tarikh-i-Rashidi*, by Miza Haidar of Kashgar. Flora Annie Steel's *Complete Indian Housekeeper*, 1892, dedicated to 'Those English Girls to whom Fate

---

[1] Some reports were briefer. An officer instructed to investigate the Mishmir area of north-east India wrote simply: 'The country is bloody and so are the people.'

may assign the task of being Housemothers in our Eastern Empire', told its readers how to build a camp-oven, how to make snipe pudding or mange ointment for dogs, how to treat cows with colic or husbands with prickly heat, the best means of keeping sparrows out of the house, the cost of hiring a bullock-drawn van in Ootacamund, the Tamil word for horse-barley, the right underwear to take to the Punjab, the proper way to load a camel and the only correct recipe for boot-dubbin (fish oil, mutton suet and resin).

Languages especially were an imperial concern. The Englishman might be notorious for his inability to learn French or German, but the most unlikely members of the imperial services, dim infantry subalterns or district officers reserved to the point of misanthropy, seemed able to master Burmese or Arabic, Nguni or Fijian. Many languages were first lexicographed by the British: some were first put into writing by them. The young Charles Bell, Indian Civil Service, posted to Darjeeling in 1900, so mastered Tibetan that in only four years he was able to publish his indispensable Tibetan Dictionary: the hooked alphabet of the Cree language was the only surviving memorial to the Welsh Wesleyan missionary James Evans, otherwise expunged from history owing to alleged misconduct with squaws.

The mastery of technique was the key to authority, whether it was knowing more about soil composition, or understanding the historical origins of Honduran folklore, and most of the imperial administrators were diligent in their specialities. It was a diligent Empire. As 'A Gentleman of Experience' wrote in his *Guide to the Native Languages of Africa* (1890), 'In the matter of language it is always better to go to a little more trouble and learn the exact equivalent if possible. "I am an Englishman and require instant attention to the damage done to my solar topee" is far better than any equivocation that may be meant well but will gain little respect.'

5

Behind this decency and conscientiousness, though, beyond the naive ardour of the general public, very different energies directed the affairs of Empire. Of the few dozen men who really ran the British

Empire in the dangerous years after the Diamond Jubilee, scarcely one or two were English gentlemen as the world knew the breed. Some were cosmopolitans. Some were eccentrics outside the genial English pattern. Some had married foreigners. Some *were* foreigners.

Lord Salisbury the Prime Minister was an aristocratic original, an amateur electrician whose hobby was riding a tricycle around his ancient estate, and who often did not recognize his own Cabinet colleagues. Joseph Chamberlain the Colonial Secretary was a Birmingham screw manufacturer, a Unitarian by creed, a dandy by pose, who had seized upon Empire as a means of political advance. Lord Cromer the ruler of Egypt was the great-grandson of a Hamburg merchant. Lord Milner the High Commissioner in South Africa was a German-trained lawyer of authoritarian principles. Lord Curzon the Viceroy of India was an intellectual landowner of almost preposterous grandeur, with a faint Midlands accent and an insatiable ambition. Lord Kitchener had spent so long abroad that he knew almost nothing of English life. Cecil Rhodes was a half-crazed visionary who wanted the whole world British—'the moon too, I often think of that.' Edward Elgar the maestro of Empire, having left school at fifteen, conducted the band of a Worcestershire lunatic asylum, married the daughter of an Indian Army general, and was a devout Catholic. Rudyard Kipling the imperial laureate, having started life as a newspaper reporter in Lahore, married a domineering American, dabbled in mysticism and wrote short stories no simple soldier could understand.

Such were the master-imperialists, the dogmatists, prophets and executives of the creed. Few of them spoke the language of Newbolt, or even of Kipling himself. Few of them had been at English public schools.[1] Most of them looked cynically upon the gaudy patriotism of the day—some indeed chose to work in the Empire because they so detested life in late Victorian England. Yet they were heroes to the masses. Country doctors and surburban solicitors nodded their

[1] Of the nine I have named, only Salisbury and Curzon, who were both at Eton. Salisbury left when he was fifteen and Curzon used to give wine parties in his ornately over-furnished room—'what struck me painfully in him', wrote Gladstone after visiting him there, 'was the absence of any sort of reverence for anything like age or tradition.'

heads in agreement, when they read Mr Chamberlain's latest speech in the *Morning Post*. Crowds flocked to quay or platform when the great Kitchener came home from the wars. The British saw in their leaders the best of themselves, the truest: they did not often know, and would not willingly believe, what excesses were sometimes committed in their name.

Take the young Mahdist commander Emir Mahmoud, captured by Kitchener at Berber in 1897. Would the proud father of *Following the Flags* like his children to know what happened to him? Chains were riveted around his ankles, an iron halter was put around his neck, his hands were bound behind him, and he was paraded in ignominy through the town. Kitchener rode in front, magnificently on a white charger: Mahmoud came behind, sometimes dragged, sometimes running. Whenever he fell, Sudanese soldiers drove him on with leather-thonged whips. The crowd hooted and pelted him as he passed: far in front, beyond the tossing cavalry, the Sirdar rode on impassive, looking neither left nor right.

Every Empire rests on force, and though the British were not habitually cruel, they were certainly ruthless on their frontiers. By any standards but their own they might be considered bullies— almost the worst category of villain, in the vocabulary of Victorianism. There were few rougher fighting men than the British soldier in revengeful mood, shouting Irish scurrilities perhaps, or sustained by the eerie wailing of the pipes, as he advanced with a gleam of that abattoir weapon, the bayonet. 'Severity always,' was an old imperial maxim, 'justice when possible.' 'Butcher and bolt', is how they described that familiar imperial exercise, the punitive expedition.

For by now an assumption of superiority was ingrained in most Britons abroad.[1] The English milord had long before travelled through Europe as though he owned the place. The British imperialist travelled through his subject lands with the same proprietorial air, and when ennobled often formalized the status by including some foreign fief or battle-ground in his title: Kitchener became Kitchener of Khartoum and Vaal in the Colony of Transvaal,

[1] This was fundamentally an Empire, so the poet W. E. Henley thought, of 'us Anglo-Normans, Empire-builders, masters of the earth . . .'

Roberts was Roberts of Kandahar, Wolseley was Wolseley of Tel-el-Kebir—for all the world as though they really were hereditary squires of those recondite properties.[1] The Englishman expected the best seat, throughout his quarter of the world. He expected to be treated, by a quarter of the world's people, with a proper respect, even with the gratitude due from a tenant to a benevolent landlord. Men of the middle classes acquired patrician pretensions, for between the castes of Empire and the castes of home there was a recognizable correspondence, while the ancient social orders of the subject nations were all too often ignored or mocked.[2]

Though these snobberies were more often a matter of habit than intent, they placed the African or Asian subject, struggling to speak English and to Anglicize himself, at a perpetual disadvantage, and on a level of high policy they could be translated into insufferable assumptions of authority. It was as though the British were gods themselves. Lord Salisbury, who hated racial bigotry, and called it the 'damned nigger attitude', nevertheless seriously considered transferring half the population of Malta to the island of Cyprus, to prevent Greek irredentism there.[3] Kitchener once suggested that all the more contumacious of the Boers, in South Africa, might conveniently be expelled to Madagascar, Fiji or the Dutch East Indies. There were no Indian natives in the Government of India, Lord Curzon once observed, because among all the 300 million people of the sub-continent, there was not a single man capable of the job. When the Khedive of Egypt, ostensibly an independent sovereign, once ventured to criticize the way the British were reorganizing his own army, his Minister of War was instantly obliged to resign, and it was even suggested that the Khedive might have to abdicate.

[1] The urbane Viceroy Lord Dufferin, having added Burma to the Empire, wanted no such grandiloquent memento, and found most Burmese names 'like something out of Offenbach or *The Mikado*': and so he devised the most elegant of all the imperial ranks, the Marquisate of Dufferin and Ava.

[2] 'The Aga Khan', the College of Heralds in London once declared, 'is held by his followers to be a direct descendant of God. English Dukes take precedence.'

[3] A proposal strongly supported, it is only fair to add, by many of the Maltese—particularly, I dare say, those who would not have to go.

Sometimes the hauteur was self-defeating. In 1900, for instance, the British in West Africa determined to seize the Golden Stool of the Ashanti. This was far more than a stool really, being the most sacred possession of the Ashanti people, delivered to them magically out of Heaven at the beginning of the eighteenth century: it rested upon its own Chair of State, was hung about with talismanic emblems, and was in fact the very repository or ark of the Ashanti nationhood. Who destroyed it, destroyed the nation; who abused it, insulted the soul of the Ashanti people. The Golden Stool was kept in an inviolable shrine, never to be seen by strangers, and around its mysterious presence revolved the entire Ashanti meaning.

The British found that meaning generally unprepossessing, for the Ashanti were addicted to human sacrifice, dishearteningly unresponsive to Christian improvement, and reluctant to pay an indemnity they owed the Queen of England. In 1900 the Governor of the Gold Coast, Sir Frederick Hodgson, accordingly travelled with his wife and a stalwart bodyguard through the rain-forests to Kumasi, the Ashanti capital, and there called a palaver of the chiefs. He was hardly one of Newbolt's cricketers, being variously described by his contemporaries as 'rotten', 'an egregious ass', and 'no gentleman', and he wasted no time on subtleties. Summoning the elders and potentates of the Ashanti around him, Lady Hodgson white-gloved at his side, he adjusted the Order of St Michael and St George around his neck, and demanded the instant surrender of the Golden Stool. 'Where is the Golden Stool? Why am I not sitting on it at this moment? Why did you not take the opportunity of my coming to Kumasi to bring the Golden Stool, and *give it to me to sit upon?*'

The people rose to arms at once, the Governor was shut up in his own fort for six weeks, and though inevitable retribution followed, imperial columns fell upon Kumasi from all directions, and the Asantahene Prempeh I was banished for twenty-four years to the Seychelles Islands, still from that day to the end of the British Empire, no white man ever set hands upon the Golden Stool of the Ashanti.[1]

[1] The Asantahene was allowed to return in 1924, as plain Mr Prempeh, but in 1935 a successor was installed as Prempeh II. When he received me at

6

Yet if there was something loathsome to this arrogance, there was often something impressive too, and even Britain's enemies begrudgingly conceded it. The implacably hostile Boers, to whom the whole British arrangement of life was inexplicable, if not actually deranged, were nevertheless moved to admiration by the certainty of it all: the unshakeable patriotism, the acceptance of hierarchy, the paradoxical brotherhood, the beauty of pageantry and tradition, the inspired courage—'I must say this for the British officer,' said one Afrikaner critic of Empire, 'that I never once saw one who was a coward.'

This intangible power, beyond morality, beyond measurement, Joseph Chamberlain the Colonial Secretary now hoped to channel into efficiency. He was the odd man out in Salisbury's Government, a modernist, a bourgeois, and he believed that the Empire needed an altogether new driving force, to pull it together and prepare it for a more difficult future. The white self-governing colonies, who now liked to be called Dominions, were gradually coalescing: Canada had been a single federation since 1867, the six Australian colonies were about to become a federal Commonwealth. Chamberlain wished to see these 'overseas Britains' supplemented by the black, brown and yellow colonies in one self-supporting, self-sufficient political unit— a new kind of super-Power, embracing supply and demand, raw materials and manufacturing ability, malleable labour and constructive capital. He saw the whole amorphous conglomeration as one enormous estate, and he wanted it run by the best principles of modern management, like his factories in Birmingham. Imperialism should not be an adjunct to national policy, as Salisbury saw it, but a national purpose in itself—'the days are for great Empires,' Chamberlain said, 'not for little States.'

Times moved fast, in those days of discovery and realization: if

---

his palace in Kumasi in 1957 he showed me many signed portraits of British Governors, but they did not include, I noticed, Sir Frederick. I also met several old men who remembered the fateful palaver of 1900, but they declined to talk about it, and looked very fierce.

Victoria's Empire had seemed the Very Latest Thing at the end of the 1880s, by the end of the 1890s it needed a new direction. Schematic solutions had always been a Victorian speciality— everything from ethics to biology had been tabulated, subdivided and classified—and now Chamberlain and his imperial reformers wanted to do the same for Empire. 'There is no article of your food,' he told the British, 'there is no raw material of your trade, there is no necessity of your lives, no luxury of your existence which cannot be produced somewhere or other in the British Empire . . . nothing of the kind has ever been known before.' Empire should be an economic system, a political solution, a modern career. It should have an altogether new kind of romance, principled but brisk, adequately defined by Lord Cromer, the pro-consul of Egypt, when he offered a philosophy of life to the boys of Leys School, Cambridge: 'Love your country, tell the truth and don't dawdle.'

How to achieve this dawdleless fulfilment was a constant pre-occupation of Chamberlain and his supporters, in the years after the Diamond Jubilee. They thought of an imperial federation, or an imperial parliament, or a Grand Council of Empire, but the shape of the thing was so inchoate, its constituent parts were so varied, and the dead weight of its backward regions was so colossal, that the self-governing colonies could never be persuaded far along the way. They thought of an imperial customs union, embracing the whole Empire in an immense free trade area, but the British themselves were too dedicated to Free Trade, while the Dominions were altogether too protectionist. They tried repeatedly to give some pattern to imperial defence, sharing its responsibilities between Britain and the Dominions: but for the most part the colonists were very content to have their security looked after by the Mother Country, and the most the British achieved was a battleship or two contributed by the colonies to the Royal Navy, and some garrisons of colonial militia.

At home Chamberlain campaigned assiduously to place his idea of Empire at the centre of British politics. He tried to develop tropical colonies with Government money, as an inducement to private investment there. He founded research institutes and schools of tropical medicine, to encourage progress in the more uncomfortable possessions. He gave hope to imperialist agencies of many kinds, from public institutions like the British Empire League to enthusiastic journals like *The Imperial and Colonial Magazine*, or private propagandists like the journalist Arnold White, who foresaw the worst if the Empire was not reformed—'There is no time to lose! What is to arrest our Gadarene rush down the steeps of inefficiency to the sea of national destruction?'

But the people of the homeland, like the colonists, responded half-heartedly. They loved their Empire dearly, but since Chamberlain's vision was strenuous, challenging, and might put prices up, it met with a mixed response. The British never did believe in taking pleasures too seriously.

7

It was too late, anyway. Though for the rest of the Empire's history there were intermittent attempts to give it logic, in fact it had gone too far already, was too odd, too heterogeneous, for centralist reforms. Its separate units were more competitive than complementary, and the only real unifying bond was the authority of London. The white colonials, far from wanting closer ties with the Motherland, only wanted more independence for themselves: the subject peoples, when if ever they achieved equality, were bound to demand control of their own resources. The Empire was essentially irrational, not to be transformed into a smooth-running joint stock company, and its truest energies were highly individualist, not to say arcane.

Chamberlain was a utilitarian. He never did see that the poetry of Empire was not merely half its point, but actually its chief support. There was an absurdity to this structure which could not survive down-to-earth Birmingham analysis, but which was a strength in itself: as the bumble-bee aerodynamically could not fly, so 40 million northern islanders patently could not rule 370 million subject peoples in the face of the world's jealousy. Yet as the bee flew, so the Empire stood. Bluff, pageantry, confidence, faith, habit, tradition, even sleight-of-hand—all these were to prove, in the end, more resilient than Chamberlain's criteria of advantage, and the least successful imperial experiments were those which relied upon common-sense.

These truths were realized most clearly in India, the most tremendous of all the imperial territories, where the spectacle of so few Britons administering so vast a country made even less sense than most of the imperial demonstrations. The British had been in India for 200 years, and they had long before adapted to the Indian taste for colour and display, partly to amaze the indigenes, partly to fortify themselves. In a country of princes, they deliberately used the mystique of monarchy as an instrument of dominion, carrying the idea to lengths which some of the more sophisticated Indians found ludicrous, but which seemed to work among the susceptible masses. Elephant parades, soirées, durbars, palaces—these were among the technical devices of British rule in India, a showy palanquin of authority beneath which those thousand civil servants proceeded workmanlike with their court hearings, their tax collections and their inspection tours.

Let us then, since this strain of splendid pretence was a true *leitmotiv* of the imperial climax, end this chapter of attitudes with a visit to one of the ceremonial durbars with which the British in India from time to time glorified their own achievement. We will choose the most glorious of them all, which Lord Curzon, Viceroy of India from 1898 to 1905, personally staged outside the walls of Delhi, the ancient seat of the Moghul Emperors and the most vital centre of Indian history.

Nothing could be much more gorgeous, much more ridiculous, much more deceptive or much more effective than this enormous fantasy. A city of tents was erected on the plain north of Delhi, and there a kind of gigantic morality play was enacted. The setting was symbolic in itself, for while the durbar ground was very old, and dusty, and Indian, the durbar camp was all new, progressive and British. It was laid out to a practical grid, it had well-paved roads and excellent water-supplies, and everywhere among its white marquees ran the drooping telegraph wires that were the threads of Empire, the electric cables which illuminated it as brightly as London itself, and the winding rails of the tramcars.

To this exhibition of progress, ablaze upon the brown plain, all the feudatories were summoned: the bewhiskered Maharajahs of Punjab, the bold soldier-princes of Rajasthan, royalty of Nepal in peculiar hats, sleek Bengalis and beautiful Tamils, Sikhs with gilded scimitars, gaunt Baluchis with ceremonial camels, Burmese and Sikkimese and Madrasis and wondering rustic potentates from Gujarat or Kerala. All were called to the tented city beyond the Lahore Gate, and there they paraded dutifully in the great Durbar Square to swear fealty to the Crown of England.

This was Empire! Here illusion mastered reality, and theatre became life. Trumpets sounded, guns fired, soldiers presented arms, plumes waved, elephants snorted, jewels glittered, cameras clicked ('nearly everyone had a kodak,' wrote one participant, 'even many of the natives themselves'). Here were aged veterans of the Indian Mutiny, led by a blind centenarian who, we are told, 'turned his sightless orbs towards the cheering and feebly saluted.' Here were standard-bearers and heralds, scrolled and tabarded, and High Court judges in their wigs. Here were the twelve State trumpeters, and the twelve military bands, and the 40,000 parading soldiers. And here upon his slender-pillared dais stood the Viceroy of India, the Crown's embodiment, George Nathaniel Curzon of Eton and Balliol, forty-three years old, half-crippled with pain in his back, an accomplished inventor of comic verse, a well-known eastern traveller, dressed in the flamboyant accoutrements of his office and accompanied by his wife, the former Miss Leiter of Washington, in a pale blue dress embroidered with gold.

The trumpets sound. The drums roll. The regiments present arms. As the vast polychromatic crowd rises thunderously to its feet, slowly, rather muffled there fall upon that dry air the first solemn notes of the British National Anthem—so dignified, so old, so far from home, so simple in that exotic setting, so touching, so profound, that the very soul of India seems to be stirred.[1]

---

[1] Or so the British thought. Gandhi says in his autobiography that some of the Maharajahs were ashamed of the ridiculous clothes they had to wear for the occasion—'we alone know the insults we have to put up with in order that we may possess our wealth and titles.' As Gibbon observed of a not dissimilar function, Elagabalus' presentation of the sacred black stone in Rome in 219, 'the gravest personages of the state and army, clothed in long Phoeni-

8

Chamberlain never visited India, and he might have found all this preposterously irrelevant to the times. Our proud paterfamilias would have adored it, and his son's schoolmasters probably pinned pictures of it on the notice-board in the corridor. Curzon himself was highly satisfied with the arrangements,[1] and that blind old hero of the Mutiny was so excited by it all that the very next day he took to his camp-bed and died. But before the end of the century, anyway, the behaviourist fallacies of Empire were to be exposed, and its high style irrevocably chastened, by a catastrophe of the imperial story, the second war against the Boers.

---

cian tunics, officiated in the meanest functions with affected zeal and secret indignation.'

The Delhi durbar ground, first used for the proclamation of Queen Victoria as Empress of India in 1877, was later the scene of a Coronation Durbar for George V, and is now a bleak and generally deserted pleasure-ground called Coronation Park. Several statues of imperial worthies have been taken there from elsewhere in the city, notably the elongated effigy of George V, in Coronation robes, which used to be a focal point of New Delhi. The equestrian Durbar statue of Edward VII, though, has been removed from India altogether, and re-erected by subscription in Queen's Park, Toronto.

[1] Having prudently forbidden the playing of 'Onward, Christian Soldiers' —'Crowns and thrones may perish, Kingdoms rise and wane . . . .'

CHAPTER FOUR

'The Life We Always Lead'

IN the immensity of the South Atlantic, blazing hot and slowly heaving, one Victorian steamship slowly overtakes another. On the liner *Dunnotar Castle* is General Sir Redvers[1] Buller, VC, one of the boldest generals of the British Army, who is on his way to assume command of the imperial forces fighting the Boers in South Africa. On the old transport *Nineveh* are Australian volunteers on their way to join his armies. Slowly the two ships close, until they are almost within hailing distance, the decks and riggings are crowded with waving men, and the general, tearing himself away from his maps and campaign plans, emerges portly and beaming from the converted ladies' dressing-room he is using as an office.

From the rigging of the *Nineveh* somebody is signalling with a flag, and word by word the message is interpreted on the liner: *Is—Sir—Redvers—Buller—On—Board?* Yes, goes back the answer, and robustly across the gap between the ships, above the swish of the waves and the pounding of the engines, come three lusty Australian cheers, to a waving of wide-brimmed hats and a ribald whistle or two. The general is much gratified, the *Dunnotar Castle* responds with a hoot of its siren, but as the ships draw apart again, and Sir Redvers returns to his calculations, a second signal is flagged from the *Nineveh*. It is harder to read this time, for the gap is widening fast, but a hundred pairs of binoculars are raised upon the liner, poop to bridge, as the letters are spelled out. Is it a message of loyalty or good luck? Is it a patriotic slogan? No, it is another inquiry, hardly less topical than the first: *What—Won—The—Cesarewitch?*

In such a spirit did the soldiers of the Empire go to war against the Boers in the autumn of 1899—cocky after a century of easy

[1] Pronounced 'Reevers'—or later in his career, 'Reverse'.

victories, secure in their tribal jokes and customs, confident in their leaders, anxious only about the racing results: but though the Empire had a population of 370 million, and there were not much more than 100,000 Boers altogether—though General Buller had 85,000 men at his disposal, and the Boers only 35,000—though the Army believed it could end the war by Christmas, and it was the ambition of every British officer to get to the front before then—still the campaign to which they were so boisterously sailing marked the beginning of the end of their Empire, and the first faltering of their pride.

Sir Redvers sailed on, chuckling and drinking champagne: but nothing would be quite the same after the Boer War, and even Queen Victoria, as if recognizing it to be the end of her era, died when it was half-way through.

2

The British and the Boers were old enemies. They first confronted each other after the Napoleonic wars, when the British acquired the Cape of Good Hope, and they had been skirmishing on and off ever since. Before the arrival of the British the Cape had been colonized, under the Dutch flag, by a community of mixed Dutch, Flemish, German and French Huguenot stock, known then as Boers ('farmers') and later as Afrikaners ('Africans'). These varied settlers had long ago coalesced into an all too recognizable unity, an African tribe in fact. Tight-knit, traditionalist, racialist, individualist, devout in a severe Calvinist style, the 'Volk' were fundamentalists in life as in faith. They were suspicious of all change, determined to live by their own ideals, and convinced of their unalterable rights under a God of absolutes.

From the start the British found them a terrible nuisance, for South Africa played an important part in the imperial thinking. It stood pivotally upon the Empire's eastern trade routes, and it contained a large population of black pagans, towards whom the best of the British felt themselves to stand in the condition of Godparents. The Boers, though, were irreconcilable. Some did settle into the British pattern of things, but many more wanted nothing of the

Empire and its liberal humbug, being perfectly convinced that the white man would be forever superior to the black, and that the Negro was divinely ordained to be a permanent hewer of wood and drawer of water. Throughout the Victorian century the most resolute of them had been withdrawing stage by stage from the British presence. By 1890 some 100,000 had out-distanced the imperial expansion, and were established in two more or less independent Boer republics, far in the South African interior—the Orange Free State, with its capital at Bloemfontein, and the Transvaal with its capital at Pretoria. Spiritually they had retreated farther still, and were living up there in *laager*, as they would say, ramparted by the wagons and earthworks of their dogma.

The British could not tolerate such anachronisms within their South African paramountcy. By 1898 they were settled themselves in three colonies down there, Cape Colony, Natal, and Rhodesia, founded eponymously by Cecil Rhodes and his Chartered South African Company; and just as they had annexed the territories of the Zulus and the Matabeles, the Ashanti and the Baganda, so inevitably they would one day subdue this most refractory tribe of all. Twenty years before they had briefly seized control of the Transvaal, only to be humiliated in the hilltop battle of Majuba, but since then they had established their own settlements so far to the north that the Boers were now surrounded on three sides. Since then, too, the Transvaal had been discovered to contain, in the highlands around Johannesburg, the world's largest deposits of gold.

In those days gold played more than a practical, almost a mystic part in the affairs of the nations: that this vast new supply should fall within a British sphere of authority seemed to the imperialists, not to speak of the City speculators, almost a divine dispensation. So by the last years of the century strategy, morality, economics, instinct and plain greed made it inevitable that the Boer Republics must be tidied up beneath the Crown—'sooner or later', as Winston Churchill wrote, 'in a righteous cause or a picked quarrel . . . for the sake of our Empire, for the sake of our honour, for the sake of the race, we must fight the Boers.'

The ostensible *casus belli* was the presence within the Transvaal of a large foreign population, much of it British. These were the men who worked the gold-mines, and they actually formed a majority in the Republic. They paid 80 per cent of the taxes, they mined the gold, but they were allowed no rights of citizenship, being treated by President Kruger of the Transvaal with a disagreeable mixture of contempt and suspicion. In 1895, just before the Jubilee, Cecil Rhodes had connived in a conspiracy to overthrow Kruger by a *coup*

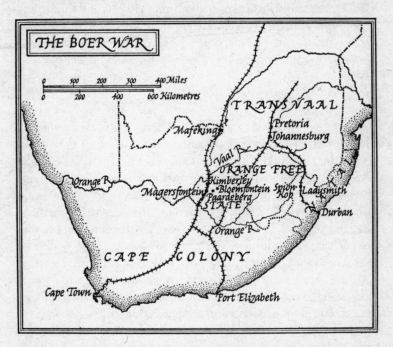

d'état, with a rising in Johannesburg and a filibustering invasion from British territory: Chamberlain the Colonial Secretary turned a blind eye, but the Jameson Raid ended anyway in fiasco and ignominy.[1] By 1899, with British ambitions all over Africa reaching a peak of energy and fulfilment, the issue had gone beyond plot or maverick, and war came about scarcely by intention at all, but in the

---

[1] 'Ah!' said the Public Orator of Oxford University, in Latin, of this sorry adventure, 'let not excessive love of country drive to rashness, and do not resort more than is proper to alliances, stratagems and plots!' He was addressing Rhodes, who was getting an honorary degree nevertheless.

natural course of events. It had to happen. The Boers were in their last encampment, the British at the apogee of their imperial advance.[1]

The Boers actually started the war, by a preventive invasion of British territory. On October 11, 1899, after presenting an impossible ultimatum, they crossed the frontiers of Natal and Cape Province to invest the railway towns of Ladysmith, Mafeking and Kimberley. Within the week Sir Redvers was on the high seas with the three divisions of his Army Corps, twice the size of Wellington's army at Waterloo, in eager expectancy of greater fame and further glory.

3

On the face of it the odds against the Boers were farcical, which is why they pinned their hopes upon a sudden attack. If they could overwhelm the weak British garrisons in South Africa, they might force the Empire to agree to their own independence before reinforcements could arrive. In fact the disparities were more of scale than of effectiveness. Two very different forces clashed in the Boer War, and they might have come from different centuries, the one looking back to Omdurman, the Indian Mutiny and even Waterloo, the other looking forward to Ypres, El Alamein and even Sinai.

The British Army in 1899 was essentially an imperial force, accustomed to colonial wars against primitive opponents, fought at great distances from home along complicated supply routes. Within its specialized limits it had been very successful, but since the Crimean War the only European enemies it had faced had been the Boers themselves, who had effortlessly trounced it at Majuba Hill, and its famous victories had mainly been over Asian and African primitives. It had no general staff, and only two intelligence officers to keep in touch with military affairs throughout the Empire. Its rigid conceptions of class made for discipline and unshakeable camaraderie, but reduced the private soldier to a willing cipher. His was just to do or die. It was an army instinctively drawn to the battle-

---

[1] There are fuller accounts of all these events—the Great Trek of the Boers, Majuba, the Jameson Raid—in the opening volume of this trilogy, *Heaven's Command* (London and New York, 1973).

square and the close-order advance, those glorious specialities of British arms since the days of the great Marlborough. It accepted change with great reluctance, and though it was now armed with machine-guns and repeater rifles, it tended to use them like cannon and muskets. It went into action with bands and pipers, had only recently abandoned the red-coat as standard battledress, and put great store upon its magnificent roll of battle-honours, fought for through many generations in every clime and country.

The Boer Army, on the other hand, was hardly an army at all, and its chief battle-honour was Majuba—Majuba Day, February 27, was one of the great secular festivals of the Boer year. It was simply the Boer manhood *in toto*, mustered in local mounted units called commandos, owning its own horses, electing its own officers, wearing its own casual interpretations of uniform, and relying heavily upon the inbred fieldcraft, horsemanship and enterprise of its individual soldiers. Its discipline, like its morale, was variable: often its soldiers drifted away from the battlefield home to the farm, or decided to try their fortunes on another front, or demanded new officers. The Boers had armed themselves, though, with the most modern equipment from European arsenals, they kept open minds on military matters if on no others, and above all they were born to the terrain. They were a nation of horsemen, hunters, trekkers, pastoralists of the open veld. Since every Boer male went to war, among their officers were men of striking intellect, too, who were quicker by far than their professional British opponents. All in all they were born irregulars, perhaps the best guerilla soldiers in the world.

These were opponents, then, different in kind. They had much in common nevertheless, and in particular they shared an emotional sense of brotherhood and purpose. Each army was secure in its convictions, and each was bound by a tribal trust and integrity. When, at the start of the war, the Transvaal commandos assembled in Pretoria to collect their arms and orders, the occasion was likened to the gathering of a huge family. From every part of the republic they came, schoolboys to grandfathers, riding their tough and shaggy ponies with a shambling ease, their saddlebags bulging with biltong, their slouch hats hard over their heads, greeting friends and relatives everywhere as they rode into the little capital, and saluting President Kruger himself, as they passed his modest frame house on Church Street, as they might greet a family patriarch on the farm.

On the other side the fellowship was just as strong, if less egalitarian. 'We'll do it, sir! We'll do it!' cried the soldiers when, early in the war, Colonel Ian Hamilton told them they would send the newsboys crying victory through the streets of London. And here is the coded message by which an approaching British column declared itself to the besieged British garrison at Mafeking: *Our numbers are the Naval and Military Club multiplied by ten; our guns, the number of sons in the Ward family; our supplies, the O/C 9th Lancers.* No Boer alive could crack this impenetrable cipher, but almost any British officer in Mafeking could interpret it. Everyone knew the Naval and Military, the 'In and Out', was No 94, Piccadilly; most people knew the Earl of Dudley, Bill Ward, had five brothers; and anybody in a decent regiment was aware that the colonel commanding the 9th was that very nice fellow Malcolm Little.[1]

4

As Buller sailed out to Africa, *Punch* published a cartoon of two London urchins discussing the war. 'The Boers will cop it now', one was saying to the other. 'Farfer's gone to South Africa, *an' tooken 'is strap!*' The British saw the war at first as a punitive campaign like so many others, in which a recalcitrant tribe of the frontiers was to be summarily brought to heel. In the event the fighting lasted three years, and required a great deal more than farfer's strap.

First the Boers, investing the three railway towns, pushed deep into Cape Colony, hoping that the Boers living there under British rule, the 'Cape Dutch', would join their cause in rebellion. The British, when they had assembled their expeditionary force, responded with two main counter-attacks: out of Natal to relieve Ladysmith, through the Orange Free State to relieve Kimberley, where

---

[1] Hardly less tribal was the message once urgently flashed by heliograph, uncoded, to the half-starved garrison of Ladysmith: SIR STAFFORD NORTHCOTE GOVERNOR OF BOMBAY HAS BEEN MADE A PEER.

Rhodes was shut up in his own diamond fields, and Mafeking. Both these thrusts, which were meant to converge upon Pretoria, disastrously failed. After a succession of defeats which became known as Black Week, 1899, Buller was replaced as Commander-in-Chief by the aged and adored Lord Roberts of Kandahar. With him as Chief of Staff there inevitably arrived, fresh from his Governor-Generalship of the Sudan, Lord Kitchener of Khartoum, by now easily the most imperial of all the imperial soldiers.

In the New Year Roberts opened the second phase of the war with a massive and skilful offensive directly up the railway line to Pretoria itself. Bloemfontein, Johannesburg and Pretoria were all captured, the three railway towns were relieved, the two republics were officially annexed to the British Empire and Kruger fled to Europe.[1] Ten thousand British soldiers sang Kipling's *Recessional*—'Lord God of Hosts, be with us yet!'—in a victory ceremony outside the Volksrad, the Parliament of the Transvaal.[2] The war seemed to be over; but instead the Boers transformed it into a protracted guerilla campaign, in which the huge British armies, repeatedly reinforced, were harassed by roaming, self-supporting commandos, while raiding columns struck deep into Cape Colony. Kitchener, assuming the command from Roberts, beat them in the end only by ruthless and laborious methods of attrition, burning their farms, herding their women and children into detention camps, and criss-crossing the entire countryside with interconnecting blockhouses.

[1] He died in Switzerland in 1904, but his body was brought home to Pretoria, and he lies now in the Old Cemetery, West Church Street, not far from Queen Victoria's grandson Prince Christian Victor of Schleswig-Holstein, who died of enteric fever in 1900 while serving with the British Army against the Volk.

[2] And at home a magazine for girls held a competition for limericks to commemorate the victory. Here is one of the winning entries, kindly sent to me by Mr C. P. Wright of Wolfville, Nova Scotia:

> *There was an old man of Pretoria*
> *Who said 'My! How I pity Victoria,*
>   *Oh, summon the ranks*
>   *And let us give thanks!'*
> *But something went wrong with the Gloria.*

When at last the Boers surrendered, in May 1902, 20,000 commandos were still in the field, but both sides were exhausted and embittered. The British had suffered terribly from heat and disease: of their 22,000 deaths, two-thirds were from cholera and enteric fever. The Boer guerillas ended the war half-starved and virtually destitute, and their families were decimated by the appalling conditions of Kitchener's detention camps: 24,000 Boers died in the war, but 20,000 of them were women and children. Before the war was over Buller's expeditionary force of 85,000 men, sailing out so confidently to their victory by Christmas, numbered 450,000 men, the largest British Army ever sent overseas.

Such were the bare bones of it: but no military summary can do justice to this lacerating and ironic war, fought against a magnificent backdrop of veld and mountain, by enemies whose dislike was often turned to admiration, and whose hostile causes were full of paradox. The Boer War came to be called 'the last of the gentlemen's wars'. This is because it was fought, in its first stages anyway, to a set of conventions, based upon the Christian ethic but shaped too by the physical and historical setting. These were white men fighting, and they were watched in all their sieges and manoeuvrings by the silent black mass of the indigenes.

To the Boers, though, the war was a climax and an ultimate challenge. To the British it was only another stage in the long march of Empire. During a local armistice on the Tugela River in February, 1900, while the dead and wounded were being recovered from the battlefield, a Boer soldier engaged a British officer in conversation. 'We've all been having a rough time', he remarked. 'Yes, I suppose so,' replied the other, 'but for us of course it's nothing. This is what we're paid for. This is the life we always lead—you understand?'

'Great God', simply said the Boer.

5

Let us peer through our field-glasses (Dollond and Aitchison, By Appointment to the Duke of Cambridge), at four of the most significant Boer War battle-grounds—each offering its own dramatic unity, each to be immortalized in legend.

Look first at Spion Kop. There it stands in our lenses now, a bulky flat-topped hill above the Tugela, in northern Natal: grander and more imposing than its neighbours along the ridge, bare, silent, and looking much higher than its 1,500 feet. On a brilliant summer day in South Africa nowhere could be more suggestive of heroic purposes—the shadows slowly moving along the flank of the Hill, the brown Tugela winding at its feet, the doves cooing gently among the shrubby trees, the scent of flowers and dry grass, the rolling mass of uplands stretching away eastward into Zululand, westward to the great massif of the Drakensberg. Here and there the beehive kraals of the black people are scattered around the landscape, but there is no sign of human life on the mountain: only the buck and the wild turkey live up there, and the bees hum among the mimosa.

General Buller, leading his southern army to the relief of Lady-smith in the last week of January, 1900, was persuaded that this hill was the key to the beleaguered town, twenty miles beyond. His reasons were vague, his plans unformulated, but handing over executive command to his subordinates, he set up his headquarters on the south side of the river, and ordered an assault. British maps of the area were rudimentary—five miles to the inch—and nobody really knew what shape the mountain was, or what lay immediately beyond it: nevertheless on the night of January 23 an assault column climbed a steep spur to the summit, and overwhelming a small Boer picket up there, raised three cheers in the darkness and dug itself in as best it could in the sloping stony plateau at the top.

A heavy mist hung around, blanketing everything in damp obscurity, but so far as they could tell the British were masters of the mountain: they settled uncomfortably in their shallow trenches to await orders and reinforcements from below. When the sun came up, though, and the mist cleared, they found to their horror that Spion Kop was not as they supposed. The small green triangle of the summit, perhaps an acre of gently sloping grass, was overlooked by two outlying knolls, and from these positions, as soon as the light broke, a terrible point-blank rifle-fire was opened upon the crouching soldiers. Boer artillery soon found the range, too, and fired pom-pom and high-explosive shells almost without pause into the British positions. There was virtually no cover on the plateau, only the odd boulder and hummock, so the British were entirely exposed. They had taken the mountain indeed, but they were trapped upon it.

Spion Kop was one of the most cruelly confined of all battles. Some 2,000 British soldiers were packed within a perimeter about a quarter of a mile round, without water and, as the day wore on, in blazing heat. They could scarcely move at all. The moment anything stirred, a head, a hand, a rifle, a marksman's bullet was there from the Boer positions north, east and west of them. Time and again the Boers pressed upon the plateau, at one end of the line or the other, to be beaten desperately back, but all the British could do was hang on—for them there was no hope of advance or retreat while daylight lasted.

It was like a high proscenium. Buller's entire army, massed in the valley below, could see the smoke and the shell-bursts far above, and through binoculars could even make out the figures of soldiers, crouched or stumbling through the shell-fumes, but the performance was allowed to run its course. Buller wavered, in his command post across the river: orders were muddled, mislaid, misinterpreted; at one time three different commanders all believed themselves to be in command on the hill; when an enterprising cavalry brigadier mounted a diversionary attack along the ridge, he was testily recalled and reprimanded. Twice reinforcements were ordered up the mountain, mule-trains took ammunition up, and through the day a straggle of individual officers, war correspondents and stretcher-bearers clambered like pilgrims up the summit track, its surface scratched and scored now by the bootnails of the soldiers. But nothing essentially happened. Pinned to their ground, exhausted by desperate attacks and counter-attacks, the British on Spion Kop simply sweated and died through the long day, holding an objective that had no meaning.

Boers and Britons were only a few yards apart up there, and the battle sometimes degenerated into rough-house, the men falling upon each other with rifle-butts, bayonets, boulders and even fists. One group of Lancashire Fusiliers, holding white handkerchiefs, rose from their trench to surrender. Others scrambled demoralized over the rim of the plateau, and wandered sheep-like in irresolute groups here and there across the mountain flanks. The rest fought gamely but helplessly on, and by evening their bodies were toppled one on another in their trenches, an obscene pile of boots, sun-helmets and khaki serge, twisted limbs and shattered faces. Half the force was killed and wounded, and the survivors fought on in a daze. When darkness fell, and the Boer fire ended, the senior surviving officer on the summit, Colonel A. W. Thorneycroft, declared that he and his men would fight no more—no argument would persuade him to spend another day on the summit: and so, trudging down again through the darkness with his long line of wounded, exhausted and appalled soldiers, on his own initiative he ended this squalid and futile engagement. The Boers had also withdrawn as the light failed, many of them too swearing that they would never go back, and it never occurred to them that the battle of Spion Kop had been won: but when in the small hours some of them climbed up again to look for a comrade's body, they found to their astonishment that the British had gone, and the mountain, littered with its dead and debris, stood deserted once more in the mist.

6

For a very different battlefield, look at Magersfontein, on the western front, a low and undramatic cluster of hills which stood astride General Lord Methuen's route to the relief of Kimberley. In December 1899 the British, pushing the Boers with difficulty across the Modder River, reached the wide flat plain before this modest redoubt. As Spion Kop was to Ladysmith, Magersfontein was to Kimberley: only a few miles beyond it lay, as the whole army knew, the beleaguered diamond town and the chance of glory. The Boers held Magersfontein in strength: but so wide was the landscape all around, so dun and featureless the plain, relieved only by the shine of water in a hollow here and there, or a bumpy tree-less kopje, that from a distance the whole veld might have been deserted. Behind, on the railway line, a plume of black smoke marked the arrival of a British supply train, following the army up from the Cape, and all around the station at Modder River was the dust and bustle of an army on the move. In front, on the night of December 10, 1899, the veld was motionless.

One of the most famous fighting units of the British Army, the Highland Brigade, was ordered to lead the assault on Magersfontein, commanded by one of the most famous of its fighting generals— 'Andy' Wauchope of Niddrie, Midlothian, whose gingery Scots face we last glimpsed beside the ruins of the Khartoum Residency. First, though, the ridge was heavily bombarded—two hours of shelling, by thirty-one guns, so overwhelming that Methuen himself thought nobody on Magersfontein could have lived through it. Then at midnight the kilted and bearded soldiers set out, led by the Black Watch. They were marshalled with guide ropes, and moved to a compass bearing, advancing as if blindfold, bayonets fixed, across the five miles of scrubby rough ground towards the ridge. It was a very black, wet night, with a driving rain from the north-west, thunder and flashes of lightning: in no time at all they were soaked to the skin, and the ground was turned to mud under their feet. None of the men had been told what they were going to do: they simply marched as always into the stormy night, led by Wauchope with an old claymore.

Beyond the ridge, playing on the low cloud, the violet shaft of a searchlight from Kimberley beckoned them on to battle. They moved in a dense square of ninety lines, 4,000 men in all, clutching each other's clothing to keep contact, tripping often over boulders or ant-heaps, squelching in muddy pools, catching their kilts on thorn-bushes. Often they stopped while their guides checked their compasses: and after three hours, still in dense close order, as the storm passed and the day began to break, half a mile before them they could make out the dim shape of Magersfontein. It was time to deploy the brigade, the guiding officer whispered to Wauchope— 'this is as far as it is safe to go.' The general disagreed. He was afraid the troops would lose themselves. 'I think we'll go a little further', he said.

So a little further they went, shoulder to shoulder: and so, just as Wauchope did give the order to deploy, and the Black Watch was struggling through an especially awkward patch of mimosa thorn, the Highland Brigade was suddenly blasted by the most violent fusillade of rifle-fire any British soldiers had ever experienced. The ground in front of them seemed ablaze with flame—'lit up', one Scottish sergeant said, 'as if someone had turned on a million electric lights'. Most of the fire came not from the ridge at all, but from a continuous line of trenches concealed and unsuspected at its foot. This was something new to the British, something their traditions and disciplines did not allow for, and for a few seconds the soldiers simply stood there, stunned with shock and deafened by the noise, before throwing themselves to the ground.

Wauchope was killed almost at once, with eighteen of his officers. Some of their soldiers turned in the half-light and ran, knocking over their own officers, scrambling back through the scrub and puddles towards Modder River and safety. One group of men did find a gap in the Boer line, and began to climb the ridge, but they were caught between the Boers and their own artillery, and those who did not die were taken prisoner. The rest lay where they were: and so the morning broke, and the day wore on, with the Highland Brigade pinned helplessly to their ground before the Boer trenches, ordered only to 'hold on till nightfall'.

The sun blazed down, and hour after hour the soldiers lay there. They had not eaten since noon the day before. Other troops went forward, and fighting continued in confusion all day. Guns were deployed, cavalry tried to turn the Boer flank, even the pipers went into action: but at last the Scottish soldiers broke, and getting to their feet spontaneously but with a strange deliberateness, almost as if they had been given an order, turned their backs on the enemy and ran—'for all their worth,' as an eyewitness said, 'officers running about with revolvers in their hands threatening to shoot them, urging on some, kicking on others, staff officers galloping about giving incoherent and impracticable orders. . . .' By sunset it was all over, and the horse-drawn ambulances went out under a brilliant moon to pick up the wounded still left lying in the veld. The British never tried to take the ridge again, and when the time came to renew the advance, went another way.

7

Two British failures: for a famous British success, let us inspect the besieged railway town of Mafeking, away to the north. It was really hardly more than a village, and the Boers did not invest it very resolutely, but it was defended so jauntily by its dapper commanding officer, Colonel Robert Baden-Powell, and projected its own self-image so successfully across the world, that its very name became synonymous with British pride and spirit. Some 1,200 Britons were shut up there, with about as many black Africans, and though they did not really suffer very greatly (except the Africans, who nearly starved), and made no real effort to get out, still they did defend the place with true panache.

It was a tiny place, a square, a church, a station, a couple of hotels, a grid of half a dozen streets, clustered in greenery around a muddy river in the heart of the high veld, and throughout the siege it retained some of the English village spirit. Baden-Powell was the undoubted squire of Mafeking, the life, soul and character of it all, around whose cheerful conceited figure everything revolved. The centre of activity was Dixon's Hotel in Market Square, with the horses at its hitching-posts and the loungers on its verandah, and from there every kind of enterprise was mounted. It might be a foray into the enemy lines. It might be a jolly ruse to deceive the Boer sentries. It might be a fancy-dress ball for Sunday evening, or a comic couplet for the *Mafeking Mail*. Numerous swells were invested in Mafeking. Baden-Powell's Chief of Staff was Lord Edward Cecil, lately of Fashoda, and ensconced in a beflagged and cushioned dug-out was Lady Sarah Wilson, daughter of the Duke of Marlborough and Winston Churchill's aunt ('Breakfast today horse sausage', she wired home in April, 1900. 'Lunch minced mule and curried locusts. All well').[1] 'B-P' dominated them all, though, from his headquarters next door to Dixon's, and he gave to the defence a perky humour that caught the fancy of the world.

He disseminated it carefully, in a flow of vivacious and not

[1] Communications were never broken, though the London *Times* correspondent in town complained that the postal service was quite deplorable.

always strictly accurate messages home. If he made things in Mafeking seem more desperate than they were, that did not detract from the tonic effect it all had upon the spirits of the people at home, or its propaganda value elsewhere; at a time when Black Week had profoundly depressed the nation, and sadly damaged British prestige in the world, Mafeking was like a breath of the old allure. 'B-P''s cocky despatches recalled the heroic eccentricity of Gordon at Khartoum. His hard-pressed garrison, hemmed in by Mausers, showed just the same grit as the heroes of Rorke's Drift, jabbed about by assegais in the Zulu War of '79. The presence of women and children recalled the tear-jerkers of the Indian Mutiny, and the attendance of patricians too, Lord Edward, Lady Sarah, Charles Fitzclarence of Munster, the Hon. Algernon Tracy and several members of the In and Out, was an assurance that British imperialism still had *class*.

So through the days of ignominy Mafeking, far away on the Bechuanaland border, brilliantly kept the legend of Empire alive. It is true that Ladysmith was besieged more fiercely, suffered more terribly and resisted just as bravely, besides being a far more important objective. Mafeking, though, did it all with style, and style at that moment the Empire badly needed. The whole world came to know the confident figure of Baden-Powell, whistling with his telescope on his precarious lookout tower beside Dixon's: and it was wonderfully true to the Mafeking myth that when, after eight months, the first men of the relieving force clattered into the outskirts of the town, they got a distinctly laconic greeting from the first citizen they met. 'Ah yes,' was all he said, 'we heard you were knocking about.'

8

For the saddest Boer humiliation of the war we must make our way along dusty veld tracks, through low hills prickly with thorn, to the infinitesimal hamlet of Paardeberg, on the Modder River in the Orange Free State. There, at the climax of Lord Roberts's campaign, the main Boer army, with all its wagons, animals, women and children, was surrounded in its *laager* in the river-bed. It was February, 1900, the height of the South African summer. The weather was hot and heavy, with thunderstorms now and then, and black clouds piled often over the southern horizon. As usual, the Boers lay there very low. To the British, drawn up north and south of the river, nothing showed in the open veld but the snaking course of the river itself, tangled with shrubbery, and the smoke of the Boer encampment buried in its green ravine. Within the *laager*, though, General Piet Cronje's army was tensely concentrated. It was the very epitome of the last-ditch stand, down there in the airless river-bed. It was an allegory of Boerness.

On the crest of the river-banks the commando marksmen were entrenched, with clear fields of fire up the gently rising scrubland to the open veld. Behind them in the shaly gorge all the paraphernalia of the army was jammed this way and that—wagons tilted on the shingle, piles of ammunition boxes in the muddy lee of the banks, gun-litters and field kitchens, hospital tents, horses tethered restless among the trees, twitching their tails against the flies. Rough shelters had been scooped out in the bluffs, and there the women in their poke bonnets, the children in their grubby prints and frayed trousers, sheltered behind awnings of old canvas. The men lived and slept in their wagons, or in bivouacs at the river's edge, or at their guns. Their *laager* was two miles long, like a trench-grave in the veld.

For ten days the Royal Artillery tried to blast the Boers out of this place. The river-bed was thick with cordite fumes, rubble, wrecked wagons, the smoke of burning wood, the stink of dead horseflesh, and sometimes the Boers could see, in the patch of sky between the trees, the round red shape of an observation balloon, like death's scrutiny. Several times the British attacked frontally across the veld, to be beaten back with fearful losses, and the women crouching in their dug-outs could hear the rifle-fire almost above their heads. All hope of relief was lost, but Cronje, a huge, tragic, shambling figure of a man, declined offers of safe conduct for his non-combatants and refused all calls to surrender—'During my life-time I will never surrender. *Dixi*.' Instead day by day the Boers fought back sullenly and despairingly, weaker each hour, shorter of food, shorter of sleep, disillusioned in their river-bed.

Paardeberg was the greatest single reverse in the history of Boer arms. When the spirit broke at last, and on February 27 Cronje, in his wide hat and shabby green frock-coat, climbed out of the ravine to surrender to Lord Roberts, he did not reply to the victor's courteous greeting—'You have made a gallant defence, sir': and when his 4,000 ragged and half-starved burghers filed into captivity with their wives and children, carrying blankets and bundles of possessions, some with umbrellas, they looked less like an army than a band of dispossessed peasants, and the British soldiers watched them go with mingled pity and amusement. It was Majuba Day, the proudest day in the Boer calendar; when Cronje himself rode away to prison camp, he travelled stormy-faced and erect in a Cape cart, Mrs Cronje implacable at his side.[1]

9

'Say, colonel,' inquired an American observer, watching a British battalion prepare yet another frontal assault upon yet another impregnable hill, 'isn't there a way *round?*' The story of the Boer War is full of such sudden pungencies. It was not war on the gargantuan twentieth-century scale, huge conscript armies pursuing inconceivable objectives. It was war recognizably between *people*, fighting for targets all could see, commanded by generals everyone knew, animated by public emotions. 'It is our country you want!' Kruger once cried to a visiting Englishman, tears falling down his

[1] Nothing has greatly changed, in these four battle-grounds. Though Dixon's Hotel has been demolished, and the African township is today the capital of the Tswana 'homeland', Mafeking remains much the same little *dorp* it was in 1900: the brass-bound steam trains still puff away to the Rand, some of the defence positions can still be traced, and in the station yard a statue of Rhodes gazes wistfully up the track in the general direction of Cairo. At Paardeberg no memorial marks Cronje's last *laager*—unlike the British, the Afrikaners prefer to forget their defeats—but people still find shell fragments and sad mementos in the river-bed: on the Magersfontein ridge a series of monuments have been erected, including one to the Highland Brigade—'Scotland is poorer in men, but richer in heroes.' Spion Kop, though, is the most memorable of them. Though it too is crowned by a clump of memorials, and though the line of the British trench is marked by whitewashed stones, it is a marvellously peaceful and gentle place, and of all the battlefields I have visited, seems the most truly regretful.

cheeks. 'I thank God,' General Sir George White told his soldiers, when Ladysmith was relieved at last, 'I thank God we have kept the flag flying.'

It was war in the open, self-explanatory. Here, for instance, in the miserable last stages of the conflict, we see through the eyes of a young Boer soldier a scene somewhere in the northern Cape, where a ragged and emaciated commando is desperately trying to evade the British net. In terrible weather the Boers are riding deeper and deeper into Cape Colony, in the hope of getting help from the Cape Dutch, but by now they are almost beaten, hungry, wearing rags and home-made sandals, driving their bony horses in endless night marches across enemy territory. On a stormy night in September they reach a railway line, and as they do so a train approaches. Shivering with cold and wet, the Boers throw themselves on the ground as it thunders by: and out of the blustery darkness, through the rain, they catch a glimpse in lighted windows of clean and well-dressed British officers, laughing over their dinners in the restaurant car. In a moment it is past, leaving only the smell of its steam and the rattle of its wheels on the lines: and so the Boers, rousing their horses, wrap their rags around them and cross the railway track into the night.

Or here a British medical officer waits at night for the ambulance wagons to reach his hospital, No 4 Stationary Field Hospital, out of the Tugela battlefields. Two white lanterns have been hung from a flagstaff outside the hospital tents, to guide the drivers, and one by one the ungainly hooded wagons, each drawn by four horses, lurch and sway out of the darkness with their loads of maimed and dying. Usually they come silently out of the veld, but once the doctor hears, as a wagon waits outside the lines for the stretcher-bearers to unload it, a repeated half-conscious cry from a man inside. 'Can you see the two white lights yet, Bill? Where's the two white lights? Can you see them yet, Bill? Can you see them white lights? . . .'

And here, almost at the end of the war, the Boer General Jan Smuts, with his harum-scarum escort of guerillas, meets his chief opponent, General Lord Kitchener of Khartoum, after months of hit-and-run warfare across the immensities of the veld. Smuts and his companions have been brought for negotiations by train from the south, in a journey that has lasted a week—by day cautiously across the veld, for the guerilla war is still being waged, at night with the train's searchlight constantly sweeping the landscape. At last they reach the little station of Kroonstad, in the Orange Free State, where Kitchener is to meet them. The train draws into the station, and stands hissing at the platform in a sudden silence. The heat is intense. The Boers, thin, tired and sun-blistered, watch expectantly out of their compartment windows.

Presently the British general arrives, just as he arrived in his gunboats off the Fashoda fort. Reddish-brown from sun and long campaigning, expressionless as always, his blue eyes cold above his famous moustache, he rides into the station yard on a black charger. Around him ride, in the stiff English manner, three or four staff officers. Behind are a couple of willowy aides. And as they clatter up the yard the Boers see with astonishment—with admiration too, perhaps, even with a touch of awe—that behind the officers rides a bodyguard of turbanned Pathans from the north-west frontiers of India, crimson-jacketed, with jackboots polished like glass and gold-mounted scimitars at their sides. The grand illusion has arrived once more, at the Kroonstad station yard.

10

It was a war of striking personalities, and in power of character the two armies were well matched. It is the British, however, that we are concerned with, and it may be said that in the Boer War, as in most of their imperial conflicts, they got the leaders they deserved. Grandiose there upon the pinnacle of their power, they were represented in South Africa by an extraordinary gallery of originals, good or bad.

Several of the great stars of Empire were there, of course, all behaving characteristically. Roberts, 'Our Bobs', brought with him from England a Union Jack sewn by his dear wife to be hoisted over Pretoria. Kitchener, 'Old K', put on weight during the campaign. Rhodes, 'The Colossus', nearly came to blows with the British military commander of Kimberley, but was gratified to get a telegram from the Kaiser congratulating him on its defence.

Buller, 'Sir Reverse', remained through all his misfortunes the most beloved of the generals—though his soldiers often laughed at him, and nicknamed him the Tugela Ferryman because he led them back and forth across that river so often, still they never let him down. He could, so one of his colonels wrote, 'by a short unintelligible address send his defeated and diminished army merry and confident back to camp', and when he was relieved of his command they apostrophized him with a song of consolation:

> Cheer up, Buller my lad,
> Don't say die.
> You've done your best for England,
> And England won't forget.
> Cheer up, Buller my lad,
> You're not dead yet.[1]

There were many soldiers less famous but no less singular. General Sir Charles Warren, for instance, had once been Commissioner of the London Metropolitan Police, in which post he became famous for failing to catch Jack the Ripper and for issuing his orders in rhyming couplets.[2] Colonel 'Sam' Steele of Strathcona's Horse had spent thirty years as an officer of the Canadian Mounted Police, and liked to reminisce in his veld bivouac about Big Bear and the Frog Lake Massacre. General the Earl of Dundonald had been entrusted by his grandfather, Admiral Lord Dundonald, with a Secret Plan for the salvation of the nation in case of extreme emergency, and was constantly troubled by the thought that the moment for its disclosure might have arrived.[3] General Sir Henry

[1] Though presently retired on half-pay, under the shadow of his failures in South Africa, Buller remained a popular national figure until his death in 1908—hot-tempered, bibulous and jolly to the last.

[2] For example: *The Commissioner has observed there are signs of wear/On the Landseer Lions in Trafalgar Square./Unauthorized persons are not to climb/On the Landseer Lions at any time.*

[3] It turned out to be, when he did reveal it fourteen years later, a system of smoke-screens much used in the First World War. Dundonald was an ingenious inventor himself, once having himself pulled across the Thames in a watertight bag of his own design, while his grandfather was better remembered as Lord Cochrane, Father of the Chilean Navy and honoured eponymously by Chilean warships ever since.

Colvile was the author of *A Ride in Petticoats and Slippers*. Colonel Ian Hamilton kept his notes in a French shorthand of his own invention. General Wauchope, who died at Magersfontein, was one of the richest men in Scotland and had been Gladstone's opponent at Midlothian in the general election of 1892. Lord Rosslyn of the Blues was the original of *The Man Who Broke the Bank at Monte Carlo*.

Lords Dunraven, Paget, Lathom, Lovat and Donoughmore all raised private regiments, whose troopers paid their own passage to South Africa, and donated their pay to the Widows and Orphans' Fund. The ubiquitous Winston Churchill covered the war for the *Daily Telegraph*, now and then fighting a spirited skirmish on the side, and once escaping with great publicity from a Boer prison camp ('He's a fine fellow', wrote Buller to Lady Londonderry. 'I wish he were leading regular troops instead of writing for a rotten paper.') The young Mahatma Gandhi, then living in Natal, was one of the stretcher-bearers who toiled up Spion Kop. Arthur Conan Doyle, creator of Sherlock Holmes, ran a field hospital. Rudyard Kipling worked on an army newspaper. Edgar Wallace the novelist scooped the world for the *Daily Mail* with the news of the final peace agreement. Mary Kingsley the great West African explorer died of fever nursing wounded Boers. Even Queen Victoria herself, though she did not visit the battlefields physically, was *certainly* there in the *spirit*, and she sent to each of her soldiers, on New Year's Day, 1900, a tin of chocolate as a souvenir, with her crowned head on the lid, and the message 'I wish you a Happy New Year.'

11

It was a bitter war, ending more bitterly than it began, but like most wars it was bitterest away from the fighting. There were bigotries on both sides. Passions were crude, abuse was often elementary. Kipling did not hesitate to attack Kruger personally in his famous anathema *The Old Issue*—

> *Sloven, sullen, savage, secret, uncontrolled,*
> *Laying on a new land evil of the old . . .*

The Boers and their supporters, in return, loved to represent Queen

Victoria as a bloated, bug-eyed and probably dissolute old harridan, while both sides were assiduous in spreading atrocity stories about each other, generally untrue and often wildly unconvincing.

It was the first of the propaganda wars. Every incident in the field, flashed across the world by the electric telegraph, was magnified or distorted to prove a point or support an ideology. The whole world joined in this new excitement: the Boer War was the Algeria or Vietnam of its time. When, in Black Week, the British armies were so disastrously defeated three battles in a row, half the world laughed or cheered at their discomfiture: six months later, when Mafeking was relieved, the other half responded with such hysterical celebrations that the name of the little town went briefly into English usage—'Mafficking, indulging in extravagant demonstrations of exultation.'

But they Mafficked far more boisterously in Piccadilly than in Mafeking itself, and the Boers celebrated their victories only with thanksgiving to God. In the field a sad and incoherent comradeship often linked the fighting men. British doctors regularly attended the Boer wounded, and British prisoners were generally treated with courtesy. When the Boer General Kroos de la Rey went to war from his home town of Lichtenburg, in the Transvaal, he left instructions that a plot in the town cemetery should be reserved for the honourable burial of British soldiers killed in the war: his orders were obeyed, and when the time came the British dead were reverently laid among their enemies.[1]

For though the propagandists might argue otherwise, the behaviour of these opponents was often much alike. The Boers were fighting for independence, which they believed to be their sacred right: the British were fighting for an Empire which they thought to represent all that was best in human progress, headed by a Queen who was virtually divine herself. On the Boer side was the sense of 'Volk', a sense of earth, livelihood and immediacy. On the

[1] They are overlooked now by a bust of de la Rey himself, who is buried close by, and by a monument to his eldest son, who died of wounds at the Modder River in 1899. When I was there in 1975 I thanked the gardener for tending the British graves with such care. 'So long as you're satisfied', he gently replied.

British there was a noble brotherhood between officers and
uncomplaining acceptance of misery, a touching devotion
the most incompetent generals—'if he's not worth follo\
another British private wrote of the Tugela Ferryman, 'I don't k.
who is.' Both sides drew a kind of magic from the past. 'They ha
taken away our Majuba Day!' cried President Kruger in anguish
when he heard the news of Paardeberg. 'Now quietly, lads,' said
Colonel 'Bobby' Gunning of the King's Royal Rifles, addressing his
NCOs on the eve of battle, 'remember Majuba, God and our
country.' Time and again we read of chivalries in the field—a
wounded enemy given free passage, a parched patrol allowed to
water its horses, the exchange of wounded, cheerful repartee by
heliograph across the lines. Sober in victory as in defeat, the Boer
soldiers never crowed over their enemies, and were frank in their
admiration of British qualities. As for the British regulars, they came
deeply to respect the best of the Boers, and cherished the very name
'commando' to use one day for themselves.

These were Christian armies, fighting each other at the end of the
Christian era; Boer and Briton shared a trust in many old truths, and
a homely familiarity with the prophets and patriarchs of their creed.
In one of the Natal battles a middle-aged British officer, Major
Charles Childe-Pemberton of the South African Light Horse, was
ordered to lead an assault upon a hill. He was portly and greying,
having retired from the Royal Horse Guards some years before, and
was a popular racing man, known always by his racecourse nickname, Monsieur L'Enfant, or The Child. Before the battle he
confided in his brother-officers that he had a premonition of death,
and asked them to see that on his grave was inscribed verse 26,
Chapter 4 of the Second Book of Kings. The attack was made; the
hill was taken; Major Childe-Pemberton, laughing at his own
presentiment, was hit in the head by a scrap of shrapnel, and died
on the spot.

They buried him nearby, 'affectionately and reverently, in his
own clothes, just as he was', and above him they wrote his chosen
epitaph: 'Is it well with the child? It is well.'[1]

[1] He came from Kinlet in Shropshire, and the Childes own land there still,
so unfailingly rooted in Englishness that when I recently inquired after them

12

The Queen died with her century, the heroic spirit faltered, squalid images of burnt farms and diseased internment camps replaced the splendours of bugle and night march. The struggle degenerated into a messy and generally inglorious manhunt, soured by recriminations and reprisals, executions in the field, arson and broken oaths. Mile after mile the countryside was left scorched and desolate: in the internment camps the unforgiving Boer women, far from the camaraderie of the front line, nursed their dying babies. The Boers thought the British were resorting to genocide, and reproached them for betraying the white man's code by arming African scouts and sentries. The British accused the Boers of treachery, fighting as they did in civilian clothes, and disregarding many conventional laws of war.

Squat and ugly blockhouses now disfigured the landscape, 8,000 of them, one every $1\frac{1}{2}$ miles through the old Boer republics. Their protruding armoured balconies gave them an ominous mediaeval appearance, and between them thousands of fortified posts divided the country into enormous stockades, into which the commandos were laboriously penned. In one of Kitchener's drives 9,000 soldiers, 12 yards apart, formed a beaters' line 54 miles long, moving 20 miles a day, while seven armoured trains patrolled the railway tracks, and another 8,000 men manned the blockhouses all around. Across this hideous chequer-board the fugitive commandos clawed their way. They were like wild animals, Kitchener said, forever running away— 'not like the Sudanese, who stood up to a fair fight'. By the end of 1901 more than sixty British columns were in the field, but more than 20,000 guerillas still eluded them, and away in the east an exiled Government of the Transvaal, setting up its nomadic capital

a villager actually referred to the present head of the family as 'the young squire'. The Major's grave has since been moved, and is now in a small military cemetery among the outbuildings of an Afrikaner farm, near the hamlet called Acton Homes: but the epitaph remains, and the old soldier still lies within sight of his one victory (for it is thought to have been the only time Major Childe-Pemberton went into action).

in farms, woods and high valleys, survived to the bitter end. Deep within the Cape Colony, where he got within sixty miles of Cape Town itself, Smuts prayed a favourite prayer of the Griqua tribespeople: 'Lord come to our help yourself, and not your son for this is no time for children.'

The British won, of course, and the Peace of Vereeniging was concluded at their dictate in May, 1902: but the protracted guerilla campaign, the sordid anticlimax of it all, the thousands of deaths by disease or neglect, robbed the victory of any grandeur. In London the treaty was greeted far less boisterously than the relief of Mafeking two years before. The Queen had died, Rhodes had died, 'Bobs' had come home long before, Salisbury retired as soon as the war ended, Kitchener's bludgeon methods had taken the fun out of following the flags. Never again did the British go to war with the old imperial éclat, or greet their victories with their frank Victorian gusto.

The peace settlement was widely greeted as generous, especially by the British. It handsomely compensated the Boers for the devastation of their country, and it eventually gave them full equality, of law as of language, within a self-governing African union of all four European colonies. It seemed a peace of reconciliation. In this as in much else, though, the Boer War was deceptive. The treaty *was* magnanimous, but by its terms the British hoped to establish a secure, British-dominated South Africa, to establish a lasting hold over the gold of the Rand, and to ensure some measure of fair play for the black peoples of the land. The Boers were no less calculating, even in defeat. They reasoned that within a constitutional union they might one day achieve mastery not only of their own former republics, but of all South Africa, with complete control of its wealth, and with the freedom to treat their Kaffir subjects just as the Old Testament suggested.

They were right. Some Afrikaners became enthusiastic supporters of the imperial cause, and the Boer generals were greeted as prodigal sons when they visited London after the war—'Welcome to the dear old flag!' said a souvenir postcard, with portraits of three fierce commando leaders nestling incongruously beneath the Union Jack. But the Boer conviction proved, in the long run, more obdurate than the British: Jehovah survived the Queen-Empress, and the Boers were to win the Boer War in the end.[1]

[1] The bitterness of the Boer War was never quite expunged, and was fostered by the more extreme of the Afrikaner nationalists. When I first went to South Africa, sixty years later, people still told me of the ground glass allegedly put in the porridge of the internment camps, and showed me horrific pictures of black men armed by the British, while the Women's Monument at Bloemfontein, commemorating those who died in Kitchener's internment camps, was a national shrine outranked only by the memorial to the Voortrekkers at Pretoria. I must add, though, as an old admirer of the Boers, that when I explored the battlefields of the war in 1975 I heard not a word of reproach, triumph or resentment from the many kind Afrikaners who showed me around, even in my Jingo moments.

CHAPTER FIVE

The Wearying Titan

THE world watched thoughtfully. 'My dear, you know I am not proud,' wrote the Tsar Nicholas II to his sister during the Boer War, 'but I *do like knowing that it lies solely with me* in the last resort to change the course of the war in Africa. The means is very simple—telegraph an order for the whole Turkestan army to mobilize and march to the Indian frontier. That's all.'

He was exaggerating in fact, for until his central Asian railway system was complete he had no way of getting the Turkestan army to the Indian frontier, but he was only expressing the instinct of the nations. The Boer War had cracked the British mirror; the Jubilee was over; the Empire had grown too big for itself. It had seemed to most of its citizens invulnerable because of its very size, but now, it seemed, it was size that made it vulnerable. Empire gave the British a finger on every pulse, a say in every conference; but at the same time it made them subject to all the world's anxieties, innately responsive not merely to the Mauser of a Boer, but to the whim of any foreign despot.

The Boer War showed that it was getting too much for them. In the 1860s Matthew Arnold had portrayed Great Britain as a weary Titan—

> with deaf
> Ears, and labour-dimmed eyes,
> Regarding neither to right
> Nor left, goes passively by,
> Staggering on to her goal;
> Bearing on shoulders immense,
> Atlantean, the load,
> Well-nigh not to be borne,
> Of the too vast orb of her fate.

This prophetic picture would have been unrecognizable only five years before, but in 1902 the world, thoughtfully watching, saw its outline dimly delineated in the aftermath of war.

2

The scale of the Empire, which so sustained the confidence of the British themselves, had bemused other nations no less. The Empire was so inescapable, seemed so old, bore itself so majestically, that it had become a universal fact of life, something natural to the world.

It spilt far beyond its own frontiers, too, for its power was tacitly present everywhere, wherever a merchantman docked, a banker checked an exchange rate, or a statesman contemplated a course of action. It was protean in its forms. To Americans the Empire was Canada, or the Caribbean, or Pacific power, or sea-blockade, or hard cash from the City. To the Russians it was India, the Mediterranean and the Eastern Question. To France it was Africa. To Germany it was the Royal Navy. To the Japanese it was an instructive model. To the Chinese it was a cultural humiliation—'I began to wonder', wrote the young revolutionary Sun Yat Sen, 'how it was that . . . Englishmen could do such things as they have done with the barren rock of Hong Kong within 70 or 80 years, while in 4,000 years China had achieved nothing like it. . . .'

Sea-captains of every nationality knew the Empire as a chain of harbours and coaling-stations, and the most ubiquitous of maritime aids. Financiers saw it as the power of the pound sterling, the king of currencies. To Argentinian commuters it was the Buenos Aires Tramways Company, British owned and operated. To Italian rail-waymen it was the familiar Adriatic specials, the trains which, speeding from Calais to the waiting India liners at Brindisi, sent the Anglo-Indians back to their labours in the east. To the pleasure cities of Europe it was bronzed but skinny tourists with money to spend, talking to each other in inexplicable jargon, and frequently meeting colleagues in the public gardens—'Dammit, Hodgson, Helen, good to see you! Well I never! Care for a spot of tiffin? Found yourselves somewhere decent to stay? Damme, *what* a long way from Jacobabad, ain't it?'

The ampleness of it all impressed foreigners despite themselves. It was sometimes hard not to be obliged by its noblesse, and some of the Empire's most vicious foreign critics were relieved, nevertheless, to cross a distant frontier and see before them, billowing on fort or hilltop, the Union Jack that promised them order, security and a cup of thick sweet tea. The Royal Navy especially, that supreme emblem of Empire, found loyalties everywhere. When the Americans began to build a new Navy of their own, in the 1890s, they adhered so closely to the British manner that the first of their new armoured ships, the *New York*, even had an admiral's walk at the stern, a direct and quite unfunctional tribute to the Nelsonic tradition.[1] As for Admiral von Tirpitz, the creator of the new German Navy, he carried his respect to still further extremes, and sent his daughter to be educated at Cheltenham Ladies' College.

3

The spectacle of the Boer War tempered all these emotions, and presented the British Empire in less flattering lights. It was clear to everybody now, as it had been clear to the Tsar, that a single colonial war, against an enemy with a population half that of Birmingham, had tried the Empire to its limits. The British admitted as much. Arthur Balfour, who succeeded Lord Salisbury as Prime Minister, said publicly that its drain upon the imperial resources had reduced Britain in effect to a third-rate Power. The Viceroy of India had been warned, at the height of the war, that the last division was mobilized, and that if Russia did attack, he might expect no help from home. Lord Kitchener said the imperial armies were incapable of fulfilling all their commitments—'we are in the position of a firm which has written cheques against a non-existent balance.'

This was a far cry from the rodomontade of the Jubilee, and the nations responded predictably. The first humiliations of Black Week, which convinced many Europeans that the British were actually

---

[1] Never repeated in the United States Navy, though British battleships were equipped with admirals' walks for another twenty-five years.

going to lose the war, had vividly revealed foreign feelings about the Empire. The British indeed had supporters in every country, from the Anglophiles of the eastern United States, who hardly felt themselves to be foreigners at all, to Greek and Italian liberals who still saw in Britain the old champion of their freedoms. But they had far more enemies. 'Splendid isolation' had been a flattering Canadian conception of Britain's lonely magnificence,[1] but in the worst days of the Boer War it acquired an uncomfortable new meaning—as though, wrote Salisbury himself, 'the large aggregations of human forces which lie around our Empire seem to draw more closely together, to assume . . . a more and more aggressive aspect.'

The German Kaiser Wilhelm II, Queen Victoria's own grandson, kindly though he had cabled Rhodes at Kimberley, had been openly pro-Boer ever since the Jameson Raid. The Boers were armed mostly with German weapons, and their artillery was actually officered by Germans. After Black Week most of the other European States declared their sympathies too, if only unofficially. Editorialists damned the British, cartoonists lampooned them, public opinion everywhere was at once shocked by Britain's policies, and entertained by her discomfitures. From many parts of the world young men volunteered to fight with the Boers: Germans, Frenchmen, Americans, and a ferocious and treasonable corps of Irishmen, Blake's Brigade. A Scandinavian corps played an important part in the defence of Magersfontein. 'Fashoda is revenged!' a Frenchman cried as he climbed to the roof of a captured British fort outside Mafeking, carrying a bottle of brandy.[2]

They were inspired partly by idealism, partly by a taste for adventure in an adventurous age, but largely by the resentment with which, beneath the unwilling respect, the world had long regarded the British Empire. It was a resentment often envious and often hypocritical (was Europe, incredulously asked the playwright Henrik Ibsen, *really* on the side of Kruger and his Bible?), but none the less profound. To the British themselves the Empire might seem

---

[1] It was coined, I learn from *Colombo's Canadian Quotations*, by George Foster, MP for North Toronto, remembered too for another properly imperialist slogan: 'No Truck Nor Trade with the Yankees!'

[2] Prematurely, for he was shot dead a moment later, and the fort recaptured.

a mighty force for good in the world, to their foreign contemporaries it was all too often an overweening, greedy and sanctimonious cabal—'a kind of octopus', as Lord George Hamilton, a percipient Under-Secretary of State at the India Office, interpreted the general feeling, 'with gigantic feelers stretching out all over the habitable globe . . . preventing foreign nations from doing that which in the past we have done ourselves'.

The artist Pablo Picasso, who was eighteen at the start of the war, expressed these antipathies aptly when, one day during the war years, he scribbled a few doodles on his writing-pad: for there, out of his subconscious perhaps, he idly portrayed a cast of British comivillains in the veld—toothy and monocled young subalterns, all drawl and languid stoop, preposterous apoplectic generals, bovine Tommies in approximate Highland dress looking at once ridiculous, brutal and half-witted. It was not at all how the British saw their brave troops, on the souvenir plates, cards, flags and albums which were lucrative by-products of the war.

4

For the first time since the Indian Mutiny people wondered how long the British could hold their Empire. 'The fact is', said Henry Campbell-Bannerman, 'we cannot provide for a fighting Empire, and nothing can give us that power.' The white colonies had staunchly supported the Mother Country in the struggle. Some 17,000 Australians, 8,500 Canadians, 8,000 New Zealanders had fought in South Africa, and there were white volunteers too from India, the Malay States, Burma, Ceylon and most of the scattered island possessions.

Black and brown volunteers, however, were not invited, and among the coloured subject peoples the loyalty was not so absolute. Though the Nation-State was still an unfamiliar concept in Africa and Asia, the Boer War gave some encouragement to those few visionary leaders who saw that the British Empire would not last for ever. It took vision indeed to see it, from the wrong side of the colour-line. Across the globe the British presence still lay apparently immovable, and so immeasurably superior was the white race in all the techniques of command, so cowed were the coloured peoples by European technology and assurance, that the Empire really did have an eternal look. Governors and Commissioners moved freely about without bodyguards, and the Viceroy of India sometimes went walking all by himself through the Calcutta slums; for the English-man was to his subject peoples, Gandhi thought, as the elephant was to the ants.

But in the years after the Boer War the ants began to stir. In Ireland, where patriotism had survived 800 years of British occupa-tion, the old undercurrent still ran, secret societies drilled and plotted, and the Irish people only needed another in their long line of heroes to inspire them into rebellion. In Burma the Young Men's Buddhist Association, building upon its Christian model, advanced from healthy sports and social work into nationalist discussion. In West Africa the Sokoto people rebelled, in South Africa the Zulus, in East Africa the black tribes of Kenya. In Egypt a wave of anger and sorrow followed the public hanging and flogging of some villagers accused of attacking British officers on a pigeon-shoot—'everyone I met', wrote the Egyptian writer Qasim Amin, 'had a broken heart and a lump in his throat.' Most tellingly of all, in India, much the greatest of all the imperial possessions, a thin tide of patriotism was beginning to flow.

Indian nationalism had already found a voice in the Indian National Congress, originally dedicated to co-operation with the Raj, later developing more militant postures, and it had an inspiring spokesman in the Hindu visionary Bal Gangadhar Tilak, who had already been imprisoned for subversion, but was irrepressible. No country, though, was less likely to coalesce in rebellion than India—fragmented into a thousand parts by race, religion, history and geography, held so long in fee to the British that the habit of obedience was deeply ingrained in the people, and most citizens could no more conceive of an end to the Raj than an end to the world itself. Since the Indian Mutiny, half a century before, India had been held severely in check, even the Indian Army, the pride of Anglo-India, being deprived of artillery and watchfully attended, wherever it was, by British Army units. It was only after the Boer War that a mass of the people first took part in a nationalist demonstration, and the most prescient of their leaders began to see how independence might one day be achieved.

In 1905 the British decreed the partition of Bengal, the most intractable province of British India, whose population was rather larger than England's. They proposed to divide it into two lesser provinces, one predominantly Hindu, one predominantly Muslim, and their intention was self-evident: divide and rule. They brushed aside Indian objections—'if we are weak enough to yield to their clamour now,' the Viceroy reported to London, 'we shall not be able to dismember or reduce Bengal again; and you will be cementing and solidifying a force already formidable and certain to be of increasing trouble in the future.'

At first the leaders of Congress objected in constitutional terms. The British were unimpressed, being for the most part thoroughly contemptuous of Congress and all it represented, but as the day of partition approached they found themselves faced by a very different kind of protest. There were none of the habitual Indian riots, which were a nuisance indeed, but were easily enough suppressed by Indian policemen with their batons, or if necessary Indian sepoys with their bayonets. Instead, at Tilak's inspiration, thousands of Bengalis protested passively, with a silent boycott of everything British. This was a new phenomenon. Shoppers would not buy British goods. Students would not do their examinations on British paper. Washerwomen would not wash British clothes. Cobblers would not mend British shoes. Children would not suck British sweets. Lancashire textiles were ceremonially burnt in the streets. As the weeks passed, boycott went further, and those who broke it were themselves boycotted. Relatives and neighbours refused to see them, tradesmen refused to serve them, even priests and physicians denied them solace.

This was the beginning of *swadeshi*, economic boycott, which was to become one of the chief weapons of Indian revolution, and even of a greater instrument still, passive resistance. The British realized its power very soon, and some of them recognized it as the first stirring of a decisive revolutionary process. So did some of the Indians. At the height of the anti-partition demonstrations 50,000 people assembled for the Bengali festival of Durga Puja, in the temple of Kali at Kalighat in south Calcutta. This was always one of the great days of the Bengali year, but now it had a special meaning, for the goddess Durga had acquired a new cult as the personification of Motherland. The Puja of 1905 was celebrated, for the first time, as a festival of Indianness.

It was an event of theatrical effect. The temple was brilliant with the lamps of the devotees, and crammed to its high railings with their bodies, pressing around the sacrificial block in the middle of its courtyard, crowding up the steps to the Natmadnir, the shrine beyond. The air was heavy with sickly smells, flowers, incences, perfumes, and loud with thousands of voices, and bells, and the rumble of the traffic in the street outside. The pilgrims entered the shrine in groups, and there before the Brahmin priests they made a solemn vow to Durga, 'before thy holy presence and in this place of sanctity', never to buy British goods, never to use British shops, never to employ Englishmen.

As the long ceremony proceeded a violent storm blew up, and the rain teemed furiously upon Calcutta, turning the temple courtyard into a quagmire and drenching the thin clothes of the devotees. The fury of the night, though, only gave the oath a deeper meaning, and through the storm the waiting crowd could hear, over and over again, the Sanskrit injunction of the priests, as they dismissed the pilgrims batch by batch: 'Swear to serve your Motherland! Offer your lives to her service! Worship the Motherland before all other deities!'

5

Even more significantly, the world observed a first hesitation of British morale at home. A tremor, a fitful doubt, passed across the nation for all to see, in the aftermath of the Boer War.

> *Let us admit it fairly, as a business people should,*
> *We have had no end of a lesson; it will do us no end of good.*
> *Not on a single issue, or in one direction or twain,*
> *But conclusively, comprehensively, and several times and again*
> *Were all our most holy illusions knocked higher than Gilderoy's kite.*
> *We've had a jolly good lesson, and it serves us jolly well right!*

This was the national mood, chastened and bewildered, expressed here by Rudyard Kipling himself. It was a deep and subtle disillusionment, and though it was overlaid still by pride, patriotism and the joy of victory, though it lay dormant or impotent for years to come, still it was a seed that grew with time, to change the nation, the Empire and the world.

On the surface Kipling was simply attacking military incompetence, and arguing for an end to the class-ridden structure of the old army, its officers so often courageous dunderheads, its men such unreasoning serfs. In this many soldiers agreed with him. They were all 'ashamed for England', reported the war correspondent G. W. Steevens after one humiliating skirmish—'not *of* her, never that! but *for* her!' 'We are only sportsmen,' one wounded officer was heard to say with a sigh, as he hauled himself crippled with fever, dysentery and loss of blood towards the chaos of a field hospital, 'only sportsmen, after all. . . .'

Before the Boer War the British, enjoying for the first and last time in their history a militaristic phase, had assumed their armies to be the best in the world, and watched their exploits with a robust pride, adequately expressed by the poet G. Flavell Hayward:

> *Hear the whizz of the shot as it flies,*
> *Hear the rush of the shell in the skies,*
> *Hear the bayonet's clash, ringing bright,*
> *See the flash of the steel as they fight,*
> *Hear the conqueror's shout*
> *As the foe's put to rout . . . .!*
> *Ah!*
> *Glory or death, for true hearts and brave,*
> *Honour in life, or rest in a grave.*[1]

This vicarious swashbuckle was now extinguished. The British had never suffered such terrible casualties before, while the run of defeats in South Africa came as a terrible psychological shock, the first of a series which progressively whittled away the British taste for glory, and even perhaps for honour. 'Please understand,' the Queen had declared during Black Week, 'there is no one depressed in

[1] Set to music by Edward Elgar, Opus 5, No 1—*fieramente*.

this house'—but who could have imagined, five or ten years before, that she could ever make such a remark, of a war fought by the flower of her armies against 35,000 untrained bucolics?

On a deeper level, though, Kipling was reflecting a more general malaise, an unease which might not affect the Jingoist masses, but already troubled more sensitive citizens. He wrote of 'obese, unchallenged old things that stifle and overlie us'. He wrote of 'flannelled fools at the wicket, muddied oafs at the goals'. The conduct of the war had hinted at fundamental flaws in the imperial assumptions: flatulent old assumptions which, unchallenged in the euphoria of success, did lie deadweight upon the nation, and suggested themselves in failure. Britain was at the end of her aristocratic period. The public was still half-educated, half-enfranchised, and by and large the upper classes still dictated the course of events. The Labour movement was in its infancy; the new electorate was easily swayed, and inhibited too by traditional loyalties; most Members of Parliament still represented landed, agricultural interests. For the most part, in the Britain of the 1900s, what the gentry said, went.

Now this old hierarchy, which lay at the root of Empire too, was vaguely but significantly discredited. The British gentleman was not, it seemed, organically constructed to command, not entitled to success as a birthright. Calvinist dame schools of the *dorps* made better generals than Eton and Sandhurst. The British soldier's traditional loyalty to his social betters was evidently no longer enough. The great regiments of the British Army, the Guards, the Highlanders, the cavalry of the line, were of all British institutions the most devoted to the old order: yet these proud brotherhoods had been seen running for their lives through the South African night, or pinned humiliatingly among the thorn-bushes with the sun blistering the backs of their kilted knees.

Before 1900, wrote the polemicist Arnold White in his *exposé* *Efficiency and Empire*,[1] the 'accepted creed of average Englishmen' included the following clauses:

[1] Chiefly notable, by the way, for the *saevo indignatio* of its section headings: 'Influence of Bad Smart Society', 'Idleness a Trade', 'Our Most Incapable Department', or 'Serious Effect on National Character of Dishonour or Untruthfulness in State-paid Ecclesiastical Teachers'.

*The British Empire is the greatest the world has ever seen, and being free
from militarism is safe against decay.*

The British Army, though small, can do anything and go anywhere.

One Englishman can beat two foreigners.

We are the most enlightened people on the face of the earth.

By the end of the war, as White commented, every one of those
propositions was disputable, and some were obvious falsehoods.
And if the system failed in war, how long would it succeed in peace?
Could this archaic society keep pace with more modern rivals, better
educated, fresher, less circumscribed by class and tradition? Or
might it be that the nation was becoming effete? Curzon certainly
thought the British were beginning to show 'a craven fear of being
great.' The preposterously imperialist Poet Laureate, Alfred Austin,
put the same thought into a different sort of verse, as he saw the
audience for his bombast so disconcertingly diminished:

> *The sophist's craft has grown a prosperous trade,*
> *And womanish tribunes hush the manly drum;*
> *The very fear of Empire strikes us numb,*
> *Fumbling with pens who flourished once the blade.*

6

There were also doubts of a yet more debilitating kind, about the
morality of the war and so of the Empire itself. The British Empire
had never been unanimous. There had been opposition always to the
imperial idea, and several imperial wars of the past had been hotly
criticized, in Press and in Parliament, as immoral or unworthy.
Gladstone's famous Midlothian election had been concerned immediately with the injustice of the Afghan campaign of 1879—
'Remember the rights of the savage! Remember that the happiness
of his humble home, remember that the sanctity of life in the hill
villages of Afghanistan . . . is as inviolable in the eye of Almighty
God as is your own!' Britons of many kinds supported him. There
were humanitarians who thought imperialism a sin, moralists who
thought it a fraud, radical politicians who thought it an error,
economists who thought it an unnecessary fiscal device and socialists

who thought with Karl Marx that it was merely an undesirable extension of capitalism. 'This has been a day of consolation', the poet Wilfrid Scawen Blunt wrote in his diary when he heard of the capture of Khartoum by the Mahdists and the death of General Gordon, 'and I could not help singing all the way down in the train.'

The Boer War, however, revealed dissensions on an altogether different scale. There had been powerful opposition to it from the start. General Sir William Butler, the outspoken Commander-in-Chief in South Africa before the war, had resigned his office rather than be involved in it, and these are the headlines with which *Reynold's Newspaper* announced its beginning:

MONARCHY v REPUBLIC

ENGLAND FORCES WAR

CAPITALISTS' CAMPAIGN

£7,000,000 ALREADY SPENT

THE PRICE OF BREAD TO RISE

CHAMBERLAIN'S VICTIMS

70,000 BRITISH FIGHT 20,000 BOERS

WHAT ABOUT TOMMY ATKINS?

There was a revulsion against the 'stock-jobbing imperialism' which had, so many people thought, dragged Britain into the conflict. The mostly Jewish 'Randlords', the mining magnates of the Transvaal, became figures of contumely and derision, and there were hints of corruption and speculation nearer home—'the more the Empire expands,' suggested *Punch* disagreeably, 'the more the Chamberlains contract.'

When the brave Lord Dundonald volunteered for the war, taking the Secret Plan with him, his aged mother did her best to dissuade him: 'I do not like your going out to fight the Boers. It is an unrighteous war and Kruger is a religious good man and reads the Bible.' Kipling's aunt, Lady Burne-Jones, hung a black flag from her window at the news of the Boer capitulation, with the text: 'We have Killed and Taken Possession.' The popular writer H. H. Munro ('Saki') passionately opposed the war in the *Westminster Gazette*, and the *Manchester Guardian* was burnt on the Stock Exchange for its pro-Boer attitudes, surviving at all only because it was commercially indispensable to Lancashire.

Emily Hobhouse, the daughter of a distinguished Liberal family, furiously denounced to the world the dreadful conditions inside Kitchener's internment camps. The general called her 'that bloody woman', but Campbell-Bannerman told the House of Commons that the camps were a 'barbaric' method of winning a war, while Field-Marshal Sir Neville Chamberlain said there had been nothing so shameful in the annals of the British Army.[1] The war brought to fame, too, a British politician of an altogether new kind, David Lloyd George, the cobbler's ward from Llanystumdwy in North Wales—a man not simply of the people, but of a fringe of the people, somebody totally alien to the English traditions of hierarchy and dominion, an iconoclast, a man who did not even believe English to be necessarily best, but surrounded himself with Welsh aides and secretaries, talked Welsh most of the time, and paid no attention to British conventions of reticence and decorum.

7

These were no more than omens. The mass of the British remained loyal to the imperial cause and the British way of things, and few people yet wondered whether it was actually wrong in principle for one nation forcibly to rule another. When Lloyd George presumed to express his heresies in Birmingham, he was so furiously attacked by a patriotic mob that he had to be whisked away in a police van, and when the Government went to the country in the Khaki Election of 1900, it was easily re-elected. The epitaphists of Empire could still write, as one did on a memorial at Wagon Hill, Ladysmith:

[1] In the First World War Emily Hobhouse became strongly pro-German: her ashes lie at the foot of the Women's Memorial at Bloemfontein.

The Grand Illusion

Tell England, ye who pass this monument,
We, who died serving her, rest here content.

But somehow it had lost its innocence. Take that celebrated imperial occasion, Mafeking Day, May 17, 1900. On the face of it this was a simple expression of joy and relief, but there was something hysterical to it, something at once self-conscious and self-delusory. It was only a little railway town that had been relieved, far away, of small importance to the course of the war: yet London in all its tremendous history had never known such scenes of celebration. It was a kind of madness. Every class of Londoner poured rambunctious and often drunken into the streets, and the capital gave itself over to three days and nights of carnival.

In retrospect it seems a very different festival, and we see it almost as a mourning dance. Just as in adult life a suggestion of nursery fire or bedtime story conjures from the past the comforts of childhood, so Mafeking seems to have satisfied a wistful yearning in the public mind, a yearning for lost years of certainty and fulfilment. The magic was leaving the British Empire. Six months later Queen Victoria died, and with her went the imperial virtue. She was like a great old oak, whose roots run deep into a parkland, whose branches shade half a meadow, and when she died some old instinct died too, the British lost some sense of favour, the world a sense of awe.[1]

[1] She is buried in the mausoleum she had herself built for the Prince Consort in 1862, in the grounds of Windsor Castle at Frogmore—a structure open to the public only on one day a year, but potentially among the great tourist spectacles of Europe. Romanesque outside, and built in a Grecian cross, inside it is a prodigy of Victorian Renaissance, inlaid with many coloured marbles, elaborately decorated with paintings, statues and stained glass, and crowned by a ribbed gold-painted dome. The great Queen is buried with her husband beneath the largest single block of flawless granite ever quarried, and lies recumbent in white marble high on top of it, portable steps being available for visitors. There are three bronze wreaths near the tomb, all presented at the time of the Queen's death: one by the Emperor Menelik of Ethiopia, one by the crew of a Brazilian cruiser visiting London at the time, and one by 'Her Native Subjects of the District of Butterworth, Transkei'.

Sir Robert Mackworth-Young, the Royal Librarian, most kindly introduced me to this fascinating building.

CHAPTER SIX

Two Grandees

MANY aspects of Victorianism died with Victoria, and many aspects of imperialism too. All of a sudden, it seemed, the giants were leaving. Gladstone had already gone, wistfully longing for death but often breaking into 'Praise to the Holiest in the height'. Salisbury soon took his last tricycle ride round the estate. Rhodes was buried in the Matopo Hills, having instructed his portrait-painter to give him 'the eyes of the sphinx looking over the desert into eternity for the future, only in my case for the Empire of the race that I believe to be the best. . . '. Ruskin, who had once offered the British Empire an ideology, died incoherent at the height of the Boer War. G. A. Henty, who had made it one long adventure, died as soon as he had completed *With Roberts to Pretoria*. W. E. Henley, who saw it as a heroic sacrament, had time before going in 1903 to hear his 'Last Post' set to music by Sir Charles Stanford, and to publish his threnody for the Queen herself—

> *Tears for her—tears! Tears and the mighty rites*
> *Of an everlasting and immense farewell. . .*

The new king, Edward VII, was not very interested in his Empire, and the Edwardian age never did recapture the flair, conviction or vulgarity of the great enterprise. Asked long before what was the secret of the imperial system, one of its practitioners had shortly replied: 'It's not the system, it's the men.' By King Edward's day, though, the individualism of Empire had inevitably waned. The paladin imperialists of Victoria's prime, who seized whole provinces, deposed kings, established regimes on their own initiative, could not survive in the twentieth century. Rhodes the Colossus left no successor, never another Raffles created a Singapore.

The Edwardian gentleman of England, though truly an ornament of western civilization, lacked the effrontery of Empire.

In the first years of the new century, nevertheless, some remarkable Britons still found in the imperial *métier* a satisfaction worthy of their gifts, and brought to their distant offices distinction by international standards (for by and large, as the century progressed, the Empire was to become more and more provincial). Lord Cromer, the *de facto* ruler of Egypt, Lord Kitchener, Lord Curzon the Viceroy of India, Lord Milner the High Commissioner in South Africa—all these were world figures. In this chapter we look at two of them, to see how they fitted themselves into the imperial structure, and what they made of it.

2

To understand the posture of Curzon in India, we must first visit Kedleston, the country home in Derbyshire from which he took his title. The Curzons had lived at Kedleston for 800 years, providing not only its squires, but often its vicars too. They were not a particularly distinguished family—none of them appeared in the *Dictionary of National Biography*—but they were evidently tenacious. LET CURZON HOLDE, ran the family motto, WHAT CURZON HELDE.

Kedleston Hall was a famous house, designed by James Paine and Robert Adams in the eighteenth century. With its wide parkland all about, its sweep of colonnade and rose garden, its tremendous Great Hall with columns of pink alabaster, it was also overpoweringly English. It was like a country seat in a novelette (though Dr Johnson, who thought it all much overdone, said it would do 'excellently well for a town hall'). Here were the deer, grouped picturesquely around the lake. Here was the parish church, like a family chapel, where Curzons of every reign lay beneath their honorifics. Here was the estate graveyard, where gardeners and gamekeepers of many generations lay under their slabs. Ancestral portraits hung thick on every landing, and ever and again one found, sculpted on monument or engraved beneath escutcheon, slogans of familial pride—'This ancient Family which has inherited

a Good name and ample Possessions in this place From before the
NORMAN CONQUEST.'[1]

Out of such a background came George Nathaniel Curzon. He
was the eldest boy in a family of two sons and nine daughters, and
among the most precocious of his contemporaries at Eton and at
Balliol, Oxford, where he was cruelly immortalized in the famous
verses:

> *My name is George Nathaniel Curzon,*
> *I am a most superior person.*
> *My cheek is pink, my hair is sleek,*
> *I dine at Blenheim once a week.*[2]

At an early age Curzon decided that only two ambitions in life were
worthy of his pursuit: to be Viceroy of India, to be Prime Minister
of England. The first of them, after twenty years of travel and lesser
politics, he achieved at the almost unexampled age of thirty-nine
when, with his graceful American wife (the first of two), he landed
on the blazing quayside at Bombay on December 30, 1898 for the
first of two viceregal terms—seven years in all. There had been
twenty-two Viceroys or Governor-Generals before him, but none
were more truly viceregal, left more striking an image behind, or
approached the task with quite such a high-flown pride and
reverence.

Curzon was an imperialist out of his class. Not many of the landed
gentry were much enthralled by the imperial mission, and it was at

[1] It is all there still, all just the same, occupied by Curzon's nephew the
6th Baron Scarsdale, with the descendants of his deer still around the lake,
another generation of gardeners in the graveyard, and Curzon himself
culminating the family honours with his sumptuous marble tomb (where he
lies in a coffin of Kedleston oak beside his first wife. His second, visiting the
family vault one day after his death in 1925, found a note on a shelf in the
Viceroy's magisterial handwriting: *Reserved for the Second Lady Curzon*).

[2] The first couplet is thought to be by J. W. Mackail, later Professor of
Poetry at Oxford and author of Mackail's *Latin Literature*, the second by
Cecil Spring-Rice, later British Ambassador to the United States and author
of 'I Vow To Thee My Country'. 'Never has more harm been done to one
single individual', Curzon wrote nearly forty years later, 'than that accursed
doggerel has done to me.'

Kedleston indeed that a perceptive young Indian maharajah, viewing the idyllic scene around him, wondered why on earth Englishmen went to India at all, when they could stay at home in such a place, 'playing the flute and watching the rabbits'. Curzon was one of the few. Having a taste for lordliness, having travelled in his youth extensively in Persia and Central Asia, being by nature an autocrat, a connoisseur, a humorist, he responded readily to the magnificence of Empire, and believed genuinely in its usefulness. He thought it, 'under Providence, the greatest instrument for good that the world has seen . . . there has never been anything so great in the world's history'.

He saw its grand offices of authority as extensions of the British aristocratic condition. The Viceroyalty, he once said, was 'the dream of my childhood, the fulfilled ambition of my manhood, and my highest conception of duty to the State'. The Governors of Empire stood towards their subject peoples as the landowners of Derbyshire stood towards their tenants, absolute but paternal: it was a true figure of affairs, he considered, that the Governor-General's great palace at Calcutta, into which he moved that winter, should have been modelled directly if inadequately upon Kedleston Hall.

At home Curzon moved in the quickest and most elegant of English circles, the club called the Souls, a happy alliance of power, intellect and social eminence which thrived in the heady years of *fin de siècle*. His closest friends ranged from Arthur Balfour, 'the last of the Athenians', to Wilfrid Blunt, the most outspoken of the anti-imperialists, who said he hoped Curzon would be the best but the last of all the Viceroys. In India he found himself among a very different kind of Briton. The men of the Indian Civil Service were, as a whole, less than scintillating. Diligent, incorruptible, highly educated, they distinctly lacked the airy spirit of the Souls. Since the Mutiny half a century before they had become increasingly institutionalized, aloof to the world outside, enthralled by their monumental task of governing a sub-continent, but growing ever more convinced that in India especially there could be nothing new under the sun. Their methods had been tested down the generations, and were by now embodied in a thousand precedents and corollaries, docketed in a million cubby-holes, and wrapped up in red tape from Dacca to Karachi.

Curzon was much depressed by these colleagues, and by the style of bureaucratic life in India. The pace of it he found unbearably ponderous—'like the diurnal revolution of the earth', he wrote, 'went the files, steady, solemn, sure and slow'—and every initiative he made seemed to be muffled, diverted or referred back. He made few friends in India, he was often in pain from his back, injured in a riding accident twenty years before, and he was sadly homesick. 'The echo of the great world', he wrote to a friend, 'hums like the voice of a seashell in one's ears.'

So he withdrew ever more haughtily into the powers of his own autocracy: and since the idea of India, India in the abstract, India as a supreme field of British enterprise and imagination, still moved him greatly, in a curious osmosis this most devotedly English of politicians became the most nearly Indian of Viceroys. We see him sphinx-like and elegant at the head of his Executive Council in Calcutta: pink-cheeked still but slightly sneering, dominating the room with an astringency, a sheathed cleverness, that inhibited all but the most self-confident.[1] We see him receiving visitors among the trophies of the palace at Calcutta, the treaties and the captured ensigns, the portraits of triumphant predecessors or defeated enemies: the sense of history that always inspired him, his vision of India itself as the noblest trophy of them all, gives him an extraordinary power of presence, so that the most disrespectful visiting MP bows despite himself, and the most determinedly republican American involuntarily curtseys. We see him riding to the manner born in the howdahs of elephants, or passing with benevolent inclinations of his head beneath arches proclaiming him the Benefactor of India, or the Success Fighter of Famine.[2] He was the shadow of the King, taking precedence in ceremonial over the heir to the Throne. He was the Viceroy and Governor-General. He it was whose will could decree events in the remote valleys of Nagaland,

[1] 'Look upon me', he once told a favoured subordinate, 'not as a Viceroy but as a friend': but as the young man said, it was beyond imagination.

[2] Or declaring, in one of the Viceroy's favourite examples, that KARACHI WANTS MORE CURZONS.

or in mud huts of Travancore. He was the successor to the Moghuls. He was the equal of Princes. He was Curzon of Kedleston.[1]

This was the romantic view of imperial responsibilities, and as such it was an anachronism. Though Curzon much admired efficiency, and spoke proudly of his own 'middle-class methods', still he came to India in the spirit of earlier times, as though there were no obstructive bureaucrats to hamper him, or daily directives from Whitehall. He was excited rather than daunted by the magnitude of the task, and by the daring of the British presence there—'a speck of foam', he called it once, 'upon a dark and unfathomable ocean'. In his ideas if not his methods he was like one of the tremendous enthusiasts who, with gun and Bible and unshakeable assurance, had established British rule in the Punjab sixty years before. Like them, he believed that no aspect of life should be beneath the notice of a great ruler. Land revenues, railway systems, universities, the growth of industry, the control of commerce, irrigation, corruption in the police, provincial administration, border control—in matters petty or incalculable, from the Forward Policy on the frontiers to the smallest detail of office management, Lord Curzon laid down a policy, instituted an inquiry, or at least made his opinions known.

He saw India as a Power in itself—'a not always friendly Power', Balfour suggested—and was constantly chafing against the interference of London in Indian affairs. In another age one might imagine him breaking away altogether, a rebellious satrap, to establish his own Empire in the East. As it was, he considered India a suzerain in its own right, with the authority to extend its discipline over lesser neighbours. He sent, as we shall later see, an aggressive expedition into Tibet, the most private of all countries, and in an imperial gesture long remembered he made a viceregal visitation to the sheikhdoms and emirates of the Persian Gulf, impressing upon that fractious and frequently murderous congeries of princelings the power of India and the Empire—the Great Government, as the Arabs gratifyingly learnt to call it.

This was a marvellous progress. Imperial suzerainty over the Gulf was for the most part unwritten, but distinct. It was the

[1] And so he remained, as baron, as earl and finally as marquis, but the title died with him, for he had no sons.

British who had charted the Gulf waters, mapped its shores, marked its reefs with the lights of the Imperial Lighthouse Service, and put down the piracy which, for hundreds of years, had been at once the plague and the mainstay of its seafarers. They had also, from time to time, backed the more useful of the sheikhs against the pretensions of their theoretical overlords, the Turks. Curzon accordingly descended upon the Gulf in seigneurial state, sailing in the brand-new Indian Marine troopship *Hardinge* (6,520 tons), with an escort of six white gunboats. He wanted to impress upon the sheikhs, still innocent, if that is the word, of the world outside the Gulf, that they must have no truck with foreign interlopers in the purlieus of the British Empire—Germans in Iraq, Russians in Persia, or the French who had for 200 years cherished designs upon Muscat.

How he enjoyed himself! Wherever he went he was greeted with elaborate and sometimes Gilbertian deference. At Muscat, 'the most Byronical place you can possibly imagine', the melancholy Sultan described Lady Curzon as a pearl, to which the Viceroy pointedly replied that the Persian Gulf was a pearl too, and one the British Empire well realized the value of. At Kuwait, there being no jetty, Curzon was carried ashore on the backs of retainers, and there being no carriages either, a victoria had been especially imported from Bombay.[1] Sometimes sheikhs were entertained beneath the awnings of the *Hardinge*, where Lady Curzon, in muslin and white flowered hats, impeccably received them.

It was a truly Curzonian tour, not because of the excursions, pageantries and ludicrous mishaps which so amused him, but because through it all he pursued a purpose which he saw idealistically and nobly expressed: the right of the Raj to impose its own peace and order upon its perimeters. 'We opened these seas to the ships of all nations', he told the sheikhs, with justice, 'and enabled their flags to fly in peace. We have not destroyed your independence but have preserved it. We are not now going to throw away this

---

[1] The Arab horses that pulled it had never been between the shafts before, and when they had conducted Curzon to the Sheikh's palace, they turned spitefully upon the vehicle and kicked it to pieces. Curzon and his host had to walk back to the embarkation point, 'very gingerly', he reported, 'over heaps of ordure'.

century of costly and triumphant enterprise. . . . The peace of these waters will be maintained; your independence will continue to be upheld and the influence of the British Government will remain supreme.'

3

Curzon responded voluptuously to the far horizons and the scented coasts. He was easily moved, to tears and to laughter, and if he distinctly lacked the common touch, he responded to style and sensitivity in men of all races—racial pride, he thought, was a lower-class attribute. Certainly he was less at home with the British bourgeoisie than he was with a delicate Maharajah, an entertaining Kurd or the wild frontier chieftains of the north-west —'gigantic', as he approvingly described them, 'bearded, instinct with loyalty, often stained with crime'.

Curzon's first love of travel had been Persia, but he responded too to the more complex fascination of India, and grew to relish its tangled history. Surveying the disturbances and fomentations of the 1900s, he wrote: 'I have observed the growing temper of the native. The new wine is beginning to ferment in him, and he is awaking to a consciousness of equality and freedom.' The movement was inevitable, he believed, and if it was not to end in violent rebellion the arrogance of the British must be held in check. He considered that arrogance vulgar, the prejudice of uneducated men, and in trying to control it he made many enemies among Anglo-Indians.[1] Once again he was consciously enacting a lofty imperial role, the role of trustee, and never more consciously than in his famous quarrel with the 9th Queen's Royal Lancers.

The 9th, who figure prominently in half the military adventures of the Victorian age, were a fashionable cavalry regiment, rich in

[1] By which I mean, as I do throughout this trilogy, British people serving in India, of which there were two kinds—those who had their homes in Britain and those who had actually settled in the sub-continent, and were gracelessly known as the Domiciled Community. One felt itself superior to the other, and vice versa.

In 1900 'Anglo-Indian' came officially to signify people of mixed blood—Eurasians.

titles as they were in battle-honours, Anglo-Welsh in origin but long since transmuted into a kind of smart socio-military club.[1] Intensely snobbish towards inferior regiments, they were clannishly bound together themselves, officers and men alike, by a fierce regimental loyalty, making them all in all, with their rich and influential officers, their illustrious record,[2] and this tight spirit of membership, a formidable organism. In 1902 some troopers of the 9th Lancers, it was alleged, brutally attacked an Indian, leaving him dying outside their camp. Who the culprits were, nobody knew. Several courts of inquiry failed to discover, and though the regiment itself offered a reward for information, nobody came forward.

Curzon was much distressed by the affair. During the previous twenty years there had been eighty-four recorded cases in which Europeans had killed Indians, yet since 1857 only two Europeans had been hanged. 'You can hardly credit', he wrote, 'the sympathy with wrong-doing that there is here—even among the highest— provided that the malefactor is an Englishman.' He was convinced that the 9th Lancers had deliberately withheld evidence, probably with the complicity of the Army command, and he ordered that, since no individual could be punished, the whole regiment must be instead. All leave was cancelled for nine months, and as an outward mark of disapprobation, throughout that period a sentry from another regiment was placed outside each of the regimental barrack blocks.

The regiment was infuriated by this despotism, and was angrily supported by most Anglo-Indians. The case became a *cause célèbre*. The Indian newspapers were full of it, society at home gossiped about it. Even the King protested at Curzon's action. Six months later, at the Delhi Durbar for his Coronation, the 9th were detailed

[1] They had played the first game of polo in England, at Hounslow Heath in 1871—they called it Hockey on Horseback, and a contemporary account found it 'more remarkable for the strength of the language used by the players than for anything else'.

[2] *Generally* illustrious: the 9th turned tail during the battle of Chillian-wallah, in the Sikh War of 1849, but the regimental historian stoutly attributes the rout to 'the gross mismanagement of their brigadier', and certainly it was the only recorded instance, in all the 145 years of their independent history (1715–1960), of their running away from the enemy.

to provide the escort for the Duke of Connaught, who was representing the Royal family. A proud escort they provided, the horses sleek and powerful, the soldiers wearing the medals of the South African campaign, with the pennants flying from their lance-heads and their spiked helmets gleaming white. When this splendid equipage rode by the saluting base, where Lord Curzon was taking the review, Anglo-India made its feeling known: for as the 9th passed a loud and pointed cheer went up from the assembled Europeans, unmistakable in its meaning and apparently unanimous—even the Viceroy's personal guests, so Curzon noted wryly, could not forbear to cheer.

This was another sort of imperialism, something much nearer the earth than Curzon's alabaster kind. He was upset by the demonstration, implying as it did that the vast majority even of educated Europeans in India regarded the Indians as less than fully human, but was intensely conscious too of his own patrician response. 'As I sat alone and unmoved on my horse, conscious of the implication of the cheers . . . I felt a certain gloomy pride in having dared to do the right.'

He dared, too, to approach the civilization of India with a respect rare among the pig-stickers and the box-wallahs, and tried to convince the Indians themselves that they should not simply wish to be brown Britons. At the Durbar—'the Curzonation'— he insisted that every detail of the decoration should be Indian, so that the whole huge camp became a display of woodworks, enamels, carpets, potteries and lovely silks. He devoted himself to Indian archaeology, neglected then by Indians and Britons alike. He it was who restored the Taj Mahal to its original perfection, cherished the half-buried glories of Fatehpur Sikri and the exquisite Pearl Mosque in the fort at Lahore, and by reviving the moribund Department of Antiquities, gave to the British Raj in India, just in time, a scholarly distinction that would remain among the more honourable imperial legacies.

4

Out of his time—in some ways too soon, in others too late. It was Kitchener, now Commander-in-Chief in India, who baulked Curzon of his Indian mission. In a historic quarrel—'The Lord of the Realm *versus* the Lord of War'—they clashed over the degree to which the civil government should control the Indian Army. Kitchener won, obliging Curzon to resign in 1905, and never afterward did a Viceroy of such difficult originality grace the administration of India.

He left behind mixed feelings: gratitude for much of what he had done, contumely for his partitioning of Bengal, widespread resentment at his high-handedness and delusions of grandeur. He had mixed feelings himself. On the one hand he was a frank imperialist, working always 'to rivet the British rule more firmly on to India and to postpone the longed-for day of emancipation'. On the other he was a man of civilized sympathies, and hoped that he had helped India towards 'the position which is bound one day to be hers—namely that of the greatest partner in the Empire'.

His disdain was remembered by many, his fun by only a few. Though he later became Foreign Secretary, he never did get to 10 Downing Street, and he looked back on his years in India, for the rest of his life, with a nostalgic pride. Out of it all he distilled as memorable a philosophy of Empire, by his own proud standards, as was ever expressed: 'Let it be your ideal', he told his countrymen in India, 'to remember that the Almighty has placed your hand on the greatest of his ploughs, in whose furrow the nations of the future are germinating and taking shape, to drive the blade a little forward in your time, and to feel that somewhere among these millions you have left a little justice or happiness or prosperity, a sense of manliness or moral dignity, a spring of patriotism, a dawn of intellectual enlightenment, or a stirring of duty where it did not exist before—that is enough. That is the Englishman's justification in India.'

5

Our second grandee of Empire stands, beside Curzon, like a searching torchlight beside a wind-flickered flare. If Curzon was quintessentially English, rooted for eight centuries in the same plot of Derbyshire countryside, Alfred Milner was hardly English at all. His philosophies owed nothing to the easy amateurism of the

English tradition, his means of expression would not have found favour with the Souls, he lacked romance and he had little humour. He was an imperialist of a very different kind: much less flamboyant than Curzon, but much more implacable, dedicated less to ideals than to systems. He was a genuine imperial technocrat, a class of statesmen of which history had time to produce, fortunately for the allure of Empire, only one or two.

Milner was born in Germany, and had a German grandmother. His father, though British, had been born and brought up in Germany too, and Milner's boyhood years there, his schooling at the gymnasium at Tübingen, marked his mind for life, and made him vulnerable always to English sneers and innuendoes. In 1870, during a walking tour in France, he saw the German Army in action during the Franco-Prussian War, and it is said that this experience further disposed him towards order, efficiency and political cohesion—not in those days the preoccupations of English gentlemen. At Oxford, where he was at Balliol a few years before Curzon, he won practically everything. He got a double first, of course. He won four of the great University scholarships. He was President of the Union. He was elected to a New College fellowship. He was a leading figure in a very different set from Curzon's dashing and irreverent circle of Etonians. Milner's group was earnest, socially-conscious, dutiful and exceedingly clever, and the friends he made in it lastingly influenced his view of public life and private duty. He was a formidable young man, and became a formidable old one.

He came to Empire obliquely, via journalism and politics, and his imperial career started as Under-Secretary to the Ministry of Finance in Egypt, under the grandiose pro-consulship of Lord Cromer. The task of imposing fiscal order upon that anthology of disorders, Egypt, was just Milner's style. 'In Egypt,' he wrote, surveying the state of the toiling fellahin around him, 'economic causes produce their theoretically correct results with a swiftness and exactitude not easily visible in other lands': and this measurable distance between cause and effect, this self-evident translation of theory into practice, suited his sober gifts—'in Egypt,' he said, 'there is no argument but "you must".'

Milner was a black and white man, a man of figures, a prophet of 'discipline, order, method, precision, punctuality'. 'If Milner does not agree with you,' Clemenceau was to write, 'he closes his eyes like a lizard and you can do nothing with him.' Intellectually, among the servants of the Empire in the early years of the new century, he stood almost alone: it was puzzling to some of his contemporaries that one of the cleverest men in England should devote his life to that essentially irrational enthusiasm, the Empire.

6

But Milner was the born public servant—or alternatively, the born dictator. He was a bachelor until late in life, believing that he had to choose between public good and private happiness ('What he needs', Beatrice Webb the Socialist once remarked, 'is God and a wife'). He lacked the *hwyl* to lead simple people or the artistry to inspire them, but he had a vocation to serve the State, and at the start of the twentieth century he believed that the most satisfying field of public service lay within the Empire. He was not much interested in the black African and Asian possessions, except as a source of wealth, but he was a consistent believer in white imperial unity—Greater Britain, a political amalgam of Britain and the self-governing Dominions. He recognized earlier than most of his contemporaries that Britain herself would presently be outclassed by the greater new Powers of Russia, Germany and the United States. Union with the Dominions, he thought, was the only way to maintain her status in the world—'the British State' (a word he was fond of, by the way) 'must follow the race, must comprehend it. . . .'

Look at his face, in the oval-framed photograph in *Men of Empire*, 1902! His mouth is set in a downward curve, not unkindly, not even unattractively, but implacably. His moustache, while not ostentatious, is emphatic and symmetrical. His eyes have a faintly bemused expression, as though he is forever astonished at the wayward nature of mankind. His forehead is high and fine, his ears are rather prominent, he sits in a posture half slouch, half admonition, like a schoolmaster lecturing a boy who, though undeniably a trouble-maker, does have recognizable talent. Only his tie, by a lifelong quirk, is slightly skew-whiff, and shows his collar-stud beneath.

Few women might be intoxicated by this severe cerebral figure.[1] It is easy to see, however, why Joseph Chamberlain, himself a forceful mixture of the impulsive and the hard-headed, surveying the growing confusion of South Africa immediately before the Boer War, invited Milner to go as High Commissioner to Cape Town. Milner called himself 'a synoptic man', a man who could make sense of things, and he hoped to fuse the two white races of South Africa into a harmonious whole under the Crown. He went there with a reasonably open mind, he learnt Afrikaans, he conscientiously met all the Boer leaders, but the synopsis did not work. After a year of hassle and impasse, logic convinced him that the only solution to the South African dilemma was war against the Boers. British and Boer independence, like the British and the Boer mentalities, were incompatible.

Milner never did cultivate the art of the possible. Another of his self-epithets was 'a man for emergencies', and he was bad at the slow slog or the compromise. Lloyd George, who should know, said he had 'no political nostril'. He despised English constitutional methods—'that mob at Westminster', he called Parliament—resented the shifts and delays of democracy, and said, as he analysed the chances for peace and war in Africa, that he didn't care twopence for the opinion of people 6,000 miles away in England.

---

[1] Though Elinor Glyn the romantic novelist, who had a brief affair with him at Carlsbad in 1903, and thought he must be the reincarnation of Socrates, recalled a playful evening when they got lost together in the woods, and she pretended there were bears coming out of the trees to eat them—'I held his hand and made him run down into the open early moonlight.' She had an affair with Curzon, too, a few years later, and described herself as 'his grovelling slave, ever ready to kiss his hands, lick his beloved toes':

> *Would you like to sin*
> *with Elinor Glyn*
> *on a tiger-skin?*
>
> *Or would you prefer*
> *to err with her*
> *an some other fur?*

In May, 1899, Milner declared his views in a much-quoted summary: 'South Africa can prosper under two, three or six governments, though the fewer the better, but not under two absolutely conflicting social and political systems.' It was not a judgement of values, only of expediencies. Relentlessly he pressed it upon President Kruger, and the successive interviews between the two men are among the most suggestive in the Empire's history: the Briton so cold, so implacable, so Balliol, the Boer so embedded in lore and faith, inherited emotions and tribal certainties—the one high-browed and formal in his winged collar and check cravat, the other whiskered and hideous with age, slumped in his drab frock-coat across the table. The talks were doomed: after a last harangue of Kruger which reduced the old patriarch to tears, Milner broke them off, and war was inevitable. It was all Milner's fault, Lord Salisbury thought. 'We have to act upon a moral field prepared for us by him and his Jingo supporters . . . and all for people whom we despise, and for territory which will bring no power and no profit to England.'

If reason led Milner to war, the synoptic mind, when once the conflict was over, led him to new conclusions. He was still a fervent exponent of imperial unity, he recognized that true unity came best by consent, and the experience of the war perhaps convinced him that the stubbornness of the Boers could not be overcome by brute force. Instead, he thought, 'stringent patient policy' might do the trick. He imported from England a team of a dozen young Oxford bachelor intellectuals, nicknamed (naturally) Milner's Kindergarten, to help with the task of reconstruction: they included a future Governor-General of South Africa, an editor of *The Times* and a British Ambassador to the United States. Their work was greatly admired, and the imperial propagandists hailed the Milner settlement as a model of foresight, enabling the two old enemies to live together in friendship at last.

But if Curzon the great romantic was always frank, Milner the logician could be devious. A policy hailed as liberal, in that it promised the defeated Boers equality with their victors, was really hardly more than a continuation of the war itself. Milner was out to create a South Africa in which, year by year, the British element would finally overcome the Boer. 'My formula for South Africa', he once told Winston Churchill, 'is very simply: 2/5ths Boers and 3/5ths Britishers—Peace, Progress and Fusion. 3/5ths Boers and 2/5ths Britishers—Stagnation and Eternal Discord.' He was not fond of the country—'I have always been unfortunate', he wrote, 'in disliking my life and surroundings here'—but he saw it as his mission to bind South Africa permanently within the imperial unity, in the interests, one feels, not so much of human happiness or prosperity, as of applied theory.

In most parts of the Empire the British imposed their ways by sheer force of example: the western culture was so obviously superior, in economic and technical terms, that the subject peoples flocked in self-interest to its schools, its counting-houses and its drapers. In South Africa it was different: only in French Canada, Gaelic Ireland and Celtic Wales did the indigenes so uncompromisingly defend their heritage against the onslaught of Englishness. Milner's methods were thorough and subtle. He brilliantly restored the ravaged country to normal life, resettling the landless families, building new schools and farms, evolving constitutions for the new Boer colonies. He imported Chinese labourers, under indenture, to bring the Rand mines back to full production. He started irrigation schemes and experimental farms, he founded municipalities in the hangdog Boer townships, he established a customs federation for all four colonies.

In short, he really did lay the foundations of a union, in which eventually the Boers would enjoy constitutional equality with the British. But it was to be a union inalienably, irrevocably of the Empire, giving the British permanent hegemony in South Africa, giving the Boers 'a far higher plane of civilization than they had ever previously attained'. When Milner built new schools for Boers, he intended that the instruction should be in English. When he founded experimental farms, he hoped that British farmers would come and settle there. When he treated the Boer leaders with magnanimity, it was because he hoped they would be converted to the imperial plan. In 1910 the four colonies of South Africa, Boer and British, were joined in the self-governing Union of South Africa, and many people at home thought Milner had succeeded. 'The grant of self-government', declared a League of Empire publication that year, 'has enabled two once hostile peoples to combine in a common ambition.' 'The Boers,' said *The British Empire in Pictures* more ingenuously, 'seem to be comfortable and contented under British Government.'

7

They were not contented at all. Though some of their leaders really did embrace the imperial ideal, the most obdurate of the Boers were as calculating as Milner himself, and had no intention of allowing the British to swamp or pervert the Volk—just the opposite, in fact. Piece by piece the policy collapsed. British settlers never did achieve Milner's algebraic serenity, while Boer national feeling only became more cohesive over the years, finding expression in political movements, secret societies, religion and an implacable cultural chauvinism.

Milner had returned to England for good in 1905, and blamed it all on too much haste—it takes time, to de-culturize a people. But he was blamed himself for disregarding the blacks of South Africa, who were denied all franchise in the new Union, but whose welfare had been, to many Englishmen, the only good reason for fighting the Boer War in the first place. 'For the first time', cried the Socialist Keir Hardie, 'we are asked to write over the portals of the British Empire, "Abandon Hope All Ye Who Enter Here." ' Milner's rash importation of Chinese labour for the Rand, something that seemed dangerously close to slavery, horrified liberal opinion hardly less, and in the end brought down the Conservative Government that had sent him to South Africa. Even Churchill, an old friend of Milner's, could only say in a celebrated Parliamentary circumlocution that the indenture system could not be 'classified as slavery in the extreme acceptance of the word without some risk of terminological inexactitude'.

Racialism, absolutism, illiberalism—all these were now seen as Milnerisms, and the radical Press fell upon him. 'Is it any wonder when an essentially German mind drives English policy that the result is not exactly what the English public look for?' demanded

The Speaker. The *Manchester Guardian* said that what used to seem exciting in Milner's character was now merely disgusting, while H. W. Massingham, who had been a war correspondent in South Africa, described Milner as 'essentially un-English'—'he had hardly any English characteristics, but was a pure bureaucrat and a pure ideologue.'

Milner never quite recovered from the taint, and ten years later a newspaper diarist could still say of him that 'rightly or wrongly, few men in the country are more distrusted'. The Liberal leader Campbell-Bannerman called him 'a most dangerous element in the national life'. He went on to be War Minister and Colonial Secretary, but his name would always be associated first with South Africa, and so with failure: for there the tribal spirit outstayed the synoptic skill, and made South Africa in the end a sad antithesis of the Empire's good and glory.[1]

8

Yet it would be wrong to leave Milner on a note of logic betrayed, for despite appearances he was far more than a mere intellectual. He had true gifts of friendship, his clever young disciples worshipped him, and behind that imperturbable façade, which made the Boers think of him as the Englishman incarnate, he burnt with lofty aspiration. He was, he said himself, 'from head to foot one glowing mass of conviction'. With a vision less poetic but more furious than Curzon's, he believed in the imperial mission almost as a revelation: he was one with the possessed expansionists of earlier times.

When Milner returned from Egypt, in 1892, he wrote a book, *England in Egypt*, in which he set out with his usual clarity the purposes of the British presence there. It was a plea for the acceptance of great responsibilities, and became one of the seminal texts of the New Imperialism. Churchill said it was 'more than a book', it was 'a trumpet-call which rallies the soldiers after the parapets are stormed, and summons them to complete the victory'. Milner's imperialism, when it came to the point, ran deeper, more religiously,

[1] He made sure of it anyway by adopting, when ennobled in 1901, the title Baron Milner of St James's and Cape Town.

than Curzon's. It possessed a conspiratorial element, and in later years the members of his Kindergarten, reconstituting themselves in London as the 'Round Table', propagated the imperial faith like cultists, given to arcane images and confidential reports, and becoming the most persistent of all political pressure groups. Milner's life was dominated, he said himself, by the dream of Empire: for years he spent every Empire Day, Queen Victoria's birthday, talking about it to Rudyard Kipling. Though he claimed to see the Empire as a means of social reform—power equalled prosperity equalled a better life for all Britons—he really believed in it as an end in itself, to which any weapon might be directed: diligence, magnanimity, neo-slavery, cunning, logic, war.

Even treason: for years later, when the Protestant patriots of Ulster, in northern Ireland, proposed to defy the authority of the Crown in the interests of Empire, Lord Milner offered them all his help. It would not really be a rebellion at all, he reasoned, though the Royal Navy was preparing even then to put it down, it would be 'an uprising of unshakeable principle and devoted patriotism—of loyalty to Empire and the Flag!' Such was the lifelong fervour of this misleading man, who looked so grave, and thought so icily, but felt with such unexpected passion.[1]

9

Two late grandees, each to end his life, in succession, as Chancellor of their old University, that last compensation of aged fame. There would never be imperialists like them again. No men of their calibre, in future generations, would see in Empire a proper arena for their talents, or satisfaction for their profoundest hopes. In the twentieth century the British Empire could no longer be the master

---

[1] Milner married, when he was seventy, the widow of our old friend Lord Edward Cecil, living happily ever after until his death in 1925—two months after Curzon. Ronald Storrs the diplomat, who met him once in Venice, reported that he expressed a particular sympathy for Carpaccio's dragon in the Scuola San Giorgio—who looks up reproachfully, almost wistfully, as St George thrusts his long spear implacably down his snout. Almost alone among the undergraduates of his day, Milner kept a cat in his rooms at Balliol.

of its own fate, still less dictate the course of world history, so that it lost its power to fascinate men of great ambition. Here, in farewell, are characteristic quotations from each of our imperialists, each illuminating a strand of Empire and a kind of man. This is Milner on his conception of imperial patriotism: 'I am a British (indeed primarily an English) Nationalist. If I am also an Imperialist, it is because the destiny of the English race, owing to its insular position and its long supremacy at sea, has been to strike fresh roots in distant parts of the world. My patriotism knows no geographical but only racial limits. I am an Imperialist and not a Little Englander because I am a British race patriot. It is not the soil of England . . . which is essential to arouse my patriotism, but the speech, the traditions, the spiritual heritage, the principles, the aspirations, of the British race. . . .'

And here, more stylish, less anxious in the end, is George Curzon's own epitaph, above his tomb behind the great iron screen at Kedleston:

> *In diverse offices and in many lands*
> *as explorer, writer, administrator*
> *and ruler of men,*
> *he sought to serve his country*
> *and add honour to an ancient name.*[1]

[1] Above the tomb there hang, side by side, a mediaeval war banner of the Curzons and the flag of the Indian Empire. Milner, who had no roots and left no heir, is buried less grandiloquently at Salehurst in Sussex, unnoticed even by the guide-books.

CHAPTER SEVEN

A Late Aggression

'BEFORE the Boer War', wrote Sir Edward Grey, 'we were spoiling for a fight. Any Government . . . could have had war by lifting a finger. The people would have shouted for it. Now this generation has had enough excitement. . . .'

It was true. The militarism, so briefly summoned, was over already, and the crude aggression had gone out of the Empire. Since Black Week the British had been recognizably on the defensive, even in attack. They had already lost the violent bravado which had been an inescapable part of Victorian imperialism, and which Lord Salisbury had once grandly described as 'but the surf that marks the edge of the wave of advancing civilization'. 'The vast majority of the Cabinet', wrote Lord George Hamilton, 'look with apprehension and dislike on any movement or any action which is likely to produce war or disturbance in any part of the British Empire.'

One last aggression of the old sort was launched, but there was no heart to it. It was conceived in India, the most traditionalist of imperial possessions, by Lord Curzon, the most historically minded of pro-consuls; its purpose was the establishment of British influence in that last stronghold of obscurantism, Tibet; yet it never rang true, and brought no glory with it.

2

For generations the British in India had been obsessed with the Russian menace. This was understandable. During the last 400 years, it was estimated, the Russians' own empire, on the other side of the Himalayan massif, had been growing at the rate of fifty-five square miles a day, and by the 1900s it was steadily advancing, with railway lines and garrisons, through the deserts of Central Asia towards the Indian frontier, besides nibbling eastward into the borderlands of China. 'We must be very careful', the Queen wrote in 1900, 'and well prepared, and have plenty of artillery.' The Russians themselves realized that their putative threat to India was the one sure way of circumventing British command of the seas, and influencing British foreign policy, and they played the Great Game of the Indian approaches assiduously, to the satisfaction of the Tsar and the frequent consternation of Calcutta.

The British fluctuated in their responses. Sometimes they retreated behind their own frontiers, leaving an undefended No-Man's-Land, sometimes they went to the opposite extreme, breaking out of India altogether to impose their authority in neighbouring States. They even toyed with the idea of countering a Russian invasion of India by an attack on Finland, then a Russian province, or the Caucasus. By the turn of the twentieth century they had achieved a mean. Their main defence lines were well within India, but they had established along their northern frontiers a series of neutral States, more or less subject to British policy, bound by treaty to allow no foreign influences within their borders, and thus providing a *cordon sanitaire* along the Himalayas.[1]

One State in particular stood outside this system, and that the most inaccessible of all: Tibet, whose capital was virtually unknown to Europeans, whose inner affairs were plumbed only by spies and hearsay, but whose frontiers marched with India's. To Curzon this exception was at once sinister and compelling. He profoundly distrusted the Russians. He watched with foreboding the slow progress of their railway lines, ever closer across the steppes, and their advance into Chinese Turkestan, and he took very seriously the incessant rumours of Russian intrigue and subversion gathered by his powerful intelligence system—that ever-present engine of suggestion and report, that sub-world of agents and informers, which Kipling immortalized in *Kim*. He believed that the Russians

[1] An economical process. All that was needed, wrote General Sir Charles Brownlow when it was proposed to establish a very small buffer State in Chitral, was 'a title, a subsidy, a couple of honeycombed guns and some bales of half-worn tunics as uniform.'

really had designs upon India, and by 1903 he thought that the chief danger came by way of Tibet.

There was no British representative at the Tibetan capital, Lhasa, and the information that reached him from inside the country was dim but disturbing.[1] Tibet's Government was arcane, for its titular head, the Dalai Lama, was a monk called to office by destiny as a reincarnation of all his predecessors. The incumbent in 1903, Nga-Wang Lobsang Tupden Hyatso, The Precious Protector, the All-Knowing Presence, was one of the few Dalai Lamas to reach his majority, most of them being eliminated in childhood for the convenience of Regency Councils, but he was theoretically subject anyway to a higher political suzerainty: for 200 years the Chinese Empire had claimed authority over Tibet, and a representative of that dilapidated Power, called the Amban, was always resident in Lhasa as a kind of Viceroy. As for the everyday Government of the State, it was altogether in the hands of the country's Buddhist lamas, who exercised despotic authority from their enormous high-walled monasteries, and were not only the priests, administrators and landlords of the country, but its policemen and soldiers too.

Tibet also marched with Russia, and it was repeatedly rumoured that the Russians had achieved some hold over Lhasa. Sometimes military missions were said to have reached the city, sometimes an ambassador was supposed to be in residence, sometimes the Russians had concluded a treaty of alliance with the Tibetans, or had definitively persuaded them that Russia was the legendary land of Shambala, the Source of Luck. The Dalai Lama's tutor, a Mongol Buddhist who was in fact born in Russian territory, was generally assumed to be a Russian agent, while the Tsar was repeatedly rumoured to have become a Buddhist himself.

Shadowy and sometimes preposterous though these reports were, Curzon was legitimately concerned. His own advances to the Dalai

---

[1] Much of it came from 'the Pundit explorers', a remarkable company of Indian, Sikkimese, Nepali, Mongolian, Tibetan and Persian agents trained by the Survey of India and sent on frequent secret missions north of the Himalayas. They were equipped with prismatic compasses disguised as prayer wheels, and were referred to only by initials.

Lama had been twice rebuffed. Besides being insufferably impertinent —'the most grotesque and indefensible thing'—this attitude seemed to suggest a deeper Tibetan hostility towards Britain. Relations between the two countries, vestigial at best, became at once remoter and testier. There were petty border disputes, between the frontier officers of the British Empire, far away in the last yak-meadows of the Raj, and the secretive priests who spoke for the Dalai Lama on his southern marches. Grazing rights were contested, trading arrangements were repudiated, and a British proposal to demarcate the frontier more exactly was murkily rejected.

But when Curzon suggested that some action should be taken over Tibet, London was unresponsive. Lord Lansdowne the Foreign Secretary was trying to improve relations with the Russians, and the last thing he wanted was any recrudescence of the Great Game. Besides, they did not take Tibetan dangers very seriously in Downing Street. Curzon was a notorious exponent of the Forward Policy—'Let me beg you as a personal favour', Sir William Harcourt had asked him when he had first gone to India, 'not to make war on Russia in my lifetime'—and they suspected that he was now exploiting trivial episodes to achieve dramatic ends.

When, therefore, in the spring of 1903, the Viceroy of India resolved to impose the will of the British Empire upon the refractory Tibetans, he did so less in obedience to Balfour's Government in London than to his own instincts in Calcutta. 'Of course we do not want their country,' he wrote, 'but it is important that no one else should seize it': and nothing could ever be done with Tibetans, he uncharacteristically added, until they were frightened.

3

He chose as his chief confidant in the scheme a man scarcely less ardent and impulsive than himself, Francis Younghusband of the King's Dragoon Guards. The son of an Indian Army general, Younghusband was an Englishman who could trace his forebears to Saxon times, but he had discovered like Curzon romantic affinities with Central Asia, and had become famous in his twenties for an astonishing solitary journey overland from China to Kashmir.

Thereafter he had become a prominent player of the Great Game, as explorer, as theorist, as Political Agent in Chitral, and so it was that in May, 1903, he was summoned to Simla, the summer capital of India, into the presence of the Viceroy.

Younghusband was a kind of mystic. Very short, conventionally moustached, with clear blue eyes and a somewhat peremptory expression, he had been at Clifton with Henry Newbolt, married the daughter of an MP ('I was not really suited to my wife', he recorded blandly in old age) and often acted as correspondent for that daily tapestry of orthodoxy, *The Times*. He was much concerned with 'face' and image, with other peoples' opinions, and he was an eager puller of strings. He was ambitious, he could be double-faced, and in his fortieth year he was aggressively imperialist—the Tibetans, he said, were 'not a fit people to be left to themselves between two great Empires'.

Behind this abrasive front, though, he was a dreamer. The landscapes and philosophies of Asia had profoundly, if hazily, affected his thought. He toyed with ideas of reconciliation and renunciation, the fusion of faiths, inner lights and brotherhoods. Not a very intellectual man, educated to practical tasks and not widely read, he approached these mysteries with an ingenuous wonder. The Himalayan presences, opaquely drifting through the high valleys of Sikkim and Chitral, expressing themselves it seemed in the marvellous snow-peaks of Pamir and Hindu Kush, deeply moved him, and when he crossed the northern frontiers he went not simply as soldier or explorer, but as slightly winsome pilgrim too.

It was to this impressionable man, part 'B.O.P.', part Light of Asia, that Lord Curzon confided his plan for the taming of Tibet. It was almost like a Curzonian Jameson Raid. Since London would sanction nothing bolder, the Viceroy proposed first to send a frontier commission to open negotiations with the Tibetans at Khamba Jong, an almost invisible settlement fifteen miles over the Tibetan frontier from Sikkim. There the British, the Tibetans and representatives of the Chinese Amban would meet, overtly to settle the frontier difficulties, tacitly to impress upon the Tibetans—'these wretched little people'—the true strength and authority of the Empire. After that—who knew? An occupation? A permanent mission? At least,

The Great War: Anzac dispatch rider under fire, 1915

both Curzon and Younghusband undoubtedly hoped, a sphere of influence once and for all.

Younghusband leaped at the chance to lead this venture, and viewed its purpose spaciously. It would, he thought, not merely keep the Russians out of Tibet, but would 'prevent the junction of any future spheres of French and Russian influence north and south across Asia'. At Simla he and Curzon discussed the scheme. It was the height of the summer season, a gymkhana was in progress, and Simla was full of ponies, girls in summer dresses, handsome young subalterns, proud mothers and more than usually indulgent administrators. Always in sight, though, beyond the tower of the Anglican church and the flagstaffs of the Viceroy's palace, beyond the last flower-potted villa, the furthest picnic ground, the highest thicket of pines, there shone the unimaginable mass of the mountains, the ramparts of Tibet.

Over there lay the minds of the two Englishmen, as they sat in their deck-chairs watching the competitions, straw hats tilted over their eyes, Curzon now and then interrupting their conversation with viceregal formalities -- 'Well done, Miss Leatherby, splendid round'—'Such a nice boy of yours, General, I'm so glad to see him doing well.' There was a rapport between the two men. They had last met in the mountains, when Younghusband had been Political Agent at Chitral, and Curzon had been his guest ('both a pleasure and a trial', he had reported home). Both loved the feel of the high places, both relished an adventure, both were active imperialists, both were convinced of Russian designs upon Tibet. An element of collusion sealed their understanding, Younghusband feeling flattered by the great man's confidence, Curzon sensing that he had found a willing instrument. Their plans were settled among the deodars, to the laughter of memsahibs and the badinage of subalterns, and Younghusband left Simla in high spirits. 'This is a really magnificent business', he wrote to his father, home in retirement in England, 'that I have dropped in for.'

4

The first part of the plan proved, as the two men perhaps hoped, abortive. The mission crossed the frontier safely, and encamped upon a grassy knoll at Khamba Jong: but though they stayed there for five months, and Younghusband spent long hours in constructive meditation—for it was, he wrote, 'high up among the loftiest mountain summits . . . that the very essence of sublimity must be sought for'—still they achieved nothing at all. After some weeks a delegation did arrive from Lhasa to see what was happening, but its lamas firmly declined to negotiate, shutting themselves up incommunicado in a nearby fort. The mission accordingly withdrew again, and in the winter of 1903 Curzon put into effect the next part of his scheme.

This was much more drastic. Now Younghusband was to occupy an entire Tibetan valley, the Chumbi, and take his mission far deeper into Tibet, some 150 miles to the town of Gyangtse. Balfour's Government sanctioned this proposal even more reluctantly. The Prime Minister, it was minuted, 'is incredulous as to the importance of Tibetan trade, and dislikes the idea of allowing ourselves to be permanently entangled in Tibet'. Sanctioned it was, nevertheless, and so a die was cast. When Younghusband went back to Simla for consultations, he asked Lord Kitchener, the Commander-in-Chief, if he could take some British soldiers with him, to reinforce the imperial message. Kitchener saw the point at once. 'All right,' he replied, 'you shall have a section of a British mountain battery and two Maxim gun detachments, all of British soldiers, and I will give orders that not a single man is to be under six feet.'

So the British Government, far away, found themselves committed to an imperial military incursion into a neighbouring independent State. Curzon had won his point. It was to be one of the most bizarre and demanding such expeditions ever mounted. The idea of crossing the Himalayan passes in the middle of winter was dismissed by most strategists as lunatic, but Younghusband was convinced it was possible, and Curzon was determined to get into Tibet as soon as possible. The Jelap La, the main pass from Sikkim into Tibet, was rough and narrow, and no wheeled vehicle could cross it: instead everything the Mission had, all its food, all its supplies, all the tentage, equipment and ammunition for 1,200 men, had to be carried on the backs of animals and humans. Some 10,000

porters, men and women, were recruited, and nearly 20,000 animals
—mules, bullocks, buffaloes, ponies, Nepalese yaks, Tibetan yaks,
6 camels and 2 'zebrules', half zebra, half donkey, which were taken
along as an experiment.[1] Even the track had to be re-made, for it was
blocked everywhere with ice and landslides and in its upper reaches
was deep in snow.

The Mission's armed escort included Gurkha infantry and Sikh
pioneers, besides the British gunners and Maxim men, and on
December 11, preceded by a man on horseback carrying a Union
Jack, they moved off through the snow towards the Tibetan frontier.
This really was like one of the classic Anglo-Indian enterprises of the
previous century: the long winding caravan along the mountain
tracks, the monotonous chanting of the Lucknow litter-bearers—

> *Mountains are steep*
> *Yes they are*
> *The road is narrow*
> *Yes it is*
> *The sahib goes up*
> *Yes he does*
> *The sahib goes down*
> *That is so*

—the old smells of dung, sweat and leather, the great woodfires
when the camp was pitched at night, the grunts of restless animals
and lascivious porters, the handful of Englishmen huddled in their
tents as they smoked their pipes and drank their last whiskies before
turning in. On the march the expedition extended for many miles,
and when its advance guard reached the head of the passes, 14,000
feet up and cruelly cold, its rearguard was only emerging from the
tree-line far below: but by December 13, 1903 every last yak and
zebrule had crossed the frontier of Tibet, and the Mission was
encamped in a pine forest beneath the Union Jack, surrounded by its
multitude of guards, bearers and beasts of burden.

[1] An unsuccessful one: everybody learnt to loathe them, and they were
never tried again. Of the other animals, 99.2 per cent of the buffaloes, 98.9
per cent of the Nepalese yaks, and all six camels perished in the campaign.

5

Tibet had been invaded, though first to last Curzon never used the word, and a British armed force lay within the Forbidden Land. This is what the Viceroy intended all along, but the British Government at home only half realized what was being done in their name. Balfour's was not a bold Cabinet, and among its least swashbuckling adherents was St John Brodrick, who became, in October 1903, Secretary of State for India. Evasive, ambitious, with piggy eyes and a determined mouth, Brodrick was hardly an imperialist in the high old style, and it was not surprising that he had a complex about Curzon. They had been at school and at Oxford together, and in the manner of the times Brodrick had often declared his changeless admiration and affection for his glittering contemporary—whenever Curzon won a prize, or was promoted, or published a book, or became Viceroy, Brodrick was among the first to congratulate him. Festering away there, though, was an inner resentment, only awaiting outlet: and just as Younghusband's little army prepared itself for its march into Tibet, conveying Curzon's mastery across the roof of the world, the moment of revelation arrived: for at the end of October Brodrick assumed the seals of his office, and thus became George Curzon's immediate superior.

Almost the first thing he did was to make clear that he was not going to be held responsible for the Tibet Expedition. 'In view of the recent conduct of the Tibetans,' he telegraphed his old friend, 'His Majesty's Government feel that it would be impossible not to take action, and they accordingly sanctioned the advance of the Mission to Gyangtse. They are, however, clearly of the opinion that this step should not be allowed to lead to occupation or to permanent intervention in Tibetan affairs in any form. The advance should be made for the sole purpose of obtaining satisfaction, and as soon as reparation is obtained a withdrawal should be effected. While His Majesty's Government consider the proposed action to be necessary, they are not prepared to establish a permanent mission in Tibet, and the question of enforcing trade facilities in that country must be considered in the light of the decisions conveyed in this telegram.'

This was a far cry from Omdurman or Fashoda: Brodrick's directive was a *caveat*, to be quoted when the need arose, or kept as a hedge against political embarrassment. The truth was that the British Government of 1903, like the British public that elected it, was not conditioned to such an imperial *coup-de-théâtre*. Balfour and his Ministers did not really know what to do about the Tibet Expedition, and hoped it would somehow come to nothing, or alternatively not take long.

6

In fact it progressed slowly. It met no opposition, but took no chances. By Christmas, 1903 the British were sixty miles into Tibet, and had occupied the awful town of Phari, where the filth of generations was piled so high against the houses that holes had to be scooped in the excrement to give access to front doors. By the New Year of 1904 they were moving across a bitter snowy plain, where wild asses ranged and skirmished, to Tuna, 15,000 feet up. There they stayed for three months, and there Younghusband again met delegates from Lhasa. The lamas were no more accommodating than before. They denied all contact with the Russians, they said the British were behaving like thieves and brigands, and they demanded that Younghusband should withdraw at once—'using forms of speech', Younghusband reported indignantly to Curzon, 'only used in addressing inferiors'.

Not until the end of March did the expedition resume its progress towards Gyangtse. Younghusband, noting that in the mornings the snow summits became 'the most ethereal blue shading into the cerulean blue of the sky above', derived only 'a sense of peace and quiet and coming joy' from the prospect: and though a horrible wind often swept off the mountains, scoring everything with grit and stinging the soldiers' faces, still everyone was glad to be on the move again. Bent half-double against the wind, coat-collars up, heads down, the little force moved off through a landscape of gravel, rock and snow. They had brought collapsible carts with them, and these, the first wheeled vehicles ever seen in Tibet, now carried their gear. Mounted infantry fanned out in front, the four guns were in the middle, and the column now numbered a thousand rifles.

Some ten miles from Tuna, a hundred miles within Tibet, they reached a hamlet called Guru. There the track led between an escarpment and a dry salt lake, overlooked by a range of stupendous snow peaks, and there on the gravel flats the Tibetans had built a barricade against them. It was a large crude rampart of stones, placed naively across the track, and through their binoculars the British could see that all around it a mass of Tibetans milled and swarmed, like brown bees against the scree. The expedition was in no mood to be delayed, and so within a few minutes, on March 31, 1904, between 600 and 700 Tibetans sacrificed their lives to the imperial purpose.

7

The barricade was nothing. It could easily be outflanked, and anyway a few shells would destroy it. The British approached it cautiously, nevertheless, and soon a group of Tibetan horsemen rode out to meet them. They came, they said, to warn the expedition to turn back, but Younghusband dismissed them abruptly. Nothing would stop the British marching to Gyangtse, he said, and if they were opposed they would fight. The guns were assembled, the troops prepared for action, the Tibetans were given fifteen minutes to clear the road: if they failed to do so, the expedition would break a way through.

The time passed. The Tibetans did not move. In open order across the gravel plain, on a grey and bitter day, the infantry moved slowly forwards, with orders not to fire unless fired upon. Silence hung over the scene. The troops moved in absolute silence towards the silent barricade. Gurkhas silently climbed the escarpment, Sikhs silently turned the flank of the barricades, the ponies of the mounted infantry trotted silently up the valley to cover the line of retreat. Before the wall the artillery was ready: on the higher ground to the east the six-foot Maxim gunners of the Norfolk Regiment, squatting behind their tripods, cocked their weapons and waited.

Below them the Tibetans were ready to be slaughtered. They were surrounded now, entirely covered from higher ground, within easy artillery range. Their only line of retreat was blocked. Yet they did nothing. Some stood to their guns at the barricade loopholes, the rest aimlessly ambled about, watching with bewilderment the methodical positioning of the British all around. Their leader, a handsome and powerfully built lama, brown-cloaked and felt-booted, sat with sword in hand actually *outside* the defences: his men, who had never seen a machine-gun before, simply gaped and speculated. These were truly strangers meeting, beside the salt lake on the road to Gyangtse. The two sides looked at each other hardly in hostility, perhaps not even in fear, but in astonishment.

It seemed best to disarm the Tibetans, before some terrible mistake occurred: so a detachment of Sikhs and Gurkhas walked impassively into the crowd, and began to seize their matchlocks, clubs and swords. Some handed them over without protest, but others refused, until all around the wall little groups of men were struggling for weapons, sometimes wrestling, sometimes pushing and hitting each other. Now the silence was broken. There were shouts, curses, blows, and at last, late in the morning, a single shot. The Tibetan commander himself fired it. Seeing the trouble fester, he had mounted his pony and ridden into the crowd to intervene. When a Sikh seized his bridle, the lama drew a pistol from his belt and shot the Indian through the jaw.

At once a battle began. The Tibetans fired their old muskets furiously, and slashed about with their swords, and kicked, and clubbed, and shouted, but it only lasted a minute or two. The sepoys withdrew, and at point-blank range from either side the massed British riflemen opened fire upon the Tibetan mass, while the Maxims sent a stream of fire into the mêlée behind the wall. There was no need to aim precisely. The target was unmissable, and could not escape. Soon the artillery opened fire. Shrapnel shells burst behind the Tibetans, scattering iron fragments, stones and gravel among the mob. In the distance the British mounted infantry prepared to fall upon the retreat.

The Tibetans did not panic. They seemed more appalled than alarmed, as though their gods had let them down. They turned away from the wall, sorrowfully, and walked very slowly along the track to the north—'with bowed heads', one eye-witness reported. In a long straggling line of brown they retreated, figures of sad reproach, to disappear behind a spur in the hill half a mile away. The soldiers fired at them sporadically as they went, and now and then one would fall: but slowly they walked away, as if they were lost in prayer or thought, until one by one they were gone behind the hill.

The British let them go. They were aghast. 'It was an awful sight,' wrote one officer. 'The slowness of their escape was horrible and loathsome to see.' Half the Tibetan force had been killed, including its commander: six of the imperial force had been wounded. The British slept that night chastened by their own victory, and Younghusband wrote one of his less visionary letters to his father next day. 'I was so absolutely sick at that so-called fight', he reported, 'that I was quite out of sorts.'[1]

8

To Lord Curzon he reported differently, as the expedition resumed its march to Gyangtse. The massacre of the Tibetans, he told the Viceroy, had been 'an exemplary punishment' which would live in the memories of the Tibetan people for ages—especially as the total British losses had been 'two fingers and an arm'.[2] Moreover the affray had given Lord Curzon the ideal pretext for sending the expedition still further into Tibet, to the palace of the Dalai Lama at Lhasa itself.

Away in London Brodrick and his colleagues were not so sure. They were at twos and threes on Tibet. That summer Curzon went home on leave, and tried to stiffen their resolve, but he failed. Some thought the country should be placed under British paramountcy, like Nepal, or obliged to accept a permanent British Agent in Lhasa, or simply be taught an old-fashioned imperial lesson—a few houses burnt down, an indemnity exacted. Others again believed the whole venture should be cancelled, and that Younghusband should be

---

[1] Three months later the Tibetan dead were still lying there, stripped of their possessions, but 168 wounded men were treated by the British in a field hospital at Tuna. 'I think', wrote Edmund Candler, of the *Daily Mail*, 'the Tibetans were really impressed with our humanity.'

[2] And the arm only belonged to Mr Candler.

withdrawn at once to India. Scared of Russian or even Chinese reactions, timid of offering hostages to fortune, hindered by inner rivalries, sensitive to public opinion on the eve of an election, wildly ignorant of Tibet and bureaucratically suspicious of the men on the spot, they lacked the élan either to send the Flag dominant to Lhasa, or to admit an error and call the Mission home.

In all this they truly demonstrated the weakening of the Empire's confidence. Younghusband had fought his way to Gyangtse, constantly harassed by the Tibetans, and by the end of July was ready to march upon Lhasa itself: but when at last he was ordered to do so, his instructions were limp. He was not to occupy Tibet, or make a protectorate of it, or even insist upon the opening of a British Agency in the capital. He was merely to conclude a trade agreement, extract an indemnity, and insist that Tibet must not deal with foreign Powers without British consent. How these demands were to be enforced His Majesty's Government did not say, but Younghusband was not perturbed. He set off for Lhasa in high spirits, eager for the climax of 'the magnificent business'.

The weather was splendid now. The valley of the Kyi Chu was fertile and inviting. The column, reinforced by four companies of the Royal Fusiliers for the sake of prestige, plodded its way along the dusty road full of expectancy. Dragon-flies and butterflies fluttered through the thin brilliant air, clematis clambered over the roadside boulders, unfamiliar birds called, cows anomalously grazed. At each bend in the road the soldiers expected to see the fabulous city, but it was masked by a succession of spurs and ridges, and by thickets of poplars and willow trees. They were almost in the outskirts of the place when the advance guard of horsemen scrambled up a bluff and saw below them, a golden jumble of roofs, a slab of white stonework, a flash of turquoise tapestries, the legendary palace of the Dalai Lama, the Potala of Lhasa.

9

'I question', wrote Edmund Candler, 'if ever in the history of the world there has been another occasion when bigotry and darkness have been exposed with such abruptness to the inroad of science, when a barrier of ignorance created by jealousy and fear as a screen between two peoples living side by side has been demolished so suddenly to admit the light of an advanced civilization.' Once the marvellous shock of the Potala was over, though, Younghusband's entry into Lhasa offered only anticlimax. It was on August 4, 1904, that he entered the capital, preceded by an escort of the Amban's bodyguard, and followed by a company of the Royal Fusiliers. He wore full-dress diplomatic uniform, with a plumed hat and braided trousers: the British infantry behind him, in their khaki drill and sun-helmets, marched with a proper pride as the first European soldiers ever to enter the capital: but when they got inside the gates, they found only squalor and disillusion.

The Dalai Lhama had fled. There were no Russians. Except for the extraordinary palace, there was scarcely a building of interest in the place. The streets were filthy, with pools of scummy water everywhere, piles of rubbish, open sewers running between the houses, scavenging dogs, pigs and birds of carrion. Swarms of ragged monks watched with surly resentment as the Fusiliers marched spick-and-span through these disagreeable surroundings, but the ordinary citizens displayed only a vacuous unconcern—'idiotic-looking people', one Briton thought them, and pictures of the scene do indeed show some less than intellectual faces, peering in perhaps syphilitic listlessness around the frames of muddy photographs.

For several weeks nothing much happened. The Dalai Lama was well on his way to Outer Mongolia, and the National Assembly of senior lamas was frightened to reach decisions without him. The British set up their main camp outside the gates of the city, and soon erected the familiar paraphernalia of Empire. The flag flew, the bugles sounded. Servants polished officers' boots in the sunshine, grooms combed ponies' manes, troops drilled on the dusty parade-ground, the gunners greased their beloved guns, whitewash and regimental crests blossomed in the scree. They organized gym-khanas, race meetings and football matches, they fished in the Pygong river, they went bird-watching, they hunted wild donkeys in the hills outside the city.[1] Far, far away into the distance from

[1] And sent one to the London Zoo.

their camp, the signallers' telegraph poles carried the cable wire all the way back to India, and so to London: when asked by curious Tibetans what it was for, they said tactfully that it was to show them the way home again.[1]

Meanwhile Younghusband patiently groped his way towards a treaty. At first the National Assembly rejected all his demands, and very reasonably suggested that far from the Tibetans paying the British reparations, it ought to be the other way round. Day by day, nevertheless, argument by argument, Younghusband won them over, sometimes by charm, sometimes by threat. The filthy lamas grew on him: he now thought them, though 'intangible' and 'un-get-at-able', still 'extraordinarily quaint and interesting', and even 'good fellows'. He himself believed that the most important requirement was the stationing of a British Agent in Lhasa: once he was there, he thought, all else would be plain sailing—in trade, in foreign relations, Tibet would fall easily enough into the pattern of the imperial buffer-States. London, though, had specifically forbidden any such demand, so that while on the one hand Younghusband was trying to persuade the Tibetans into compliance, on the other he was hoping to evade the direct instructions of Whitehall.

He managed it in the end, if only briefly, and the scene of the signing of the Anglo-Tibetan Convention, though it had been won at a terrible cost of innocent lives, against the instincts of the British people themselves, and despite the agonized opposition of the Tibetans, remains among the jollier spectacles of the imperial pageant. Early in the afternoon of September 7, 1904, Younghusband and his staff, covered by a battery of artillery, escorted by mounted infantry, and dressed in their fullest fineries, rode across Lhasa to the long sloping causeway which, striking diagonally across the face of the Potala hill, gave access to the palace. Here, dismounting,

---

[1] Only one of many imperial anecdotes about the electric telegraph, by now essential to the mystique of Empire. Others concerned the Jameson Raiders, who were supposed to cut the telegraph wire to Johannesburg, but got so drunk they cut a barbed-wire fence instead; the Ashanti, who assumed the wires to be a potent ju-ju, and criss-crossed the forest with emulative cables of knotted creeper; and the Egyptian who was alleged to think that all the cables were connected to Queen Victoria's personal bell-pull.

they found their aplomb put to the test, for the cobbles of the causeway were so slippery that in their nailed boots they could hardly get up it. They were like men bewitched in a fairy tale. Slithering all over the place, their plumes erratically waving, their Sam Brownes and decorations pushed awry, the representatives of advanced civilization clambered farcically up the hill of bigotry, to disappear at last, panting with the altitude and hastily straightening their accoutrements, into the dim and dirty corridors of the palace, where monks with butter-lamps guided them to their appointment.

The treaty was signed in the Dalai Lama's audience hall, a large pillared room already crowded, as the British delegates entered it, with dignitaries—the Regent of Tibet, the Amban, the Abbots of the three greatest monasteries, the Tongsa Penlop of Bhutan, the Nepalese Resident—and with monks, British officers, Sikhs, Gurkhas and Royal Fusiliers. The British had arranged this peculiar Durbar didactically. 'Those who have lived among Asiatics', Younghusband later wrote, 'know that the fact of signing the treaty in the Potala was of as much value as the treaty itself.' When it came to the point, though, the ceremony was more boisterous than instructive. The dead of the long road were forgotten, the brutal intrusion of Empire was apparently forgiven, and the bonhomie of the Tibetans presently infected the British too, so that they themselves became, as one of them recorded, 'rather a rowdy lot, and made a beastly noise'.

Tea was served, with dried fruits, and the Tibetans handed the several copies of the agreement around with jokes and pleasantries. It took an hour for everyone to sign them, and then Colonel Younghusband made an amiable if schoolmasterly speech, reproaching the Tibetans for their former disrespect, but hoping that friendship and peace between the two countries would now last for ever. Everybody clapped, and smiled, and laughed, and shook hands, and drank more tea, and ate more fruit, and the proceedings broke up in high good humour. The British had been disconcerted, clambering so ludicrously up the causeway for the start of the ceremony, but going down again they found hilarious. So did the Tibetans. As the departing officers scrabbled and slid away down the ramp, almost convulsed themselves with the ridiculousness of it, they heard guffaws and high-pitched laughter somewhere above them, and looking up to the great face of the palace-monastery, saw the monks of the Potala, high on their serried terraces, merrily observing their predicament.

10

Younghusband had not abided by Brodrick's prim telegram of the preceding year. His treaty obliged the Tibetans to accept a British Agent at Gyangtse, with the right to visit Lhasa when necessary, and to deal with no foreign Powers without British consent. They were to pay an indemnity of £50,000, in seventy-five annual instalments, and until it was paid—until, that is, the year 1979—the British would occupy the whole Chumbi Valley. The British Empire in return promised nothing whatever.

This harsh agreement apparently satisfied the Tibetans, who had found the British at least more pleasant to deal with than the Chinese—'when one has known the scorpion', as one of their proverbs had it, 'the frog seems divine.' It did not, however, please the British Government in London. They had been at pains to convince the Powers, and especially Russia, that the British had no acquisitive ambitions in Tibet, but were merely concerned to regularize existing arrangements. Younghusband, in the manner of earlier imperialists, had boldly exceeded his orders, but the British of 1904 were not excited by his initiative. The days of splendid isolation were over, Britain could no longer behave as though she held the world in fief, and when the terms of the convention became known the Russian, German, American, French and Italian Ambassadors all called at the Foreign Office to protest. King Edward himself was shocked, and thought Younghusband had acted in an 'extraordinary' way, 'in direct and deliberate defiance of instructions'.

Having sanctioned, however reluctantly, an expedition which, after a year of danger and difficulty, had succeeded in reaching an agreement with the most evasive Power on earth, the British Government now repudiated it. The indemnity was reduced by two-thirds, and the Chumbi Valley was to be evacuated in 1908.

When the British left Lhasa, six weeks later, the Chinese resumed their old ascendancy, and not until 1936 did the British have their own Agent in the capital. None of it mattered anyway, as it turned out, and the course of imperial history would have been altogether unaffected if the Mission to Tibet had never been invented.

The expedition itself had gone exactly as Curzon had planned, in its gradual development from a negotiating body to a military campaign, from talks on the Sikkim frontier to the occupation of Lhasa itself. It returned safely to India like so many imperial columns before it, its soldiers receiving a handsome medal engraved with a picture of the Potala, and a campaign bar, 'Tibet 1903-4', to add to the battle-honours of the regiments. A couple of thousand men went home, briefly to enrich the folklore of Norfolk or Punjab with their tales of Tibet.

But it was an anachronism already, so soon after the excitements of *Following the Flags*, or the frenzies of Mafeking night. Even the thrills of penetrating the unknown, even the satisfactions of bringing enlightenment to the barbarians, had lost their appeal for the British. Only a Power of supreme strength and confidence could embark upon such ventures with the requisite assurance, strong enough to ignore or defy all rivals, confident enough to be sure that the deaths of poor peasants, innocent tribesmen and unquestioning soldiers could be justified by results. Ten years before Britain had been such a Power, but no longer, and the invasion of Tibet never attracted much attention in the school history books of England.

As for its participants, only Younghusband, perhaps, paradoxically found what he wanted in Tibet. He was blamed for the diplomatic embarrassments of Lhasa, being awarded only the lowest order of knighthood for his achievement, and always around his name there was to hang the suggestion, growing dimmer over the years, of an obscure disgrace. His control of the expedition, though, had been admirable, and the journey itself had been one of the most remarkable in military history, involving the highest battle ever fought—at 19,000 feet on the pass of Karu-la.

More importantly, he really did encounter in the Forbidden Land that sublime essence he pined for. When, on September 23, the

British left Lhasa and began the long march home to India, they pitched their tents on the first night within sight of the Potala. As the camp settled down to sleep, Younghusband climbed a hill nearby and looked back at the towering mass of the palace, serene now in the evening light, and bathed for him in an aura of success and reconciliation. Alone there on the hill he experienced a moment of ecstasy. An intoxicating happiness overcame him, like a transfiguring drug injected into the bloodstream. 'This exhilaration of the moment grew and grew until it thrilled through me with over-powering intensity. Never again could I think evil, or even again be at enmity with any man. All nature and all humanity were bathed in a rosy glowing radiancy; and life for the future seemed naught but buoyancy and light.'

Sure enough, presently he abandoned the imperial service, and never went to war again. He devoted the rest of his life to mysticism, religion and peaceful geography, and was buried when the time came in a Dorsetshire churchyard, beneath a bas-relief of Lhasa.[1]

[1] Losing its clarity rather now, and fuzzed with lichen, but still recogniz-able on his gravestone at Lytchett Minster, near Poole. On his coffin was placed an image of the Buddha, given him on the morning he left Lhasa in 1904, and on the footstone of his grave is remembered his only son: 'A Tiny Child, Francis Charles, 1898'.

CHAPTER EIGHT

On Power

IN 1905 a Liberal Government came to power in London, under Henry Campbell-Bannerman. With its advent, the New Imperialism died: it was the end of Jingo—what the new Prime Minister had once called 'the vulgar and bastard imperialism of irritation and provocation and aggression . . . of grabbing everything even if we have no use for it ourselves'. The British people, Edward Grey said, were back to normal. This was fortunate, for by now Salisbury's threatening 'aggregation of human forces' seemed more menacing every year. There was war in the air, not simply the running colonial war in which the British had been engaged for a century or more, but the greater international conflict which their own dominant power had so long prevented. All the symptoms were brewing; economic rivalries, patriotic frustrations, the ambitions of leaders, dynastic squabbles, the general sense that an epoch was disintegrating and could only be cleared away by violence.

The British could not escape this gathering maelstrom, however pacific their new mood. They were traditionally the regulating Power of the world. Their navy made them militarily significant beyond their size, their command of raw materials gave them unique economic leverage, their possessions everywhere gave them a universal stake, while the immense accumulated wealth of the City of London was a hidden factor even in the banks and chanceries of Europe. The Pax Britannica really had kept the peace of the world for three generations. Now, for the first time since the end of the Napoleonic wars, the task seemed beyond it.

So we see the British, so recently all-confident, apprehensively looking for allies. The newly federated German Empire, under Wilhelm II, was emerging as the potential enemy: almost anybody would do as a friend. Nation by nation the British patched up their relationships, in agreements, concessions or full-blown alliances: with Japan, with France, by 1907 even with Russia, when the Great Game was ended at last in an Anglo-Russian convention. Within the Empire, too, they repaired their friendships. Lord Kitchener toured the Dominions to enlist their military potential. An Imperial General Staff was established. Churchill devised a scheme for an Imperial Squadron of warships from all the major colonies, to be based upon Gibraltar and sent wherever it was needed.

If there was a slight air of desperation to these preparations, it was because time was running short. The Edwardian age, that mellow epilogue of Victorianism, died with its eponymous patron in 1910, and the world became more urgent. There were repeated naval scares, when tales of the vast new German battle-fleet cast a chill over England. Louis Blériot flew the English Channel, making the islanders feel a little less insular, and the liner *Titanic* sank on her maiden voyage, making them feel rather less titanesque. The trade deficit grew bigger every year: by 1910 the profits Britain made from invisible exports, insurance, banking, services of many kinds, no longer covered it, so that for the first time the British could not pay their way in the world by their own skills, but depended upon investment income from abroad.

In this shifting world only the Empire itself seemed to stand firm, still maintaining in its strongpoints and bases across the continents a posture of unassailable power: and this magnificent façade exerted a profound psychological effect upon its enemies, as well as upon itself. Just as the British economy now depended upon investments made by a previous generation, so the British reputation rested upon the constructions of earlier imperialists. For a century the imperial outposts had been an inescapable reminder of the British purpose, from Esquimalt at one end of the Empire to Hong Kong at another, and between them they kept an ostentatious watch upon the world: Bermuda covered the Atlantic, Simonstown and Trincomalee commanded the Indian Ocean, Singapore looked out to Japan and Australia, Gibraltar and Alexandria were sentinels of the Mediterranean. The very sight of them on the map, let alone from the bridge of a reconnoitring warship, had been an inhibition to the Powers. As Thackeray had written long before,

> *Jerusalem and Madagascar,*
> *And North and South Amerikee,*
> *There's the British flag a'riding at anchor,*
> *With Admiral Napier, K.C.B.*

2

Far in the east the jealous foreigner, sailing up the China coast, would discover the Crown Colony of Hong Kong, by which the British staked their authority throughout the further East. Not so long before it had been supposed that one day Queen Victoria might add China itself to her dominions, and it was said of the British in Hong Kong that they had cut a notch in China as a woodsman marks a tree—'to mark it for felling at a convenient opportunity'. In the event the British had stopped short of an empire in China, and now shared Chinese Spheres of Influence with several European Powers: when in 1900 the Chinese rose against foreign interference in the Boxer Riots, Britain was only one of eight nations which sent troops to put them down. Hong Kong, nevertheless, kept a watching brief upon the affairs of the Chinese land-mass, and regularly sent its gunboats up the Chinese rivers to protect the vast British interests there. Further north the British possessed another base on the China coast, the naval anchorage of Wei-hai-wei: but Hong Kong was the centre of their power in the Far East, and seemed to stand there four-square and perpetual as an earnest of British stability. Every Englishman must feel a thrill of pride, Curzon once wrote, to visit this 'furthermost link in the chain of fortresses which . . . girdles half the globe': and every foreigner, the intention was, should feel a *frisson* of respect.

Its nerve-centre was the Peak, the rocky promontory which crowned Victoria Island, and upon whose slopes were concentrated the military, economic and social power of the colony. There the flag pointedly flew, there the morning bugles sounded, and there up the elevated railway the tired rich Englishmen went home in the afternoon, to their delectable garden villas above the heat. Hong

Kong was overwhelmingly Chinese in the mass: besides the island, it included a slice of territory, leased for ninety-nine years in 1898, which was actually part of the Chinese mainland, and looked it, with its walled villages and its shuttered temples, the willow-patterns of its country scenes and the immemorial postures of the workers in its paddy-fields. In the particular, though, nowhere in the Empire was more unquenchably British, for an element of defiance heightened its imperial attitudes, and made of it, so unsympathetic visitors sometimes thought, almost a caricature.[1]

3

The great men of Hong Kong were the taipans, the British merchants who had been in the colony from the beginning, and had never relaxed their grip upon its trade. They lived very grandly, sometimes in Hong Kong, sometimes in country houses at home in England, and their firms were household names throughout the east—Jardine Matheson, Butterfield Swire, names which spanned the generations, and seemed changeless constituents of eastern life. The taipans were opportunists by definition, quick, clever, rich men, who had absorbed some of the Chinese qualities themselves in so many years upon the coast, and who were there not for the sake of the Empire, still less to undertake a share of the White Man's Burden, but to maintain their annual profits. Their attitudes were imperial nevertheless, and often dazzled foreign visitors by sheer hubris. The colour bar in Hong Kong was absolute, and the British held themselves almost psychotically aloof from the swarming Chinese who were their work-force. They seldom spoke Chinese,

[1] *Its leading characters are wise and witty,*
Their suits well-tailored, and they wear them well,
Have many a polished parable to tell
About the mores of a trading city.

Only the servants enter unexpected,
Their silent movements make dramatic news;
Here in the East our bankers have erected
A worthy temple to the Comic Muse.
*   —W. H. Auden, 'Hong Kong', Collected Poems.*

and to heighten the lordly effect their servants talked to them in pidgin English, a literal translation from the Cantonese which gave their employees an air of child-like dependency. The British residents of Hong Kong genuinely, without affectation, thought of the Chinese as foreigners in the colony, and themselves as true natives.

The colony's government was, of course, altogether British. The Governor lived in lavish style on the slopes of the Peak, guarded by soldiers at his gate.[1] The administrators and generals and admirals of the island lived hardly less consequentially. They ran it with a paternal authoritarianism, making few concessions to liberal thought, and almost none to Chinese susceptibilities. Everywhere the symptoms of Empire showed: the ships always steaming in from India, Australia or Britain itself, the Indian soldiers who often formed its garrison, the Sikh policemen and hotel doormen, the Australian jockeys who won all the races at Happy Valley, above all the British mesh of the place, the webs of money, style and sovereignty which bound the colony so unmistakably to the imperial capital far away. English equity was the basis of Hong Kong's trade, with British cash to keep it flowing, and British power to back it.[2]

4

There was a caravanserai feeling to Hong Kong, a feeling of

[1] And almost certain to be commemorated, when the time came, in a street name, a quarter or an institution. Today Pottinger Peak, Mount Davis, Bonham Strand, Bowring Town, Blake Pier, Peel Rise, Northcote Hospital, Grantham College, Sir Cecil's Ride and Robinson, Nathan, Macdonnell, Kennedy, Hennessy, Bowen, Des Voeux, Lugard, May and Stubbs Roads are all dedicated to gubernatorial memories. In 1973 academics at the University of Hong Kong analysed the nouns, verbs and adjectives most frequently used in the *South China Morning Post:* 'Governor' came third.

[2] British condescension, too. When in 1910 the powerful new German cruisers *Scharnhorst* and *Gneisenau* arrived in Chinese waters, the officers of the British China Station promised that if it ever came to a battle, their own armoured cruisers would give the Germans a sporting chance, and use only nine of their ten 7.5-inch guns. Both German ships were to be sunk by the Royal Navy, using all its weapons, at the Battle of the Falkland Islands in 1914.

movement and opportunity and intelligence, as a ceaseless flow of traders, scholars, refugees and wanderers moved through it in and out of Asia. This gave a cogent impression of British omniscience. The colony's newspapers were always full of Chinese news, commercial information from Shanghai, trade reports from up the Yangtze, rumours from Tibet or Mongolia, lists of travellers arriving from Canton or Vladivostock. The Peninsula Hotel, one of the supreme imperial hostelries, was associated with the Grand Hotel des Wagons Lits in Peking, and the red brick and mahogany railway station, down by the water-front at Kowloon, was linked by daily train with Canton and the Chinese capital. It was a caravanserai, though, conducted strictly to the imperial rules: undesirables were briskly deported, and the very first thing the new arrival saw, when he stepped out of the train from Shanghai, Peking or even Moscow, was a trimly uniformed British policeman, in pith helmet and khaki at the end of the platform.

It was very beautiful. At first most visitors were intoxicated by the setting of the place, and did not grasp its full meaning. The colony lay there all a'bustle, with the ferry-boats passing and re-passing across the crowded harbour, the steam-launches sweeping away from the quay outside the Peninsula, the ships steaming in and out, the flags, the trams and the rickshaws, the carriages crawling up the flowered Peak, the deep tireless hum of life from the Chinese tenements. Away down the coast lay lesser islands, green and silent in the sun, and in crystal bays the junks lay anchored, awkward but elegant, looking as though they never moved from their moorings at all, but lay there for ever as upon a china plate. Outside their barracks the sentries stamped and strutted: in a thousand workshops the Chinese diligently worked their crafts, the apothecaries and the silversmiths, the tailors and the magicians, the shipwrights and the bookbinders and the calligraphers, bringing to the island some profounder stability from China itself.

But the beauty was misleading, for the real purposes of Hong Kong were brutal. The colony had been acquired to facilitate the selling of opium to the Chinese, and even in 1912 a quarter of its revenue came from the official monopoly in the opium trade. It was there to make money, by means fair or relatively foul—and by being there, by seeming to bristle with bayonets, riggings, wealth and self-assurance, to advertise the Law of Empire.[1]

5

Far away, and even more explicit, lay Malta. This had been one of the great British naval bases since its acquisition in 1814, and its sole purpose now was to serve and display the Royal Navy's Mediterranean Fleet. Half-way between Gibraltar and the Suez Canal, it was at once a defensive station on the imperial route to India, and a watch-post off the southern flank of Europe, and no fortress on earth looked more satisfyingly commanding. The ramparts of the Grand Harbour, built by the Knights of St John during the two centuries of their regime, were colossal in themselves, but doubly so when they sheltered, mirrored in the still waters of the haven, the towered and turreted battleships of His Majesty's Fleet.

The Navy was dominated still by its capital ships. Some of them were always based at Malta, for everyone to see, and their tremendous shapes were inescapable in Valletta, the capital, glimpsed at the end of city streets, or basking in the sun, like sea-monsters, below the public gardens. In 1910 there were six on station, a sixth of the Royal Navy's battlefleet. Four were vessels of the *Duncan* class, looking stolid, almost ecclesiastical, with their close-set funnels like pince-nez and their bridges like church chancels. Two others were the striking *Triumph* and *Swiftsure*, built originally for the Chilean fleet, and still curiously exotic—tall, slim ships, with slender funnels and ornamental prows, and large goose-neck cranes, between the masts, which made their silhouettes unmistakable to the

---

[1] Of all the imperial possessions Hong Kong was to retain its posture longest, for to this day it is not only still a Crown Colony, living still by wits and paternalism, but it is larger in population than all the rest of the surviving Empire put together. There is a garrison of Gurkhas, and the taipans on the Peak are richer than ever, being active in shipping, trading, banking, insurance, speculation and agency all over the Far East—'I'm so sorry,' I was told once when I inquired after the friend of a friend, 'but Mrs W— is away in Japan, launching a ship.'

sailors of any Navy.[1] Flotillas of cruisers, destroyers, torpedo boats and submarines attended these great ships, and the Commander-in-Chief, Mediterranean, flew his flag sometimes on his flagship, *Cornwallis,* and sometimes from the yardarm of his comfortable Admiralty House ashore.

In British eyes Malta was an extension of this fleet. They saw the island specifically in terms of firepower, fuel and repair works. Here there were no hinterlands to be defended, spheres of influence to be disputed, squabbling chiefs to be pacified, or even taipans to pursue alternative priorities. Malta was a fortress, no more, no less, and nobody who went ashore on the island could escape the fact. Malta felt both powerful and permanent, for the British were of the opinion that the base was absolutely essential to the security of the Empire, and therefore destined to be British for ever and ever.

As it happened, too, this was one place in the Empire where the British had a true respect for their environment, or at least for its history. Towards the contemporary Maltese they had adopted a benevolent, tolerant, but superior pose, erecting a kind of imaginary colour bar (Maltese being not much swarthier than Welshmen), excluding Maltese from the best clubs, and seldom inviting them to dinner-parties. Towards the previous rulers of the island, though, they could scarcely be patronizing. Not even a British sub-lieutenant could sneer at the Knights of the Order of St John of Jerusalem. In many ways they had behaved just as the British did themselves, and honoured similar values. Fighting capacity, a taste for splendour, the team spirit and the family tradition—all these characteristics of the Order exactly matched the British imperial ethos, and made the Knights in retrospect seem a very decent lot.

So a towering sense of continuity dignified the British presence in Malta, and made it seem more formidable still. The magnificent architecture of the Knights, the white ramparts above the Grand Harbour, the great gates and the stepped city streets, had become by symbiosis part of the British imperial tradition. The house of the Commander of Galleys was now the house of the Dockyard Captain. The Fort of St Angelo flew the White Ensign as naval

[1] Their offices were still marked in Spanish, and they were known to the Navy as *Occupado* and *Vacante.*

headquarters, and its slave dungeons were used as ammunition stores. Where the galleys had careened the cruisers refitted. The narrow alley called Strait Street, a byword among the Knights for sex and skulduggery, had long been adopted by the bravos of the Navy too, and was known throughout the Fleet as 'The Gut'. And in the Palace of the Grand Masters themselves, where generations of soldier-priests had ruled the island, a British Governor now sat in similar authority, distributing the pomp of Empire in a very knightly manner.

6

It all looked down to the Fleet. All the sailors' bars, the Happy Return, the Cricketers' Arms—all the whorehouses and cafés and cheap souvenir shops—the Union Club, the Opera House, the bosomy Methodist Church, the Anglican Cathedral spired above the harbour—the dock-workers streaming home when the evening hooter sounded, the sentries with their bayonets fixed at the Main Guard in Palace Square—all the mass and life of Malta, stacked there so vividly above the sea, looked always towards the warships in its lee. A visit to Malta was less an experience than an indoctrination, and among the grandest historical spectacles of the day was the return of the Mediterranean Fleet, after exercises at sea, to its incomparable haven.

For its commanders this was a supreme professional moment, and they handled it theatrically, hoping foreigners were watching. Some Admirals preferred the approach stately, the grey ships treading steadily and regally to their moorings; some liked to do it at maximum speed, with froth, hiss, split timing and brilliant displays of seamanship. Either way, the impact was tremendous. Crowds hastened to the quays to watch the fleet come in, flags ran up poles and yardarms, children hopped about in excitement, wives chatted happily beneath their most fetching parasols. The distant thump of a band, the muffled thudding of engines—and there they were! First the destroyers, smoke streaming, swept past the harbour mouth to their Sliema moorings. Then the big ships approached in line ahead, their decks lined with ratings, their lamps flashing, their Marine bands overlapping in rich discord as, one by one, they entered Grand Harbour to the strains of 'Rule, Britannia' or 'Hearts of Oak'. Presently, in an atmosphere of festive splendour, they were at their moorings, and the Commander-in-Chief, piped overboard from his flagship in the sunshine, was scudding across the water in his steam-barge to pay his respects to the Governor.[1]

All the ships were moored facing out to sea, for it was a maxim of Malta that the Mediterranean Fleet was ready for instant action at all times, against all comers.[2]

7

Such, many times multiplied, was the show of power. The majesty of it all, and especially the arcane and gorgeous ritualism of the Navy, certainly helped to overawe potential enemies of the Crown, even as the century entered its second decade, for it gave an impression of strength more than merely military or economic, but actually organic. The Empire provided a kind of talismanic screen, like the mystery of ikons and war-banners behind which mediaeval armies went into battle.

But it was only half true. In many ways the Empire made the British weaker rather than stronger—it presented, John Morley the Liberal had said in 1906, 'more vulnerable surface than any empire the world ever saw'. Now the British reluctantly attended to their less recondite defences. In 1908 Campbell-Bannerman was succeeded as Prime Minister by Herbert Asquith, a rather less liberal Liberal. In 1910 Edward VII was succeeded as King by George V, a more imperial Emperor. A succession of reformers

[1] Unless he was Sir John Fisher, commander of the Mediterranean Fleet from 1899 to 1902, who preferred to enter harbour first and walk up to the Barracca Gardens to watch the Fleet moor. Once the flagship of his second-in-command, Admiral Lord Charles Beresford, was so ineptly handled that Fisher sent him a malignant signal of public rebuke, observed throughout the Fleet: 'Your flagship is to proceed to sea and come in again in a seamanlike manner.'

[2] In 1972 I went to Malta, by then an independent State within the British Commonwealth, to see what was left of the Mediterranean Fleet. All I could find was the wooden minesweeper *Stubbington* (360 tons).

worked to implement in the British Army the lessons of Spion Kop and Magersfontein: its structure was altered top to bottom, its tactics and weaponry were drastically revised, at last the hidebound traditions of the parade-ground and the open square were jettisoned in favour of field-craft and initiative. By 1914 it was no longer an imperial army in the old sense, trained specifically for the imperial purposes. Its core was now a highly professional expeditionary force ready for service anywhere, but particularly against European enemies.

At the same time Admiral Fisher, now a volcanic and visionary First Sea Lord, prepared the Royal Navy for duties more demanding than patrolling the China coast, showing the flag in South America, or entertaining sheikhs to wardroom dinner parties. Ruthlessly he cut away the dead wood of the Fleet, most of it the rot of Empire: ships that were good only for showing the flag, drills that were performed only for exhibitionism, unnecessary stations, irrelevant exercises. Oil power was introduced, and the British Government acquired a controlling interest in the oilfields of southern Persia, specifically for the fuelling of the new warships. The focus of the Fleet was shifted, from the blue waters and the far horizons to greyer waters nearer home: the commander of the home fleet, Fisher used to say, was the only man who could lose a war in an afternoon. By the summer of 1914 almost all the great ships were home, ready to face a European enemy in northern waters—of all the Royal Navy's battleships, only seven were on the imperial stations.

8

They were just in time. In 1897 the Kaiser, already in possession of Europe's most powerful army, had ordered the construction of a German High Seas Fleet. He had seen it from the start as a deliberate challenge to Britain's command of the seas, and so to the established order of things. One day, he told his admirals and constructors, it would be God's Instrument of Justice—'until then, silence and work'. By 1914 the work was done, the silence broken, and as Europe burst like an abscess into war, Queen Victoria's Empire found itself challenged by equal force of arms for the first time since she had succeeded to the throne, almost eighty years before. The grand illusion was collapsing. The Powers no longer waited, straightening their ties or adjusting their ribbons, upon Admiral Napier, KCB.

CHAPTER NINE

The First War

THE Pax Britannica came to an end in August 1914, and the British Empire entered its first general war. It was not really an imperial war, but many an old imperial preoccupation, the Eastern Question, the Overland Route, Cape-to-Cairo, Freedom of the Seas, the Supremacy of Race, now came home to roost. From distant seas and perfumed stations the cruisers came hurrying back, and men whose most formidable enemies had been the fugitive commandos of the veld armed themselves to face the greatest army in Europe. The recruiting offices of the Empire were besieged once more by volunteers, as they had been fifteen years before at the start of the Boer War: but now the patriots offered themselves in a different spirit, a spirit of willing sacrifice, almost of sacrament, knowing that this was a conflict separate in scale, in kind and in consequence. They went to South Africa to the jingles of Alfred Austin and the music-hall balladeers. They went to the world war with a very different Poet Laureate, Robert Bridges:

> *For Peace thou art armed, thy Freedom to hold:*
> *Thy Courage as iron, thy Good-faith as gold.*
> *Thro' Fire, Air and Water, thy trial must be:*
> *But they that love life best die gladly for thee.*
>
> *The Love of their mothers is strong to command;*
> *The fame of their fathers is might to their hand.*
> *Much suff'ring shall cleanse thee, but thou through the flood*
> *Shall win to salvation, to Beauty through blood.*
>
> *Thou careless awake! Thou peacemaker, fight!*
> *Stand, England, for honour, and God guard the right!*[1]

[1] Published in *The Times*, on August 8, 1914, four days after the declaration

To Beauty through blood. The entire British Empire went to war with Germany and her allies that August, all 450 million subjects of the Crown being bound by a single declaration from the King-Emperor. The imperial mobilization was presided over by the most famous of all the imperial fighting men, Lord Kitchener, who was immediately appointed War Minister, and the Empire's response surprised even the British themselves. 'Our duty is quite clear,' announced the Prime Minister of Australia, 'to gird up our loins and remember that we are Britons.' Within ten days New Zealand had despatched an expeditionary force of 8,000 men. Within two months 31,000 Canadians had been recruited, drilled and sent to Europe. The South Africans, led by the conciliatory Smuts and Louis Botha, the victor of Spion Kop, not only sent soldiers to Europe, but also took on the task of evicting the Germans from their colonies in south-west Africa, while from every last island, promontory or protectorate came volunteers, offers of money or at least flowery messages of support.

There were dissenters. In Ireland, as we shall presently see, old enemies of the Empire obeyed an old Irish dictum—'England's trouble is Ireland's chance.' In India and in black Africa a few premature nationalists chose this moment of crisis to rebel. In South Africa 10,000 'last-ditch' veterans of the Boer War rose to arms rather than support the war, and had to be put down by their old comrades-in-arms. In Canada many French-Canadians opposed the war as a matter of sectarian principle. Generally, though, the Empire attained a unity in conflict which it had never achieved in peace, and the fact of war in Europe gave an unexpected boost to the imperial ideal. The war propaganda of the British was rich in Sons of Empire, Bonds Across the Sea, posters of stalwart Sikhs or devoted Sudanese, cartoons of Britannia attended by her statesmen, her young men of the frontiers, or her familiar imperial menagerie—lion, tiger, emu, springbok, kangaroo and beaver.[1]

---

of war. Before the year was out it was set to music by Arthur M. Goodhart, and dedicated to 'LORD ROBERTS and to all Etonians serving in the War'. 5,687 were to serve: 1,160 were killed.

[1] Or sometimes, when a more stately Canadianism was required, moose.

2

The bugles of England (wrote J. D. Burns of Melbourne)[1] *are blowing*
   o'er the sea,
Calling out across the years, calling now to me.
They woke me from dreaming in the dawning of the day,
The bugles of England: and how could I stay?

To the wife of a volunteer in Winnipeg or Alice Springs, to the
Sikh rifleman from Amritsar or the illiterate Gold Coast askari, the
bugles must have sounded distant indeed. Why were they going?
What were they fighting for? The first Gurkha detachments for
Europe, approaching Calcutta for embarkation, sharpened their
kukris as the train drew into the city, supposing that they were
nearing the battlefront. The epicentre of the war was western
Europe, its issue was essentially the balance of power in Europe, its
deepest causes lay not in the overseas enterprises of the Powers,
but in cloudier and more complex differences within their own
cramped neighbourhoods.

The imperial soldiers found themselves transported, all too
often, direct from their own sunlit spaces to the mud and drizzle of
Flanders and France, where they floundered and died, were gassed
or mined or mutilated, shivered in the unaccustomed cold or
miserably ate their alien rations, year after year, trench after trench,
sadness after sadness to the end. But the war reached out to the
imperial territories, too. German ships were hunted down in atolls
of the Pacific or in African creeks, campaigns were fought to capture
the German colonies of Africa and the Pacific, and at the back of
British strategic thinking there often lay an imperial aim or instinct.
The more imaginative of the British strategists cast their eyes
beyond the confined butchery of western Europe, towards the
greater landscapes and grander chances of their imperial tradition.
Kitchener himself was baffled by the trench fighting in France—
'this isn't *war*', he used to say—and his planners were divided into
rival schools, 'westerners' who believed the whole effort should be
directed towards victory in Europe, 'easterners' who responded to

[1] Killed in France, 1915, aged twenty.

these more liberated impulses. The latter looked in particular towards an old arena of imperial intrigue and aspirations, the Ottoman Empire.

The Turks entered the war on the German side in October 1914. Theoretically they were suzerains of almost all the Arab World, even including Egypt, and their emergence now as enemies conveniently clarified the anomalous British position in the Middle

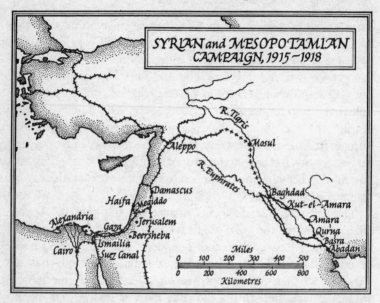

East. Traditionally the British had supported Turkey—whatever happened to the Ottoman Empire, Lord Salisbury had decreed, could only be for the worse, 'and therefore it is in our interest that as little should happen as possible'. Also the Sultan of Turkey, in his capacity as Caliph of the Faithful, commanded the spiritual allegiance of some 70 million Indian Muslims, and was not lightly to be antagonized by the temporal rulers of India. The British already possessed, though, a shadow-empire of their own in the countries that lay between India and the Mediterranean. They ruled Egypt and Aden, they controlled the Persian Gulf, as the greatest Muslim Power they felt themselves to have special interests in Arabia, the Holy Land of Islam. Their oil supplies came

The first rebels: Easter Rising, Dublin, 1916

from southern Persia, which they had declared a British sphere of influence, and all over the Arab countries they had agents and emissaries, from the diplomats of the Levant Consular Service to the young men who, from time to time, struck out from Kuwait or Bahrein to make contact with the Bedouin of the Arabian interior.

Now all was clearer. By April, 1915, an official committee was discussing what ought to be done with the Ottoman Empire after the war—divided among the victorious allies, partitioned into zones of influence, or set up as some kind of independent State. Whatever happened, the committee agreed, British interests must first be safeguarded, in an area particularly important to the Empire: and the best way to safeguard interests, as everyone knew, was to control the place yourself, whether you did it frankly or covertly. The war against Turkey *was* an imperial war. The Turks, whose military reputation was dim but whose armies were powerfully stiffened by German generals, officers and men, presently called for a *jihad*, a Muslim holy war against infidels.[1] Unsuccessful attacks were made upon Aden and the Suez Canal. The British responded in kind, and embarked upon three offensives against the Turks. One was a miserable kind of victory. One was a resplendent success. One was a great defeat, and gave to the imperial annals the most poignant of all their tragedies.[2]

[1] 'You are enjoined to make war against the infidels even though this may displease you: for you may well be displeased by that which is to your good, as it may be that you like a thing which instead is harmful to you: but God knoweth and you know not': *Koran*, 11: 216.

'Oh Muslims! Embrace ye the foot of the Caliph's throne and join in the *jihad*, the Holy War. Warfare is ordained for you. Your enemies will not cease until they have made you renegades from your religion if they can. Drive them out! If they attack you, slay them! Such is the reward of unbelievers': *Sultan Mehmed V, the Caliph, November 1914.*

The title of Caliph was abolished after the war, and has never been revived.

[2] I have disarranged the order of them, in telling the story. Here is the correct chronology:

	1914	1915	1916	1917	1918
Mesopotamia	······· —				·
Gallipoli	·······	····	··········	··········	··········
Syria	············	············	············	········· —	

The half-naked fakir: Gandhi in London, 1931

3

The Turkish possession called Mesopotamia, the Land of the Two Rivers, now Iraq, had been a subject of British anxiety for years. It was one way to get to India. British explorers, surveyors and spies knew it well, and a British company, Lynch Brothers, had a monopoly of steam navigation on its rivers. In recent years the Germans had been active there, building their Berlin to Baghdad Railway, and the imperial strategists had long been agitated by the thought of an enemy moving down the line of the Tigris and Euphrates to the Persian Gulf. By 1914 they were concerned too for the security of the Persian oilfields. The oil was pumped to a refinery on the island of Abadan, at the head of the Gulf, which lay directly across a narrow channel from the Mesopotamian port of Basra. Once Turkey came into the war, the British assumed, it might only be a matter of hours before the oil was cut off, the refinery destroyed and half the Royal Navy immobilized.

The invasion of Mesopotamia—'Mespot' to the British Army— began then as an operation to seal off the Persian Gulf, and protect Abadan by seizing Basra. The Gulf had always been supervised from India rather than Britain, so it was an Indian Army expedition, commanded by an elderly Anglo-Indian general, equipped from Indian sources, escorted by the battleship *Ocean* from the East Indies Station, which in November 1914 sailed up the Gulf in four transports and successfully disembarked in the port of Basra. The Turkish authorities, such as they were, fled. The local Turkish forces, consisting mainly of Arab conscripts, did not long resist. The rambling mud city behind its waterside palms readily submitted, and in no time at all there was a British Political Office there, and an indefatigable Political Officer, Percy Cox, Indian Political Service, brisk behind his files in his upstairs office.

Well and good: Turkish counter-attacks were beaten off: but a new general, Sir John Nixon, assumed the command in the spring of 1915, and he had plans to go further, and to move his troops northward up the line of the rivers. He was partly concerned to push his defence lines further inland, by the truest principles of the

Forward Policy, but he was also impelled by profounder imperial instincts. In this he was not alone. In London the War Cabinet felt that, with Abadan safe, the expedition's task was done: but the activists of the Indian Empire had their eyes on greater prizes, on Baghdad, everyone's epitome of an oriental city, and on Mesopotamia itself, which to some strategists, as well as many theologians, was truly the heartland of the world. Cox had already proposed an advance to Baghdad. Nixon had been ordered to prepare plans. 'So far as we can see', minuted the Indian General Staff in Simla, 'all advantages, political and strategical, point to as early a move on Baghdad as possible.'

Southern Mesopotamia, where the Tigris and the Euphrates flowed, first separately, then united, towards the Persian Gulf, was more beguiling in history than in fact. Here were Babylon and Nineveh, here Sennacherib had fought his battles, here indeed, some said, had been the Garden of Eden at the start of the world. But it was a fearful country now. Much of it was empty desert, inhabited by lawless predatory Arabs who loathed nearly everyone, the rest wide and foetid fen, inhabited by amphibious marshmen who detested everyone else. The irrigation works of the ancients had long since crumbled, and the long years of Turkish rule had left only decay and depression. There were no paved roads, no railways. Such towns as existed were hardly more than excretions of mud, like piles of rubbish in the wasteland, relieved only by the minarets of shabby mosques, or the lugubrious walls of forts. In the summer it was indescribably hot, in the winter unbearably cold. In the dry season everything was baked like leather, in the wet season 10,000 square miles were flooded, the waters gradually oozing away to leave malodorous wastes of marsh. Fleas, sand-flies and mosquitoes tormented the place, and its inhabitants lived lives of ignorant poverty, enlivened only by sporadic excitements of crime or brigandage, the illusions of religion and the consolations of sex.

Is this the land of dear old Adam (one British soldier wondered),
And beautiful Mother Eve?
If so dear reader small blame to them
For sinning and having to leave.

There was almost nowhere in India as forbidding as this. It offered none of the boyish stimulations of frontier war, upon which the Indian Army had been raised: yet intuitively perhaps, led on by the highway of the great rivers, and by the promise of great achievement, General Nixon deployed one of his two divisions northwards, towards Amara, Kut and Ctesiphon, and the distant prospect of Baghdad. In April, 1915, the Viceroy of India, Lord Hardinge, arrived himself to inspect the situation[1]: at the end of May the imperial forces set off for their first objective to the north, Amara.

4

The British field commander was Major-General Charles Townshend, of the 6th (Poona) Division, of whom we caught a sentimental glimpse when, as a promising captain, he stood at Kitchener's side in the ruins of Gordon's Residency. Now fifty-four, he was a moody, contradictory, eccentric man. The heir presumptive to a marquisate, he had high social aspirations, besides being intensely ambitious. He relished publicity, loved to strike poses with his banjo and fluent French, but was a soldier of mercurial gifts—a soldier out of the common mould, who studied Napoleon assiduously, and perhaps thought himself a successor. Townshend was much liked by his British soldiers, but not by his Indians: and this was perhaps because he had no high regard for Africans and Asians in general, having beaten them in battle from Chitral to Omdurman, and proposing to do it again now.

The country south of Amara was flooded, and the main Turkish defence position was a series of islands in the flats, protected by mines. To attack it Townshend put one of his brigades upon the water. His infantry paddled out in 500 commandeered Tigris boats, called bellums. His machine-guns were on rafts, his field guns in tugs and barges, and there were launches to sweep the minefields, and three sloops of the Indian Marine to give supporting fire. It was all an exhilarating success. All among the flooded palms this

---

[1] Still seriously incapacitated by an attempt on his life in Delhi two years before. It is not true, Mr John Bowle the historian assures me, that when the bomb exploded in his howdah the Viceroy cried 'Save the elephant!'

amphibious army sailed, the launches first, the sloops next, the troops in their armada of quaint craft punting themselves forwards with poles. When they came to a Turkish position, every gun was turned upon it until, the frightful mess of mud and explosives having subsided, the infantry waded ashore with fixed bayonets to capture it. The Turks did not long resist this unnerving process. By noon on the day of the attack they could be seen scurrying away in boats from every remaining island; by nightfall the whole Turkish force was in retreat, helter-skelter up the Tigris to Amara.

First thing in the morning Townshend was off, personally, in pursuit. His brigade was embarked upon paddle-steamers, but the general himself, with his staff and a few soldiers, boarded the sloop *Espiègle* (1,070 tons), and with her two sister ships, *Clio* and *Odin*, and four steam-launches, charged up the river at full speed after the retreating enemy. This was 'Townshend's Regatta', 'the Mespot Navy', and it was one of the most exuberant little actions of the First World War. The three ships looked more like yachts than warships, with their bowsprits and elegant prows, and their crews treating the operation like a sporting event, wildly navigating a river which swung about in violent curves, had dangerous shallows everywhere, was mined in many places and had never been properly charted.

The current was fierce. Time and again they bumped the muddy banks, and they generally had no notion what lay around the next bend. Far ahead of them raced the Turks: two steamers towing barge-loads of troops, and trailed all about by a fleet of river dhows, tacking frenziedly back and forth against the current. In the evening, when the British spotted their distant sails and opened fire, the Turkish steamers hastily slipped their tows and left the soldiers to their fate: but Townshend, detailing the *Odin* to take them prisoner, swept past the confused flotilla of barges and riverboats, and pressed on excitedly for Amara.

Next day they caught the two steamers, one abandoned, one flying a white flag, and it was time for the sloops to turn back. The water was only a few feet deep now. Townshend, though, had the bit between his teeth. With a handful of soldiers he transferred to the paddle-launch *Comet*, which had once been the official yacht of the British representative in Baghdad, and with the three other launches, each towing a barge with a 4.7-inch gun on it, sailed impulsively on. By now his brigade was far behind, but as his boats raced up the river, White Ensigns fluttering, rifles bristling, white flags appeared in one astonished riverain village after another, and all along the banks Arabs bowed low in submission.

Early the next afternoon, June 3, 1915, they saw Amara, a big brown sprawl beside the river. It was swarming with Turkish troops, but when the first of the launches approached its quay hundreds of them walked down to the bank with their hands above their heads. Townshend's brigade was 100 miles behind him. His total force on the spot comprised one brigadier-general, one naval captain, 100 soldiers and the crews of the launches. He sent ashore a corporal and twelve men, and to them the Turkish commander surrendered the town and all its garrison. Two thousand men gave themselves up, and all afternoon the Turkish officers queued on the quayside beside the *Comet* to hand over their swords. 'Safely captured', one of them is supposed to have cabled his wife that evening.

'How much I enjoyed the whole thing', the general wrote to his French wife. 'I told you, darling, that I only wanted my chance! I have only known the 6th Division for six months, and they'd storm the gates of hell if I told them to. . . .'

5

Yet before another year was over they were all to follow his orders, diseased and defeated, into the appalling prison camps of Turkey, where half of them died. Elated though Townshend was by the success of his Regatta, after so long in Mesopotamia his army was tired, ill-equipped and under strength. It was debilitated by dysentery and paratyphoid, it had no fresh vegetables, no ice and a pitiful supply of medical equipment. Townshend himself was hit by so severe a virus that he was shipped back to India for sick leave. It was known that 30,000 Turkish reinforcements were on their way, including divisions of a higher quality, German-trained and probably German-commanded. By now General Nixon, though, had his eyes firmly upon Baghdad. Though London was still lukewarm about further advances, Simla had doubts, and Townshend himself was cautious about taking inadequate forces too deep into enemy territory, still by the beginning of September, 1915, the 6th Division, Townshend back in command, was ready to move north again.

At first all went as brilliantly as before. Townshend captured the next river town, Kut-el-Amara, in a *tour de force* of surprise and deception, and seemed both to his own men and to the Turks one of those generals who cannot lose. General Nixon reported to Simla and to London that he felt strong enough for an immediate attack on Baghdad itself, and more by default than by initiative, for the Cabinet was obsessed with the greater war elsewhere, and desperate for victories anywhere, the advance was sanctioned. Among its purposes was, so Asquith told the House of Commons, 'generally to maintain the authority of our Flag in the East'.

Townshend prepared the new advance with misgivings. He thought Kut was as far as they could go—his forces were inadequate, his lines of communications were insecure, and the Turks were entrenched astride the Tigris and the road to Baghdad. Sure enough at Ctesiphon, only eighteen miles from the capital, where the great mud-brick arch of the Sassanian emperors rose gaunt and lonely in the desert, he suffered his first rebuff. There stood a Turkish army of very different calibre, powerfully reinforced by veterans from other fronts, skilfully positioned behind wire entanglements in the presence of the Arch itself.[1] Exhausted and thirsty in the terrible heat, encumbered by sick, wounded and thousands of prisoners, the 6th Division drove the Turks out, but could go no further themselves. The force was isolated and vulnerable, and Townshend decided to withdraw: sending his non-combatants down the river to Basra, he retreated himself with his army back into the walls of Kut-el-Amara. There within a week he

[1] Which the British gunners were forbidden to shell, even if the Turks used it as an observation post. The Arch is crumbling slowly anyway— Victorian travellers reported far more of it—and is nested in nowadays by rollers, greenish-blue birds like big jays, who find its brick construction convenient.

was besieged. The Turks completely surrounded the town, and on December 8, 1915 a bombardment began.

Some 13,000 British and Indian soldiers, with thirty-nine guns, formed the garrison of Kut. They still had faith in their general, 'Our Charlie', who had led them with such racy skill all the way north from Basra, and he himself exuded confidence at first. His communiqués to his men were frank, breezy and full of spirit. 'Reinforcements are being sent at once from Basra to relieve us', he said in the first of them. 'The honour of our mother country and the Empire demands that we all work heart and soul in the defence of this place.' It was a dreadful enough place to hold. Built within a loop of the Tigris, Kut was a very dirty, very dejected, very apathetic town. Surrounded on all sides by the flat and featureless desert, its only green the palm groves and gardens on the outskirts of town, it was a brown muddle of baked mud buildings, with a mesh of narrow streets meandering through it, the pimple-domed arcade of a covered bazaar, and the minaret of a mosque protruding above its roofs like a marker, to announce its presence in the waste. Its only industry was a liquorice factory, reached by a bridge of boats across the river: for the rest it lay there all but defeated, it seemed, by the heat, the emptiness, the poverty, the tedium of life. Some 6,000 Iraqis were immured in Kut too. The sanitation was frightful, and dysentery was endemic.

Kut was the *dingiest* of the famous sieges that brought drama and often despair to the story of the British Empire. Perhaps this is because for once there were no Englishwomen there, depriving the soldiers of the challenges to chivalry that ennobled or enlivened Lucknow and Ladysmith. They were alone in Kut, and lived in a sad man's world of obscenities, comradeship, bad food and bawdy. Townshend himself, though he put a bold face on the situation, knew that he depended entirely upon relief from the south. 'In military history, the history of entrenched camps is bound up with capitulations . . . history presents very few examples of the self-deliverance of an army once invested.' But weeks and months passed, summer turned to autumn, autumn to the bitter winter, and no relief came. Down at Basra, where Nixon saw his ambitions in disarray, all was confusion. The port was a shambles of ill-organization, the command was enmeshed in red tape and rivalry, the medical services were a disgrace.[1] When General Sir Fenton Aylmer's Tigris Relief Force moved off, with a fulsome preliminary message to Townshend on the radio—'Heartiest congratulations on brilliant deeds of yourself and your command'—it moved so slowly, was led so badly, and faced such tough opposition, that it lost half its own men before it got within 100 miles of Kut.

The siege itself was a squalid affair. The Turks sniped and shelled the town incessantly, and gradually it crumbled. The rations shrank, the sick list grew, soon no more Arab looters were shot, because there was nothing left to loot. Sometimes there were moments of heroism, or at least excitement. During the first few weeks the Turks repeatedly attacked the town frontally, advancing in dense grey mass across the open countryside, to be decimated by machine-gun fire, or thrown back in hand-to-hand fighting. Once two brave young officers dashed out of town and blew up a bridge of boats. Sometimes aeroplanes flew over and dropped small bundles of supplies. The King-Emperor sent a message—'I, together with your fellow countrymen, continue to follow with admiration the gallant fighting of the troops under your command against great odds, and every effort is being made to support your splendid resistance': the knowledge that the garrison was serving the Crown, Townshend gallantly replied, was 'the sheet-anchor of our defence'. Sometimes, as the months passed, they heard distantly from the south the gunfire of the relieving force, and saw its flashes in the sky, far away over the wastes and marshes.

Mostly, though, it was boredom, and dirt, and hunger, and sickness. Morale began to sag. A few Indians tried to desert. A trace of self-pity entered the general's thoughts. 'What worries, what trials, what anxieties!' he wrote in his diary, and in one of his

---

[1] 'When the *Mejidieh* was about 300 yards off', wrote one medical officer, describing the arrival of a hospital ship at Basra, 'it looked as if she was festooned with ropes. The stench when she was close was quite definite, and I found that what I mistook for ropes were dried stalactites of human faeces . . . the whole of the ship's side was covered.' When he complained about conditions, he was described by Nixon's supply chief, General Cowper, as 'an interfering faddist'.

communiqués to his troops he even blamed the higher command for their troubles—'I speak straight from the heart,' he told them, 'and you see I have thrown all officialdom overboard.' By February, 1916, the garrison was on half rations, and five or six men were dying every day from sickness and debilitation. In March an aircraft dropped a letter from General Aylmer, not very encouragingly letting the garrison know that he had been removed from his command—'Goodbye and God bless you all,' concluded this comfortless communication, 'and may you be more fortunate than myself.' By mid-March people were talking of surrender. Townshend himself decided that April 17 was the longest he could hold out, and now General Khalil Bey, the Turkish commander, sent him a courteous but chill suggestion. 'You have heroically fulfilled your military duty,' it said. 'From henceforth I see no likelihood that you will be relieved. . . . You are free to continue your resistance at Kut or to surrender to my forces, which are growing larger and larger. Receive, General, the assurance of our highest consideration.'

By mid-April life at Kut had reached its limit, so Townshend reported to Basra, but there was one last tantalizing excitement. A Royal Navy gunboat, the commandeered Lynch paddle-steamer *Julnar*, tried to break through to the town with 250 tons of supplies. She was navigated by a former pilot of the Euphrates and Tigris Navigation Company, and she set off on her desperate voyage in bright moonlight on the night of April 24. There was no hope of surprise, and as she sailed upstream she was greeted by continuous blasts of fire from both banks. On she went nevertheless, probing her way through the shallows, until only a few miles from Kut she struck a steel hawser stretched diagonally across the stream, and floundering there under a storm of machine-gun fire, was swept across the river and run aground. The Kut garrison had heard all the noise with gathering hope, but when dawn came, and they looked eagerly downstream, they could see the silent shape of the gunboat stranded on the bank, a last broken pledge of their hopes.[1]

[1] The Turks, refloating the boat and taking possession of her cargo, renamed her *The Gift*. They shot her navigator, who was awarded a posthumous VC.

'Whatever has happened, my comrades,' said Townshend in his last communiqué, 'you can only be proud of yourselves. We have done our duty to King and Empire: the whole world knows that we have done our duty.' The surrender, on April 29, 1916, was formal and full of ceremony, in the sudden silence that succeeded the incessant rumble of the guns. Khalil, who was thirty-five and very sure of himself, shook hands with the British officers, as each surrendered his sword; when it was Townshend's turn the Turk made a short flowery speech, expressing regret that such a brave soldier should be in so grievous a situation, and handing back his sword with the phrase 'It is as much yours by right as it has ever been.'

But it was all a charade. Townshend indeed was taken away to a comfortable exile in Constantinople, where he spent the rest of the war in a pleasant island villa, attended by an aide and an orderly, and living a busy social life. His men were not so lucky. After this, the most abject British capitulation in British military history, they entered upon two years of appalling captivity—first in a terrible march northwards through the desiccated provinces of the Turkish Empire, from Kut to Baghdad, from Baghdad to Mosul, from Mosul into Turkey: and then into pestilential prison camps, where they lay sick, half-starved and ill-used until, in the last month of the war, they were released at last. More than half of them died, and of those that survived, many were never healthy again, but bore for the rest of their lives the cruel stigmata of Kut.[1]

6

London took over, Nixon was removed, the army was reorganized, by the end of 1916 they were ready to try again: and in a campaign of cautious thoroughness, backed by overwhelmingly superior forces, General Sir Stanley Maude pushed his way once more up the

---

[1] Stigma of another kind attached itself to Townshend, for he was blamed for neglecting his troops in captivity. The Turks released him shortly before the end of the war to arrange terms of surrender, but though he was knighted and elected to Parliament, he received no further military appointments, and died disgruntled in 1924.

two rivers, and on March 10, 1917, captured Baghdad at last.[1] But later that year, on the western flank of the Turkish Empire, a very different offensive was launched. Mesopotamia was invaded from India, and the muddled and laborious campaign bore all the heavy hallmarks, Curzon's 'diurnal revolution', of Anglo-Indian method. The invasion of Syria was organized from Egypt, and its style was set by a commander of forcible decision, Sir Edmund Allenby, 'The Bull'.

Allenby, who arrived in Egypt from France in June 1917, was an Englishman through and through. He was laterally descended from Oliver Cromwell, had been conventionally educated at Haileybury and Sandhurst, and was commissioned into the cavalry. He was a very big man, big-featured and heavily-boned, and he struck strangers often as terse and austere: but he had a gentle private side to him, thoughtful and imaginative. Blissfully married for twenty-five years (he had lost his only son in France), he loved children, birds and books, and was a competent classicist. Almost alone among senior British generals, he had never served in India, and he heard his first shots fired in anger during the Boer War.

Allenby had not been at his best in France, but in Syria he seldom put a foot wrong. He was one of those men who needed the lonely freedom of absolute command: in the presence of superiors he could be awkward and flannelly, but given complete authority he revealed an unsuspected brilliance. He had a terrible temper, and sometimes lost it unforgivably, but those who liked him loved him, and thought him truly a great man—a 'giant', he was variously described by doting subordinates, 'a true leader', 'one who could move mountains', with 'a dreamlike confidence, decision and kindness'. He was not an attractive man to look at, being as bovine as

[1] Nixon died in eclipse, in 1921, having been blamed by a Commission of Inquiry for the catastrophe of Kut ('his confident optimism was the main cause'). He outlived the beloved and victorious Maude, though, who died of cholera in Baghdad in November, 1917, and was buried there. This is what Colonel Dickson, Maude's South African Director of Local Resources, wrote of the event:

> *Batteries have told the listening town this day*
> *That through her ancient gate to his last resting-place*
> *Maude has gone north.*

his nickname suggested: but there was an English strength and fairness to him which made men trust him—and when a man knows he is trusted, as Allenby said himself, 'he can do things'.

The Syrian war had begun in January, 1915, with a Turkish offensive against the Suez Canal. It was easily beaten off, but when in the following year the British counter-attacked across Sinai, laying a railway and a water pipeline as they went, they were held at Gaza, on the southern edge of the fertile Palestine plain, and severely repulsed. It was a shoddy performance—'nobody could have saved the Turks from complete collapse', said Lloyd George, 'but our General Staff'—and in June, 1917 Allenby was sent from France to redeem it. By now Lloyd George was Prime Minister, heading a Coalition Government that included such imperialist stalwarts as Curzon, Balfour, Churchill and Milner, and employed several members of the South African 'Kindergarten'. More even than the march upon Baghdad, the invasion of Palestine was to be an imperial campaign, for in Cairo as in London prescient minds saw in the control of Syria the key to the command of the whole Middle East. As we shall presently discover, the British tangled themselves in their own subtleties, in their devices to control the Arab countries after the war, but the first essential step was to establish a right of conquest. To General Allenby thus fell the task of staking out a new imperial province, almost a new Empire. Within four months of the fall of Baghdad, he was preparing his armies for a march to Jerusalem and Damascus.

7

They were devious war aims, and Allenby's campaign was fought with a maximum of snare and subterfuge. He surrounded himself with staff officers of high intelligence, sometimes of scholarly learning, and moved his headquarters from Cairo into the Sinai. There he studied every aspect of Syria, its history, its geography, its flora and fauna, its diseases and its resources. Lloyd George sent him a copy of George Adam Smith's Historical Atlas of the Holy Land,

[1] 'I was convinced', the statesman later wrote, 'that this work was a better guide to a military leader whose task was to reach Jerusalem than any survey

to use as a campaign aid.[1] He also read Herodotus and Strabo, and pored over the Old Testament. Around him he assembled an army of astonishing complexity, like a crusade: it included soldiers from Britain, Australia, New Zealand, India, South Africa, Egypt, Singapore, Hong Kong and the West Indies, besides three battalions of Jews enlisted in the Royal Fusiliers.[1] In the Arabian peninsula, to the south, British agents had been encouraging Arab tribal leaders to launch their own rebellion against Turkish suzerainty, and raiding posses of Arab camel-men, often led by British officers, were already active blowing up Turkish railway lines and harassing isolated Turkish garrisons.

Allenby launched his offensive in October, with the help of a famous (and much-publicized) ruse. One day that month Captain Richard Meinertzhagen, a clever, conceited and highly opinionated young intelligence officer, rode into the desert no-man's-land that separated the British and Turkish lines around Beersheba, some thirty miles south-east of Gaza. He was spotted and chased by a Turkish cavalry patrol, but after a mile or so dismounted and opened fire upon them. They galloped on, shooting from the saddle. Meinertzhagen hastily remounted and rode on, but as he did so he suddenly lurched sideways, as though he had been hit. Desperately he galloped away towards the British lines, and the Turks saw that he had dropped his binoculars, his rifle, his waterbottle and his haversack, which was bloodstained.

When the Turks returned to camp and handed the haversack to their intelligence officers, it was found to contain some money, a letter, a notebook, the agenda for a meeting, orders for an attack on Gaza and a telegram reporting preparations for a reconnaissance around Beersheba. So was accomplished one of the most admired of all intelligence deceits, whose effect, the official historian of the campaign was later to write, 'was hardly to be matched in the

---

to be found in the pigeon-holes of the War Office.' Lloyd George's own copy of the atlas is now mine, and I have used it in writing this chapter.

[1] Among them volunteers from the United States, Canada and the Argentine. David Ben Gurion, future Prime Minister of Israel, and Jacob Epstein the sculptor both served in this force, whose motto was said to be 'No Advance Without Security.'

history of modern war'. The Turks, and their German commanders, took the captured information seriously, standing ready for a reconnaissance at Beersheba, a major assault on Gaza. Allenby did precisely the opposite, capturing Beersheba after a diversionary bombardment of Gaza, outflanking the Turkish positions, and then switching his main thrust back to the coast again. So the pace and tone of the campaign was set from the start: Allenby was away, and within two weeks his divisions had broken through the Gaza line and were streaming up the Palestine coast to Jaffa.[1]

Allenby's invasion was the last great cavalry campaign in history, and he fought it as a cavalryman, making sweeping use of his 12,000 British, Dominion and Indian horsemen. He experimented with every permutation of bluff, feint and innovation—aircraft, armoured cars, launches on the Dead Sea, offshore torpedo boats, Arab raiders, propaganda leaflets, even a few tanks, the first to be used outside Europe. Time and again he outflanked the Turks, and though he repeatedly halted for rest and consolidation, and was sometimes rebuffed, and delayed for several months in central Palestine, still the impression his offensive leaves is one of swarming khaki hordes pouring, like Mongols or Huns, irresistibly through the passes, up the valleys, into the plains of northern Syria. It took exactly a year, to the day, and its climax was the devastating victory of Megiddo—the Armageddon of the ancients—in September, 1918. There on the Plain of Esdraelon, in one of the most absolute victories of the entire imperial record, the British destroyed the Turkish armies in Syria, effectively putting an end to the Ottoman Empire, and leaving the Middle East vacant for new suzerains. Allenby professed himself 'almost aghast' at the completeness of his success.

A more triumphant moment still, though, for any *aficionado* of imperialism, occurred ten months earlier. Then the first imperial troops, advancing into the hill country of Judaea, reached the

---

[1] Meinertzhagen survived to write the standard work on the birds of Arabia, dying aged eighty-nine in 1967. Though not Jewish, he became a passionate Zionist. In this he was following the example of his great-grandmother, who had launched a Return of her own, setting off for Palestine with a white donkey and a group of Jewish adherents, but getting no further than Calais.

heights of Nebi Samuel, traditionally the home of the prophet
Samuel. The Crusaders had called this place Mons Gaudii, Mountain of Joy, because from here they caught their first sight of
Jerusalem: now the British too, through the mist and rain of a
December morning, saw the Holy City golden-domed below them.
To capture Jerusalem must be the dream of any general: to a general
of the British Army it was the summit of fulfilment. This was the
capital of capitals, last entered by a Christian army 700 years
before, and even in 1917, when the vocation of Empire was faltering
rather, the Holy City remained perhaps the ultimate objective.

It had been agreed between the combatants that there would be
no fighting in Jerusalem, almost as holy to Muslims as it was to
Christians. The Turks resisted strongly in its outskirts, the Welsh
Division having to fight hard for the Mount of Olives, but on
December 8 they withdrew to the north and left the city undefended. Next morning, as the British advance troops waited uncer-
tainly outside the walls, they saw approaching them from the city
a dignitary in a long black coat, striped trousers and a tarboosh,
accompanied by a man with a white flag. It was the Mayor of
Jerusalem, clutching in his hand the keys of the Holy City. The
soldiers did not quite know what to do. Two sergeants greeted him
first, offered him a cigarette and had their pictures taken. Then he
was passed politely from post to post, headquarters to headquarters, still holding his keys, still attended by the man with the flag,
until at last General J. S. O'Shea, commander of the 60th Division,
seized history's opportunity and accepted the keys of Jerusalem.

In those days the walled city of Jerusalem stood almost un-
cluttered by suburbs on its rocky site. Its seven gates were still its
everyday entrances and exits, its ramparts, undulating with the lie
of the land, were as complete as they had been in the time of Saladin.
The army which was now encamped outside it, bivouacing among
the olive-groves, lay there just as it might have lain in mediaeval
times, the smoke from its fires rising all around the city, the dust of
its vehicles in clouds or plumes across the landscape. Three days
later, when General Allenby took possession of the city, he honoured
this sense of timelessness, and entered Jerusalem simply and quietly,
on foot.

He went through the Jaffa Gate, 'the Gate of the Friend' in Arabic, which was traditionally the entrance of the foreigner and the travelling merchant. The last foreign visitor of such eminence had been the Kaiser, who arrived there in 1898 in a ceremonial entry of preposterous pomp, preceded by brass bands, white-liveried Uhlans, Arab cavalry and German sailors, being himself dressed in a white silk cloak shot with silver, and wearing on his head a helmet surmounted by a gigantic golden eagle.[1] Allenby's entry was staged in deliberate contrast. He went more as a pilgrim than as a conqueror. The troops who lined the dusty road to the walls were dressed in their battle-frayed khaki, and so was the general himself. The citizens of Jerusalem jostled all about, as though it were not a moment of war at all, but the celebration of some perennial festival. Bearded black-hatted priests leant like schoolboys over the ramparts of the gate above, and all the heady mixture of the Jerusalem populace, its Greeks and its Jews, its Armenians and its Arabs, its Christian nuns and its Muslim imams, pressed behind the line of soldiers, or jammed the balconies of houses. No guns were fired, no flags were flown. Only the bells of Jerusalem rang.

Behind Allenby walked the American, French and Italian military attachés serving with his armies, and a group of British staff officers, some looking elated, some looking reverent, as though they were going to church, one in particular wearing a crumpled and ill-fitting uniform and moving with an air of distracted and rather donnish detachment. All alone at their head walked the general, and to most people there he undoubtedly represented not so much Christianity, or even the Allied cause, as the British Empire. He walked briskly and expressionlessly, without sword or stick, his boots dusty, his braided cap slightly tilted. Up the sunny incline he marched, in silence, and into the deep shadow cast by the gate, his footsteps echoing as he walked beneath its arch: and then, emerging into the bright sunshine on the other side, Cromwell's descendant, at the head of the armies of the Empire, entered the Holy City. He is said to have remembered the moment always as the climax of his life.

[1] The Kaiser did not actually pass through the gate, though, for a neighbouring section of the city wall had been pulled down, and a roadway laid, to allow him a more ample access.

On the steps of the Citadel—St David's Tower, where some scholars believed Pilate passed his judgement on Jesus—Allenby read a proclamation, declaring Jerusalem to be under the jurisdiction of the British Empire. Then, in the open square outside, between the Pool of Hezekiah and the Church of St Thomas, he gave audience to the dignitaries of Jerusalem—the Mayor still in his black coat, the hereditary keepers of the Holy Sepulchre, the Chief Rabbi furred and thickly bearded, the Orthodox Archbishop in his tall hat with a cross on top, the hooded Russian prelate. One by one they came forward to exchange greetings with the conqueror, and to accept his assurances of peace, tolerance and freedom of worship. Then General Allenby and his officers walked back along the ancient streets, through the Gate of Friends once more, to their waiting horses on the Jaffa Road. Already a British military administration, settling into the seat of Pilate, was reorganizing the affairs of the city upon imperial lines.[1]

8

But of the three imperial campaigns the one most truly instinct with the grandeur and the hollowness of Empire was the one that ended in defeat. The British would presently forget they ever captured Baghdad or Jerusalem; Kut and Megiddo would mean nothing to the sons or grandsons of the soldiers; but long after the Empire had

[1] So many armies have fought over this ground, before and after Allenby's, that few traces of his campaign are left. When I soldiered in Palestine, thirty years later, every blue-eyed Arab was said to have had an Australian father, there was a bust of Allenby beside the wells at Beersheba, and the main crossing over the River Jordan was Allenby Bridge, erected by the Royal Engineers in 1918. Even then, though, the most vivid memorial of the campaign was the inscription cut by the imperial army in the rocks of Nahr el Kelb, north of Beirut, where passing conquerors since the days of the Pharaohs have left their graffiti: it was a fine thing, I thought, to see Allenby commemorated there among such company—Rameses II, Nebuchadnezzar, Tiglath-Pileser III, Sennacherib, Marcus Aurelius and Selim the Turk.

Allenby himself became a viscount (Lord Allenby of Megiddo and Felixstowe), a field-marshal, Gold Stick in Waiting and High Commissioner in Egypt, before going to his grave in Westminster Abbey in 1936.

ended altogether, Britons would remember the names of Gallipoli and the Dardanelles.

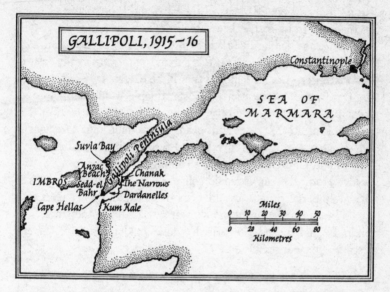

Pallid between their hills, linking the wine-dark waters of the Grecian world with the Sea of Marmara, the Bosphorus and the Black Sea, the straits called the Dardanelles had always been close to the British imperial consciousness. They lay at the core of the Eastern Question, that prolonged inquiry into the balance of authority. Only through this single channel, a mile wide at its narrowest point, could Russian ships gain access to the Mediterranean, and immediately beyond it lay Constantinople, the very crossroads of international power, where Asia and Europe faced each other, and the cold waters met the warm. What was more, the Dardanelles figured so largely in the classical education upon which English gentlemen were habitually raised, that every educated man knew of the myths and dramas that surrounded them, the heroes who frequented them, Xerxes and Leander, Priam and Ulysses. The southern gate of the Dardanelles was the Hellespont, where Leander drowned; nearby stood Troy itself, and all around lay the islands of the ancients, Lemnos and Imbros and Samothrace of the Gods.

In January, 1915, the British evolved a plan to force these famous straits, and send a fleet through the Sea of Marmara to Constantinople.[1] Ostensibly its first purpose was to give help to the Russian armies fighting the Germans in the east, and thus reduce pressure on the western front; but it was really to be a coup in the old style. Winston Churchill, at forty a dashing First Lord of the Admiralty, was its chief begetter, and he was a man brought up to the éclat of imperial enterprise. He believed that if the Royal Navy got through the Dardanelles fortifications, Constantinople would fall before its sheer presence. For a century, after all, the Navy had been the invincible arbiter. Even to Europeans it was daunting: to Asians it was incalculable, almost mystical. Faced with this intimidating instrument, Churchill reasoned, steaming admonitory towards the Golden Horn, the Turks would collapse in panic and possibly revolution. Imperial armies would occupy Constantinople, Germany would be threatened from the rear, at a bold stroke the whole war might be ended, and the British would hold in their hands the destinies of the nearer East.

9

The Navy itself had doubts about its mission. For all its prestige, it was no longer the close-hauled force of Nelson's day, addicted to risk, blind eye and heart-on-sleeve. It had fought no equal enemy for a century, and had grown stately over the years. Though Fisher had pruned, fittened and toughened it, still an attack upon so narrow and strong a strait was against all the service's instincts. Ship pitted against fort was bad enough at any time: ships committed at such a moment to a sideshow, far from the war's centre, verged upon the irresponsible. Heavy guns would be needed to force a passage, and the commanders of British battleships were no longer men of piratical boldness. Their ships were holy to them: not merely their homes and their prides, but their trust and faith too—their cathedrals so to speak, in whose turrets and superstructures were embodied almost religious responsibilities. The right place for a

[1] They had done it once before. In 1807 Admiral Sir John Duckworth forced a passage with seven ships of the line, but beat a rather too hasty retreat while the going was good, and became Governor of Newfoundland.

battleship, the Navy thought, when it was not fighting a fleet action, was safely in base behind torpedo nets, where the very fact of its existence was worth a couple of divisions.

Fisher, recalled at seventy-four to be First Sea Lord again, had misgivings himself. He had inspected the Dardanelles in 1900, when he had commanded in the Mediterranean, and he knew how strong were the forts, castles and emplacements, some modern, some mediaeval, which were embedded on each side of the Dardanelles from one end to the other. So it was mostly older battleships that he agreed, almost against his better judgement, to commit to the operation, reinforced for the sake of the alliance by a squadron of equally elderly French battleships. Most of them were due to be scrapped within the next fifteen months, and they were manned chiefly by crews left over from the Grand Fleet at home. Fired though by Churchill's enthusiasm, for the two men loved each other, Fisher agreed to add two modern capital ships: the battle-cruiser *Inflexible*, fresh from victory over the Germans in the battle of the Falkland Islands, and the *Queen Elizabeth*, the latest, fastest, most powerful and most beautiful battleship in the Royal Navy, so new that she would actually do her gunnery trials in action.

Ship by ship—some came from China—this fleet was gathered in the eastern Mediterranean. With its twenty-two capital ships, and its supporting cruisers, destroyers, minesweepers and seaplane carrier, it was the biggest naval force ever seen in those waters. After a series of preliminary bombardments and raids, like solemn warnings or declarations of intent, on March 18, 1915, this argosy of Empire, White Ensigns hugely billowing, captains princely on their flying bridges, began its direct assault upon the Dardanelles. It was like the return of the Fleet to Malta, played now in earnest: a last demonstration of naval war in the high Victorian idiom.

We are told that few who went in with the Royal Navy that day ever regretted the experience, for in all the history of warfare there was never a more momentous spectacle. 'The proud feelings that possessed us', a petty officer wrote, 'can scarcely be described.' This was the imperial theatre at its most colossal. The Straits did not look impressive from the open sea, only a narrow slit of water between the low green headlands, but on that brilliant March morning they did look ominously expectant. Everyone in the Fleet knew they were heavily defended; successive lines of minefields were covered by fixed guns on both shores, by mobile howitzers, by searchlights, submarine nets, and fixed torpedo tubes. Against these formidable defences the Royal Navy sailed head-on, fair and square, in the old style. There was no element of surprise. The objective was clear to all, the method self-explanatory. The Navy would destroy the Turkish guns, sweep the minefields and blast a way through to Constantinople.

When the Turkish gunners looked down the strait that morning, to the open sea between Cape Hellas and Kum Kale, they saw approaching them out of the south the towering grey forms of sixteen capital ships. In the van was the splendid *Queen Elizabeth*, the largest warship ever to enter the Mediterranean, her unprecedented 15-inch guns cleared for action. Around and behind her sailed a chivalry of warships, old perhaps, actually Victorian most of them, but all the more threatening for their sense of age and grandeur: the British tall and castle-like, the four French ships squatter, uglier, like huge armoured toads or turtles.[1] There were the unmistakable *Triumph* and *Swiftsure*, which we last saw in Grand Harbour at Valletta; there was the wolfish *Inflexible*, whose guns three months before had destroyed the *Scharnhorst* and *Gneisenau*, fresh from the China Station, on the other side of the world; there was the old *Ocean*, which had convoyed General Nixon's expeditionary force to Basra; there were *Agamemnon* and *Lord Nelson*, *Majestic* and *Vengeance*, *Albion* and *Prince George* and *Irresistible*. Their combined firepower was enormous, but their air of history was scarcely less imposing.

They failed, *Inflexible*, *Irresistible*, *Vengeance* and all, and instantly a myth was shattered. The Royal Navy was not omnipotent, and gunboat diplomacy, here carried to its ultimate expression, was no longer sufficient to discipline the natives. The battle lasted all day, nine in the morning to five in the evening, but the battleships never penetrated the Narrows, the strip of water between Kilid Bahr and

---

[1] The 'fierce-face' look peculiar to French battleships since the 1870s, and defined by Sir Oscar Parkes as 'piled-up superstructure, preposterous masts, uncouth funnels, tumble-home sides and long ram bows'.

Chanak which was the neck of the Dardanelles, and the minefields were never cleared. All day long they bombarded the forts, trying to batter their way through. The noise of the guns echoed and re-echoed, up and down the narrow waterway, clouds of cordite drifted between the headlands, the shores flashed with the fire of the Turkish guns. Among the forts the shells exploded in sprouts of black and brown, among the warships, hour after hour, the Turkish projectiles plunged with huge columns of spray. The sun shone all day, but it was obscured for hours at a time by the smoke, the dust and the spray of the battle.

The Turks showed no sign of surrender: it was the British who faltered. In the afternoon the minesweepers were sent into the Narrows. They were trawlers manned by civilian crews, and caught there as in a pool, in a storm of fire from both banks, shells lobbed from the howitzers on the bluffs, direct fire from the guns so close on either bank, they lost their nerve and turned tail. In that moment the battle was lost. When the French battleship *Bouvet* suddenly exploded, capsized and disappeared before everyone's eyes in a couple of minutes—when the *Inflexible* hit a mine and withdrew, listing heavily, between the other ships to the open sea—when the *Irresistible* was mined too, abandoned, and left drifting towards the Asiatic shore—when the *Ocean*'s steering gear was hit, and she steamed in helpless circles before the Turkish guns—when the barrage of fire from the Turkish forts did not, like the guns of Alexandria thirty years before, wilter and die—when it proved that the mystique of the Royal Navy could not force a passage through the Dardanelles, the great ships, turning heavily in the narrow waters of the straits, abandoned the assault and disappeared to sea, where presently the Turkish lookouts on the heights could see their dark silhouettes and billowing smoke-trails scattered among the islands of the archipelago.

Of the 176 guns that defended the Dardanelles, only four had been put out of action; of the 392 mines, not one had been cleared; but the Fleet had lost 700 lives and three great ships. The Turks fearfully awaited a renewal of the attack next day, but it never came. An age was over, and the Royal Navy never again tried to win a war by sheer *superbia*.

10

Fisher resigned—'Damn the Dardanelles! They will be our grave!'[1] Churchill wanted to try again, but he was overruled, and instead the commanders on the spot determined to land an army on the peninsula of Gallipoli, forming the western shore of the Dardanelles, and so open the way to Constantinople. One legend died, another was born. 'A wonder', is how one participant described the invasion of Gallipoli in 1915. 'Yes, that is the word for those days. The scenes, the men, the actions, the great ships, the smell of thyme mixed with cordite, the knowledge that immortal history has been made before one's eyes. I do not praise war, but there I saw deeds rise fully to the heights of a great issue, in a noble setting, giving a quality to those days, with all their suffering, that aeons of grey evolution can hardly attain.'

Everything about Gallipoli conspired to haunt men with this sense of tragic nobility. The peninsula itself was haunted, not by ghosts but by their absence. It was an arid, empty place, like a blank slate awaiting a message. Its hills, from whose summits one could see the straits on one side, the Aegean on the other, were covered with scented scrub, and it lay there sparse and aromatic, a long pile of land above the sea, ribbed everywhere, like an old skin, with gulleys and ravines. In summer it could be beautiful: the sunbaked downland of the peninsula, over whose expanses hawks hovered, through whose shrubbery lizards flickered; the silent strait below; the deep blue of the Aegean humped with its islands; the distant mountains of Anatolia grey-blue to the east. In winter it could be terrible: bitterly cold, eerie in some lights, above all sterile. Within sight across the water was the mound of Troy, with all its bright and high-flown memories. Below were the Dardanelles, through whose channel down the centuries had passed so many warriors,

---

[1] King George V thought he ought to be hanged at the yardarm for desertion, but the old admiral, growing rather grotesque in his last years, remained a national figure until his death in 1920. He was entered for the Navy by the last of Nelson's captains, his first ship was HMS *Victory*, and he died in the presence of the Duchess of Hamilton.

kings and pilgrims. But nothing had ever happened on the Gallipoli ridge, and the hills with their severe Turkish names, Anafarta, Keretch Tepe, Sari Bair, stood there loveless in the sunshine, sinister in the shade.

To this place there came an imperial army. It was the first such force in the Empire's history, and it was charismatic. Its soldiers bore themselves, so the poet John Masefield thought, 'like kings in a pageant', and its commander was a courteous and cultivated Scottish gentleman. Ian Hamilton, born in Corfu in the days when the British ruled the Ionian Islands, first entered the footnotes of history at Majuba Hill, where he was wounded, and since then had three times been recommended for the Victoria Cross. He loved the 'storm and wild joy of battle', and had seen more action than any other senior officer, certainly in Britain, probably in the whole of Europe: his left hand had been shrivelled since Majuba, his left leg had been shortened by a wound on the North-West Frontier of India, he had fought in Afghanistan and Burma and the Sudan and the second Boer War, and had been a military observer with the Japanese armies in their war against Russia in 1904. Yet he was not at all a belligerent figure, and he approached the military art sensitively, almost delicately, less like a general than an artist, or a stage director. Brave, imaginative, sixty-three years old, he went to Gallipoli in a spirit of grateful dedication, assured by Kitchener that if he won he might be winning not simply a campaign, but a war. As he sailed to his battle station the Aegean seemed to him 'like a carpet of blue velvet outspread for Aphrodite', and he observed the subsequent campaign with the same lyric response, compassionately, without ferocity or savagery: like a gentleman, in fact.

It was before he embarked for Gallipoli that Rupert Brooke wrote his elegiac poem 'The Soldier':

> *If I should die, think only this of me:*
> *That there's some corner of a foreign field*
> *That is for ever England. There shall be*
> *In that rich earth a richer dust concealed;*
> *A dust whom England bore, shaped, made aware,*
> *Gave, once, her flowers to love, her ways to roam,*

The Grand Illusion

A body of England's, breathing English air,
Washed by the rivers, blest by suns of home.

Englishry there was, cut to the finest bone, at Gallipoli. The 29th Division was one of the best formations of the Regular Army, and around its core of professionals, in the Royal Naval Divisions, in staff appointments and elegant ancillaries, some of England's brightest spirits eagerly awaited the battle. There were young poets and writers—Masefield, A. P. Herbert, Compton Mackenzie, and Patrick Shaw-Stewart, perhaps the most gifted man of his generation, who was at twenty-five not only a distinguished poet and a Fellow of All Souls, Oxford, but a managing director of Baring Brothers the bankers. There were sons of famous families, an Asquith here, a Herbert or a Napier there. Three members of Parliament fought at Gallipoli, and four future Field Marshals, and Governor-Generals of New Zealand and Australia, and a future Prime Minister of England, and Eric Partridge the lexicographer, and W. G. Grace's son, and J. H. Patterson, author of *The Man-Eaters of Tsavo*. Young Staveley, whom we saw hoisting the Union Jack in Khartoum, we meet again now as a landing officer at Gallipoli; some of the landing craft were commanded by midshipmen, fifteen years old and direct from Dartmouth; nothing could be more eternally English than the massive shire horses, shaggy-hoofed and imperturbable, who were shipped to Gallipoli to drag the army's heavy howitzers up the beaches to their firing pits.

Then there were, like auxiliaries called to the service of Rome, the imperial contingents. They included Sikhs, Punjabis, Gurkhas, the Ceylon Planters' Rifles, and the Assyrian Jewish Refugee Mule Corps, recruited in Egypt, commanded by Patterson the lion-hunter, whose badge was the shield of David, whose orders were given in Hebrew and in English, and who were said to be the first Jewish military unit to go into action since the fall of Jerusalem in AD 70.

Above all they included the Anzacs, the Australian and New Zealand Army Corps, and these men more than anyone gave the campaign its epic allure. Some 30,000 strong, the Anzac divisions had spent some time in Egypt, and they were tanned from the

Mediterranean sun, and elated by the adventure of foreign travel so far from home. Nobody had seen such soldiers before. They were truly like men from a new world, or survivors from an older one. Tall, lean, powerful, cocky, their beauty was not merely physical, but sprang from their air of easy freedom. Their discipline was lax by British standards; they made terrible fun of British officers, and regarded the British other ranks with a mixture of pity and affectionate condescension; but they brought to Hamilton's army a loose-limbed authority all their own, as though they were not the subjects of events, but their sardonic masters.

Curiously thrown together under the command of upper-class Britons, in the spring of 1915 this imperial army, with a French division attached, was assembled in scores of transports in the waters around Mudros, guarded by the warships of the Fleet. Never, perhaps, had an army been so exalted by the prospect of action. 'Oh God!' wrote Brooke, on his way to the Dardanelles, 'I've never been so happy in my life, I think. . . .'

11

It was to be the most ambitious amphibious operation in the annals of war, but it sailed to the peninsula unprepared. Its intelligence was out of date, its maps were inaccurate, it had insufficient shells, no dentists and very few mosquito nets. Two hospital ships were considered adequate for the campaign, and the Army made its own grenades out of old jam tins. Some officers, feeling themselves insufficiently briefed, took along Baedekers of Asia Minor, picked up in the second-hand bookshops of Egypt. Nobody had any idea how many Turks were defending the peninsula, and the entire staff of the Principal Naval Transport Officer consisted of a steward, a cook and a coxswain. Security was appallingly slipshod, and every stevedore in Alexandria knew the army was going to Gallipoli.

Hamilton's plan, nevertheless, was bold. He would assault Gallipoli bullishly from the south and west, and fight his way up it to command the Dardanelles from end to end—'take a good run at the peninsula and jump plump on, both feet together'. The first objectives would be the commanding heights of the peninsula, Achi

Baba[1] in the south, Sari Bair in the centre, and the main striking
force would be the 29th Division, which would be landed on five
separate beaches around Cape Helles, the southern tip of the peninsula. At the same time the Anzacs would land some thirteen miles
up the coast, to strike across the peninsula for the central hills. The
Fleet, with its terrific gunpower, would provide artillery support;
the French would make a diversion on the Asiatic shore; Hamilton
hoped that within three days the lower half of the peninsula would
be captured, the Narrows would be cleared of their mines, and the
Navy could pass through the Dardanelles into the Sea of Marmara.
Handing over tactical command to his subordinates, as was his
practice, Hamilton set the assault in motion and transferred himself
to the *Queen Elizabeth*: and in that magnificent vessel, surrounded by
the transports of the army, the battleships, cruisers and destroyers
of the Fleet, 200 ships in all, he set sail for Gallipoli on the night of
April 24, 1915.

There would be, General Hunter-Weston assured the men of his
29th Division, 'heavy losses by bullets, by shells, by mines and by
drowning'. Still the army landed on Gallipoli confident and excited,
a tremendous naval bombardment having preceded it. The Anzacs,
as 'enthusiastic amateurs', had been given what was supposedly the
easiest role: though they were landed in the wrong place, and found
their maps quite useless, they got ashore with few losses, shouting
scurrilities in pidgin Arabic, and struck inland with such gusto that
by dawn that morning a few soldiers had actually reached the
central ridge of the peninsula. On three of the British beaches, too,
around the tip of the peninsula, there was little opposition. At Y
beach there were no Turks at all: at S and X beaches there were only
a few, and officers keyed up for blood and fire found themselves
helped off their landing craft by solicitous sailors, in case they got
their feet wet.

At two beaches only was the assault as bloody as Hunter-Weston
had feared. At W beach the Lancashire Fusiliers ran into such violent
resistance, from Turks hidden in trenches in the commanding bluffs,
that in a matter of minutes 190 men were killed and 279 wounded,
before the British could dig themselves in. It was the landing at V

[1] So called because of a map error—it is really Achi Tepe.

beach, though, the southernmost beach and the most crucial, that was to provide in the first moments of the Gallipoli campaign a paradigm of the whole enterprise.

There the landing was to be made immediately below the village of Sedd-el-Bahr, where a mediaeval castle stood at the water's edge like a memorial to more ancient battles. A collier, the *River Clyde* (3,900 tons), was to be beached to act as a large landing-craft, and from its hull, it was hoped, 2,000 men of the Munster Fusiliers and the Hampshire Regiment would move across lighters to the beach, and so up the bluffs that rose, steep but not high, immediately behind. The assault went in silently at 6.20 a.m. The naval bombardment had ended, only a cloud of smoke and dust hung over the cape, and there was no sign of life at Sedd-el-Bahr. The sea was calm, the morning sunny, and this was one beach the British knew —Royal Marines had raided the place two months before. Gently and quietly, in perfect silence, the *River Clyde* ran herself ashore beside the castle, towing her lighters, and at the same time a flotilla of boats approached the beach with a battalion of the Royal Dublin Fusiliers. Everything was silent. The place seemed deserted, or stunned by the awful bombardment.

But it was Magersfontein again, or perhaps the massacre of the British on the river at Cawnpore long before. The moment the boats grounded a vicious fusillade of machine-gun and rifle fire fell upon them, from hidden positions in the escarpment. The beach was an almost symmetrical crescent, like an amphitheatre, and the Irishmen scrambling ashore were as unprotected as actors on a stage. Boat after boat was riddled with fire, the soldiers jumping overboard, slumping over the gunwales, screaming or leaping terrified into the water. Boats full of dead men drifted away from the beach, or lay slowly tilting in the water, and a slow crimson stain of blood spread out to sea. Only thirty or forty survivors, scrambling up the beach, reached the cover of a ridge of sand, where they huddled helplessly beneath the bullets raging over their heads.

Meanwhile the captain of the *River Clyde*, finding nothing to moor the lighters to, had leapt into the water with an able seaman and was holding the bridge of boats in position by his own muscles, crouching in the water with only his head and shoulders showing.

A few moments later, when the sally-ports of the collier were flung open, and the Munsters and Hampshires sprang out, they were met with a blast of fire like the smack of heat on a tropical day. They died almost as fast as they appeared, blocking the doors and gangplanks, falling into the sea: only a handful floundered ashore and took shelter with the Dubliners in the lee of the escarpment. The whole beach now was littered with corpses—'like a shoal of fish', said the Turkish commander—and through the noise of the battle one could hear always the cries of the wounded men, spread-eagled on the beach wire, or helpless in the shattered hulks of boats. When General Napier, the brigade commander, approached in a cutter to take command, the men on the *River Clyde* shouted at him through the din to go back—'Go back, go back! You can't land!' 'I'll have a damn good try', the general shouted back: and almost at once he and his officers were slaughtered like the rest.

The hours dragged on in stalemate. At midday the vast form of the *Queen Elizabeth* loomed inshore, and through a billowing cloud of green, black and yellow smoke, poured salvos into the bluffs above the beach. The little village was a ruin, the escarpment was pock-marked and crumbled with shell-holes, but still the Turks raked the beach with their fire. It was not until night fell that the men trapped in the *River Clyde* could clamber ashore over the dead and wounded—and even then, when the moon came up a little later, the Turkish machine-gunners opened fire again. V beach lay that night in a confusion of anguish and disillusion. It started to drizzle later, and the troops slept there damply among their own dead and dying, while the guns chattered spasmodically all night long, and the great ships stood offshore in an ironic blaze of lights.

The imperial armies were ashore at Gallipoli, but the experience of V Beach was to be the true index of their enterprise, from which the romantic dedication was presently to depart, leaving only a reproach of muddled waste and heroism. Almost as the campaign began, news reached the armies that Rupert Brooke, their exemplar and their laureate, had not even reached the peninsula, but had died of blood-poisoning at sea, on St George's Day, and had been buried on the island of Skyros. Hamilton was greatly moved. 'Death grins at my elbow', he wrote. 'I cannot get him out of my thoughts. He is fed up with the old and sick—only the flower of the flock will serve him now. . . .'

12

The Gallipoli campaign lasted 259 days, April 1915 to January 1916. In all half a million men were landed on the peninsula. Far from capturing their objectives by the third day, the British never captured them at all, but were confined first to last to footholds on the shore. Within forty-eight hours of the landings the two allotted hospital ships were on their way to Egypt, full of wounded; even the Anzacs had been driven off the crest of the hills, almost back to their cramped beach at Anzac Cove, and their commander was recommending evacuation at once. So another legacy of the imperial years turned sour upon the British, for even after the failure of the naval assault they had supposed this to be another species of colonial war, against superior but demoralized Asiatics. It was well known, said a statement from Hamilton's headquarters shortly before the attack, that many Turks 'looked with envy on the prosperity which Egypt enjoys under British rule', and anyway, as a staff officer wrote, the Turk had never shown himself as good a fighter as the white man. 'Who could stop us?' wrote an Australian private exuberantly before the landing. 'Not the bloody Turks!'

Four months after the first assault a second invasion was launched, the landing this time being at Suvla Bay in the north, so that at the climax of the campaign there were three separate bridgeheads, with British forces north and south, Anzacs in the centre. But the three never joined up, and what began as a campaign in the imperial kind, a war of sweep and movement, degenerated into trench warfare, just as static, just as dispiriting, as the fighting in France. Only the setting was different, for behind the backs of the Gallipoli soldiers there lay always the tantalizing sea. Serene on the horizon lay the islands, and all around the peninsula were the warships and transports, always there, dowager-like among their torpedo nets, or moving majestically along the coast for another bombardment. Sometimes the soldiers awoke to see some famous ocean liner, the

Aquitania or the *Mauritania*, standing offshore like a visitor from another world: at night the lights of all the warships, their search-lights playing, their signal lamps winking, suggested a great float-ing city, friendly and reassuring, and officers were sometimes taken out there, direct from their squalid dug-outs to the armchairs and starched linens of battleship wardrooms. The sea was always there, and always at the back of the soldiers' minds, no doubt, was the thought that if the worst came to the worst in their long fight for the peninsula, the Navy could always snatch them off.

On May 25, though, the beloved and familiar *Triumph*, bombard-ing the Turkish positions from a station off the Anzac beach, was torpedoed by a German submarine. In full view of the soldiers she capsized with a deep metallic rumble, floated upside-down for half an hour, and sank. It was a traumatic shock. The Australians watched appalled, some cursing and crying with the horror of the spectacle, and from the hills above they could faintly hear the exultant cheer-ing of the Turks. Within hours all the big ships were withdrawn to Imbros, and the soldiers, looking forlornly out to sea, saw them retreating fast into the evening, led by the battleships, with the smoke from their funnels trailing behind them. There was a momentary hush over the peninsula, as every man, British or Turk, watched them go. By nightfall they were out of sight. 'All the ships had disappeared', wrote a German officer, 'as if God had taken a broom and swept the sea . . . the joy of the brave Turks can scarcely be described.' The British felt a chill sense of abandonment, even betrayal, as darkness fell upon Gallipoli that night.

13

In hideous attack and counter-attack, interspersed with exhausted lulls, they passed the rest of 1915. Reinforcements arrived, ships came and went, twelve gunboats were built especially for service in the Danube when Constantinople was taken; British and Australian submarines, in feats of prodigious daring, passed through the Nar-rows and roamed the Sea of Marmara, sinking Turkish ships and sometimes bombarding roads—the submarine E11 actually reached Constantinople, torpedoed a freighter berthed beside the arsenal, and started a panic in the capital.[1] But on the peninsula nothing was gained. The battlefronts were often no more than a few hundred yards wide, and the salients never more than a few miles deep. The British at Cape Helles won the whole tip of the peninsula, but never got further than five miles inland: the dour mass of Achi Baba, so close across the rolling downland, was never any closer, and the soldiers were never out of sight of the very beaches where they landed. At Suvla Bay, in the north, they achieved even less, but floundered impotently about the flat lands near the beaches, losing 8,000 men and never reaching the high ground at all.[2] As for the Anzacs, though in a thousand skirmishes they hacked their way up the cliffs above their beaches, they never captured the crest, but were immured there in the end like troglodytes or fossickers, in burrows and trenches scattered over the hillsides, and straggling squalidly down to the beach. Often the Turkish and British trenches were only a few feet apart, and the enemies could easily hear each other talking; by the winter the fronts were labyrinths of trenches, and every sap or redoubt had its familiar name—Dublin Castle, Half Moon Street, Courtney's Post.

The beauty of the place, which entranced many of the soldiers when they first landed on Gallipoli, turned sour with time, and an overwhelming sense of decay fell upon the peninsula. No longer did the soldiers write with such delight of the glorious sunsets, the hyacinths and the heather. Now the landscape became terribly oppressive, as the spring gave way to the ferocious summer, and then to a wet raw winter. Flies swarmed everywhere over the

---

[1] The E11 survived the war, to be broken up at Malta in 1921. The gunboats were presently dispersed all over the Empire, many of them lasting until the *next* world war, when four were destroyed by enemy action in the Mediterranean and the China Sea. The *River Clyde* outsailed them all: refloated in 1919, she was bought by Spanish owners, and sailed the Mediterranean first as *Angela*, then as *Maruja y Aurora*, until she was sold for scrap in 1966.

[2] The general in command, Sir Frederick Stopford, had been Military Secretary to Redvers Buller in South Africa, and was said to have been the only officer in his confidence at the battle of Spion Kop. Stopford was relieved of his Gallipoli command after only ten days, and though he lived until 1929, that was professionally the end of him.

bridgeheads, day by day the wounded went away to the hospital ships queuing up offshore, the men grew dirtier, thinner, more unkempt, plagued by dysentery, septic sores, frostbite. 'The beautiful battalions of April 25th', Hamilton wrote, 'are wasted skeletons.' Corpses lay everywhere, blackened and unburied between the lines, or lost in inaccessible ravines, and their smell was inescapable: off the beaches the Navy tried to sink the floating bodies of horses and mules by churning them up with their propellers. The British infantrymen, patient as ever, grumbled their dispassionate obscenities: the magnificent Australians almost gave up being soldiers at all, fighting like brigands or guerillas, and sauntering among their dug-outs with their shirt-tails hanging out, or wearing nothing at all—'like flies', one New Zealander thought, 'wandering about like aimless men'.

The beaches were a terrible mess, clogged with supplies, littered with makeshift jetties, tumbled about with debris, broken crates, half-sunken boats. Anzac Cove, it was said, looked as though everything had been washed ashore in a shipwreck: at Sidd-el-Bahr the scarred hulk of the *River Clyde* still lay among the wreckage of her lighters. Along the tracks that led from beach to trenches, now dusty, now deep in mud, teams of mule-drawn wagons toiled; often groups of soldiers, escaping from their dug-outs, splashed about in the water; often too, in a desultory way, the Turkish gunners lobbed a shell down, to explode almost unnoticed in the sea, or splinter a few more boxes on the foreshore.

Always the noise of the battle continued on the heights above. It was never far away; it was seldom suspended; it was often savagely intense; it achieved nothing whatsoever. Of the 500,000 men who landed on the peninsula, first to last, rather more than half were killed or wounded, and though on several occasions they came heartbreakingly close to success, and the Turks suffered at least as severely as they did, still they might just as well never have gone to Gallipoli at all. *Was it hard, Achilles?* asked Patrick Shaw-Stewart, looking across the Hellespont to the plain of Troy—

> *Was it so hard, Achilles,*
> *So very hard to die?*

Thou knowest, and I know not—
So much the happier I.

I will go back this morning,
From Imbros, o'er the sea;
Stand in the trench, Achilles,
Flame-capped, and shout for me.

14

General Hamilton, one of the bravest and most experienced officers of the British Army, lacked one quality of generalship: fury. He was an optimist, but not a zealot. Considerately refraining from undue interference with his subordinates, he seems never quite to have grasped the whole momentum of the action into his own hands. His contact with the battle was more advisory than decisive. He felt for his soldiers, he was thoughtful to his commanders, he responded like an artist to the beauty and the tragedy of it all: but he was not a man to fall upon his enemies with a criminal hail of fire, steel and explosive, and he flatly refused, despite pressure from London, to use poison gas.

It was his tragedy that the Gallipoli campaign needed just such a man of blood, especially as some of his senior subordinates were abysmally inept. Risky at the best, Gallipoli was an action that could succeed only by outrage and audacity. A few more old ships sunk, and the Royal Navy might have burst through the Narrows. An instant advance from Suvla Bay, rammed home despite all dangers, and the whole peninsula might have been captured in a day. They were terrible chances to take, involving thousands of human lives, the immediate success or failure of an entire campaign, perhaps even of a war. Hamilton, a man of his time, an Edwardian gentleman, lacked the cruelty to take them. He failed by the narrowest of margins, for by the end of 1915 the Turks were almost at breaking-point: but in the conduct of great affairs, nothing fails like failure.

Gallipoli was the greatest reverse to British arms since the American Revolution, and if it was launched in a resurgence of the imperial bravado, it was lost in the deadweight of the imperial tradition. Its senior commanders had all been nurtured in the colonial wars, a debilitating legacy, and the old burden of class, which Kipling had so anathematized after the Boer War, contributed again to the débâcle: all too often generals were remote from their men, if not in courage or *esprit de corps*, at least in everyday experience —or actually in physical distance, for sometimes they preferred to conduct their battles entirely from warships out at sea. The soldiers, though they fought on bitterly to the end, lost faith in their leadership. 'Are we down-hearted?' shouted a shipload of new arrivals, approaching the peninsula that summer. 'You bloody soon will be', came the mordant reply from a departing hospital ship.

For in the end a heavy pall of sadness hung over the exhausted armies, mourning so many friends. 'For God's sake,' one officer wrote in his diary, 'get me away from the Dardanelles!' By the autumn of 1915 the British War Cabinet, looking bleakly out at the tragedy across the death-fields of France, had lost hope for the venture. Churchill was no longer at the Admiralty, and when Lord Kitchener came out to Gallipoli to see for himself, he recommended withdrawal. Game to the last and ready for another offensive, Hamilton was replaced by a very different general, the bluff and practical Sir Charles Monro: and by Christmas, silently and secretly, most of the army had been withdrawn from the peninsula.[1]

The British public was encouraged to think of the withdrawal from Gallipoli as a compensating triumph, like Rorke's Drift after the defeat at Isandhlwana, or Mafeking after Black Week. Official accounts of the tragedy always ended with images of its success— the stealthy withdrawal from the forward trenches, the skilful assembly of guns, stores, horses, the boats stealing away muffled through the night, the watchful warships standing by, and finally, as the transports sailed off beneath the guns of the Fleet, the fires of burning stores and abandoned ammunition which at last revealed the truth to the Turks. 'In that marvellous evacuation', wrote Sir Julian Corbett, the official naval historian, 'we see the national

---

[1] Hamilton never commanded again, but he lived to an honoured old age, dying in 1947. It was to Monro, who became Governor of Gibraltar and died in 1929, that Churchill bitterly attributed the apothegm: 'I came, I saw, I capitulated.'

genius for amphibious warfare raised to its highest manifestation.'
The Turks awoke on the morning of January 9, 1916, to find that
not a British soldier was left in the crannies and hidden valleys of the
peninsula. 'I hope *they* don't hear us go', one Australian is supposed
to have murmured, as his battalion stole through the graves of their
comrades down the cliff-tracks to the boats.

But it is truer to the nature of the Gallipoli story, fairer to its
soldiers, to end with a glimpse not of success in defeat, but of
tantalizing failure in victory. On August 9, 1915, half-way through
the campaign, the front element of British troops on the Anzac
front had fought their way almost to the crest of Sari Bair, the
central ridge of the peninsula and perhaps the key to all else. The
ground up there was stony and serrated, rough at any time, now
horribly cut about with shell-holes, gun-pits and lines of trenches:
for weeks the opposing armies had struggled on those unhappy
heights, now losing a vantage point, now winning a trench.

Soon after five o'clock that morning a small group of British and
Gurkha soldiers fought their way, in savage hand-to-hand fighting,
to the summit of the ridge. The Turks fled down the hill the other
side, the British pursued them over the top: and suddenly they saw
before them, for the first and only time, the object of their battle.
There below them down the eastern slopes, only five miles away
across the dun rolling countryside, blue in the morning sunshine
they saw the Narrows. There was their objective. There were the
old grey forts at the water's edge, there the cluster of Chanak with
its castle and its minaret, and only a few miles upstream the channel,
swinging between its hills, broke away into the Sea of Marmara.
They looked at it with awe and elation, wiping the sweat from their
eyes, torn and breathless and bleeding from the fight, and for a
moment felt the campaign had been won.

In a few minutes they were forced up the hill again, and leaving
their dead behind them in the sunshine, retreated once more over
the ridge.[1]

---

[1] I felt their presence still, when I stood on the very same spot in 1975,
looking through a morning mist to the straits below, for Gallipoli is the most
truly haunted place I know. Long after the war much of it was planted with
trees, but they were destroyed in a fire in 1974, and when I was there the

15

Many imperial instincts had found their epitome, or their disillusionment, in these several campaigns, so far from the crux of the world conflict. Gallipoli ended in total failure, the Middle Eastern campaigns, co-ordinated in their final phases, ended in absolute success, with the British armies sweeping far to the north, to Damascus, Mosul, Aleppo and the frontiers of Turkey. When the Turks sued for peace the British controlled the whole of the former Turkish Empire, except only the Arabian interior, and upon this achievement they would erect the last of their great imperial structures, an empire among the Arabs.

In a wider spectrum, too, the war at first seemed only to have strengthened the Empire. 'The British flag', Lord Curzon told the House of Lords after the Armistice in November, 1918, 'has never flown over a more powerful and united Empire.' For a time it had seemed that Joseph Chamberlain's vision of imperial federation might after all be realized. The Imperial War Cabinet was described by Sir Robert Borden, Prime Minister of Canada, as a 'Cabinet of Governments', and was the nearest the Empire had ever come to the projected Grand Council. Jan Smuts, so recently a defeated enemy, became one of the most influential men in London. The victory celebrations in London were almost like the Diamond Jubilee again,

---

peninsula must have looked much as it did sixty years before. Until 1964 it remained a closed military area, and not many people go there even now. The landing-beaches are unchanged, many of the trenches may still be traced, and cemeteries scattered over the hills mark the sites of old redoubts, Lone Pine, Twelve Tree Copse, Dunn's Post or The Nek. To this day the bones of dead soldiers are still found in the more remote ravines.

At Cape Helles the British erected a commanding memorial, magnificently greeting the visitor with the roster of the great ships whose heyday ended here. At the Narrows the Turks have made two more. One is simply a date, marked out in white stones on a hillside above the strait: 18:3:15. The other is a figure of a soldier, on the bluffs above Kilid Bahr, with a quotation beside it from the poet Mehmet Akif Ersoy: 'Stop, passer-by! The earth you have just unknowingly trodden is the spot where an era ended and where the heart of a nation beats.'

as soldiers from the four corners of the King's dominions marched through the adulating crowds. The Empire seemed more than ever a band of brothers: men from Fiji and Egypt had served in France, men from Trinidad at Gallipoli, men from Belize and Hong Kong in Mesopotamia. Messages of congratulation flashed across the globe. Colonial Premiers basked in the benevolence of the Court, or were welcomed effusively at victory banquets. Emblems of Empire embellished the margins of Special Editions, or were woven into the sentiments of picture postcards. 'Never while men speak our tongue', wrote *The Times*, 'can the blood spent by the Canadians at Ypres and by the Australians and New Zealanders at Anzac be forgotten. That rich tribute of love and loyalty to the highest ideals of our race has not been wasted. . . .'

It was true, and it was false. The Empire really had gone to war united, and it fought together to the end. Even India provided an army of 1½ million men, most of them volunteering for the money or the honour, but many out of loyalty too. As for the 25 million people of the 'white' Empire, they had sent 857,000 of their men overseas, and 141,000 of them were killed. A sense of common sacrifice and accomplishment really did give to the Empire's scattered peoples a new and triumphant brotherhood—'What remains to us?' cried William Morris Hughes, the spectacular Welsh-born Prime Minister of Australia. 'We are like so many Alexanders. What other worlds have we to conquer?' The Empire was more powerful than ever, possessing at the end of the war not only the greatest fleet, but also the greatest air force in the world, and from the conflict it was to win great prizes: new territories in Africa and the Pacific, a whole new paramountcy in the Middle East.

Yet it was false, for behind the triumph, the illusion was spent. After so many miseries in its name, glory was discredited in the hearts of the people, and war, which had given the British such vicarious satisfactions in the past, was recognized now in its true obscenity. The imperial peoples had gone to war in 1914 in a mood of brave, even happy idealism: by 1916, when conscription was first enforced in Britain, the spirit was lost, and 'Tipperary', with which they had marched so guilelessly to their early battles, was tacitly dropped from the Army's musical repertoire. After the war they found among the papers of Wilfred Owen the poet, killed in France in the last week of conflict, a fragment called 'An Imperial Elegy'. This is all it said:

> *Not one corner of a foreign field*
> *But a span as wide as Europe,*
> *Deep as (   ).*
> *I looked and saw.*
> *An appearance of a titan's grave,*
> *And the length thereof a thousand miles.*
> *It crossed all Europe like a mystic road,*
> *Or as the Spirits' Pathway lieth on the night.*
> *And I heard a voice crying,*
> *This is the Path of Glory.*

The Empire stood wiser but more cynical for the experience of holocaust. Elgar sang no more of the bayonet's clash, and was exploring sadder and profounder themes. Kipling, having lost his only son in the fighting, never again wrote a lay of Empire. Kitchener had shrunken in stature, as the conflict extended beyond the scale of any Paardeberg, until shipped off on a mission to Russia in 1916 he died far from his imperial exploits, drowned in the cold north sea. Through the Empire a new kind of memorial arose, not to the men who died bringing the flag to the distant dependencies, but to those who returned across the oceans to defend it. 'Sleep on, dear Howard,' said a memorial tablet in Mapleton, Manitoba, to Private Howard Pruden, killed in France—

> *Sleep on, dear Howard, in your foreign grave,*
> *Your life to your country you nobly gave,*
> *Though we did not see you to say goodbye,*
> *Now in God's keeping you safely lie.*

'In Memory of the Brave Sons of Smith's Parish', said a slab in the little country church of St Mark's, Bermuda, framed among its trees and crab-grass lawn beside the sea, 'who Risked their Lives in Defence of the Empire against the unscrupulous German Foe.' And in one of the lonely cemeteries in which, buried where they died, the Anzacs lay lost among the Gallipoli ravines, the parents of one young soldier wrote their own epitaph to their son, killed so far away, so bravely we need not doubt, in so obscure a purpose:

> *God Took Our Norman,*
> *It Was His Will.*
> *Forget Him, No,*
> *We Never Will.*

'I hope *they* can't hear us', said the Australian soldier of Our Norman and his mates, and one hopes they could not: for all too often the sacrifices of the Great War, as its contemporaries called it, were given to a cause that was already receding into history, like those discredited grey battleships, their smoke-pall filling the sky, hull-down on the Aegean horizon.[1]

---

[1] The Empire's central memorial to its dead was the Cenotaph, erected in Whitehall in 1920, the first of more than fifty war memorials designed by Edwin Lutyens. Sir Fabian Ware, who devoted his later life to the Imperial War Graves Commission, calculated that if the dead of the Empire were to march four abreast through London, they would take three and a half days to pass this monument, and one of the most vivid memories of my own first visit to London in the 1930s is that of the Londoners respectfully and unselfconsciously removing their hats when they walked by it.

PART TWO

The Purpose Falters
1918–1939

CHAPTER TEN

Into the New World

SO the British Empire moved out of the old order, which it had dominated and to some degree moulded, into a new and unfamiliar world. The Kaiser had lost the war, but had achieved at least one of his purposes—to shift the pattern of power. Everybody recognized that *la belle époque* was over, but not everyone realized that Britain's hegemony had gone, too, for the victory had been so complete in the end, the Empire's part in it had been so tremendous, and the imperial spirit seemed to have been so rejuvenated by the comradeship of conflict, that if anything the Flag appeared to fly across the world more masterfully than ever. The doubts of the 1900s were momentarily expunged by the defeat of the great enemy, and after so many triumphs and sacrifices the British Empire's status seemed uniquely privileged. As Leopold Amery observed to Lloyd George at the end of the war, if out of such heroic effort the British Empire grew stronger and greater than before, 'who has the right to complain?'

If it had not been an imperial war, it had been an imperial victory, for Britain's fundamental weapon had remained that oldest instrument of Empire, the Royal Navy. The Navy might have failed to intimidate the Turks, but it had succeeded in inhibiting its greater opponent, the German High Seas Fleet, and won the war in the end simply by existing—the profoundest use of sea-power. The Germans gave at least as good as they got in the great naval battle of Jutland, distinctly a Trafalgar *manqué* for the British, but for most of the war their magnificent surface ships stayed uselessly in harbour, blockaded by an idea. 'The surrender of the German Fleet', the Admiralty signalled to the commander of the British Grand Fleet, Admiral Sir David Beatty, *quondam* captain of the gunboat

Fateh, 'will remain for all time the example of the wonderful silence and sureness with which sea power attains its ends.'

That surrender took place on November 21, 1918, stage-managed in the classic Spithead style, and pictured like a sombre regatta in magazines and newsreels across the world. Everyone knew its meaning. It was a victory of Order over Anarchy, of the Real Thing over Upstarts, of permanent, organic values over petty ambitions and impertinences. The surrendering German fleet was very large—14 capital ships, 56 cruisers and destroyers—but it was led into Scottish waters by a single British cruiser, HMS *Cardiff* (4,290 tons), flying a huge Blue Ensign, the flag of the Royal Naval Reserve, as a recognition signal at her foremast. In line ahead the German warships, bedraggled from the demoralizing last months of the war, approached the mouth of the Firth of Forth: and there the Royal Navy was waiting for them. Admiral Ludwig von Reuter, the German commander, had hoped for foggy weather to obscure the British triumph, but in fact the day was sunny, and the Royal Navy at its most complacent. The entire Grand Fleet was assembled there (370 ships, 20 admirals, 90,000 men), flying from every mast and staff, as was the Navy's custom when going into action, all available White Ensigns. The warships formed themselves into two parallel lines, and taking up position on each side of the Germans, proceeded towards the Firth with all their crews at action stations, guns trained fore and aft on the surrendered enemy.

Thirteen squadrons of British capital ships and cruisers escorted their defeated rivals into captivity. The *Queen Elizabeth* was flagship, nearly four years after her debut at the Dardanelles, and the forty-one battleships and battlecruisers included the *Canada*, the *Australia*, the *Malaya*, the *New Zealand* and the *Emperor of India*. Three miles behind steamed a host of destroyers, their black smoke clouding the horizon, as though they were bringing the curtain down. As the German ships dropped anchor at the entrance to the Firth, at three o'clock that afternoon, Beatty sent a lordly signal to Admiral von Reuter. 'The German flag', it said, 'will be hauled down at sunset today, Thursday, and will not be hoisted again without permission.' By evening hundreds of yachts, motor-boats, skiffs and pleasure-steamers were milling festively about the humiliated High Seas

Fleet, as the sightseers had swarmed around HMS *Bellerophon*, bearing the captive Napoleon, in Torbay long before.

What could be more absolute? The Germans themselves bitterly recognized the style of it, the style of Empire itself. Was this not the Navy of Lord Nelson? Had it not been continuously at sea, manned often by the same families, from the same towns, since the Middle Ages? Was it not officered by dukes and princes, and directed from stately panelled palaces? Was its flag not honoured and familiar in every creek, channel and dockyard of the world? On the battleship *Royal Oak*, it was said, a mysterious drum-beat was heard to sound as the German Fleet sailed in. Twice messengers were sent from the bridge to investigate, but it was unexplained, and continued to sound from nowhere until anchor was dropped and the enemy was safely in captivity.

Drake's Drum, the popular newspapers romantically suggested, beating a last tattoo as England's danger ended: but at least one German captain found the Royal Navy, when a boarding party came to inspect his cruiser, less daunting in the detail than in the grand display. Half a dozen ragamuffin ratings formed the party, and they were led by a scruffy and distinctly plebeian lieutenant, patently not a duke at all, smoking a fag-end.[1]

2

The British Empire had more than survived the war, it had sizeably grown. Nothing had been lost, territorially, and much had been gained. Convinced imperialists had been influential in the conduct

---

[1] Seven months later the surrendered fleet, taken to Scapa Flow in Orkney, scuttled themselves on a pre-arranged signal from Admiral von Reuter, watched by an astonished party of schoolchildren on an excursion trip from Stromness, some of whom thought it was a show arranged especially for them. As for the *Royal Oak*, between whose decks Drake's Drum had sounded, twenty-one years later she was one of the first British warships lost in the next war against the Germans, when the submarine U47 found a gap in the Scapa Flow defences and torpedoed her twice: her shadow may still be seen there, marked by a memorial buoy, and ships of the Royal Navy salute it as they pass.

of the war, and had their say in the shaping of the peace; the Coalition Government formed in 1919 included Curzon as Foreign Secretary, Milner as Colonial Secretary, and as the Government's spokesman on colonial affairs in the House of Commons, Leopold Amery, to whom the freedom to develop and expand the Empire had been the 'first and foremost' war aim.

They had to work more subtly than before. The straightforward annexation of colonies was unacceptable now, as distasteful to the mass of the British people as it was to the world at large, and the prevailing orthodoxy was President Wilson's concept of 'self-determination'—the right of every people to decide its own future. 'Peoples may now be dominated or governed', Wilson optimistically told Congress in February 1918, 'only by their own consent'. His Fourteen Points, the basis of the peace settlement, did indeed theoretically end the imperialist age, for they specified that the interests of the subject peoples should have equal weight with those of the imperial powers, and when the League of Nations was formed its Covenant declared that the 'well-being of peoples not yet able to stand by themselves under the conditions of the modern world . . . forms a sacred trust of civilization'.

In practice the British Empire took shrewd advantage of the peace terms to extend its power and safeguard its security. Under American inspiration the victorious States devised the system of Mandates, trusteeships over former enemy territories awarded by the hopeful League to liberally-minded Powers, generally, as it happened, those which had overrun the territories in war—'the crudity of conquest,' suggested the historian H. A. L. Fisher, 'draped in the veil of morality'. This concept served the Empire usefully. In theory the League of Nations retained supervisory rights over the territories: in effect the British ruled their Mandated acquisitions as parts of the Empire, administering them like any other Crown Colonies, and colouring them red, or at least red-*hatched*, on the map of the world.

Most of the imperialists' war aims were satisfied. Nearly a million square miles was added to the Empire, with 13 million new subjects, and several old dreams seemed to be coming true. In the Pacific most of the former German colonies went to Australia and New

Zealand, as antipodean expansionists had been hoping for years.[1] In Africa the Empire gained control not only of South-West Africa, satisfactorily rounding off Imperial South Africa, but also of Tanganyika, at last fulfilling the vision of an all-red Cape to Cairo corridor. In the Middle East Iraq, Transjordan and Palestine became British Mandates, and Persia was virtually a British protectorate, so that India was linked with Egypt and the Mediterranean by a continuous slab of British-controlled territory, and one could travel overland from Cape Town to Rangoon without once leaving the shelter of British authority. The Empire seemed, on the face of it, safe and solid as never before. The Dominions had proved their loyalty. The subject races had remained mostly subject. These great new acquisitions seemed to make the whole structure complete, its gaps filled, its weak points reinforced. What other worlds, as 'Billy' Hughes had cried, had they to conquer?

3

But it was only a spasm of the old energy. The euphoria of victory made the British feel they were still masters of their own destiny, and encouraged the ageing imperialists to revive their fading dreams. They were soon disillusioned in their hopes. The nation had lost the panache of Empire, and the mass of the people resisted all attempts to make them imperialists again. Among the social hazards of the post-war years was the conversation of a Milnerite, droning once more over a brandy about the chances of imperial federation, or the insular indifference of the electorate. (*'Thank God that's over darling. We had old D. sermonizing again'. 'Not the beastly old Empire again darling'. 'That's the one! My sainted aunt, you'd think we were still fighting the fuzzy-wuzzies. . . .'*)

The truth was that Britain was changed for ever by the war, and was in no mood or condition for a revival of the New Imperialism. How could it be otherwise, when 700,000 young men had died?

[1] President Wilson was shocked. Was Australia really proposing, he asked Hughes, to flout the opinion of the civilized world by profiting from Germany's defeat and extending her sovereignty as far north as the equator? 'That's about it, Mr President,' the Prime Minister replied.

Even the imperialist balladeers were muted now, and it was Sir
Henry Newbolt himself, author of *Play the Game!*, who spoke for
them all in his poem *The War Films*:

> *O living pictures of the dead,*
> *O songs without a sound,*
> *O fellowship whose phantom tread*
> *Hallows a phantom ground—*
> *How in a gleam have these revealed*
> *The faith we had not found.*
> *Brother of men, when now I see*
> *The lads go forth in line,*
> *Thou knowest my heart is hungry in me*
> *As for thy bread and wine:*
> *Thou knowest my heart is bowed in me*
> *To take their death for mine.*

Here was a silence more terrible than the breathless hush in the
close that night, and though the British soon recovered their natural
jollity, and entered the 1920s in a spirit of resolute escapism, still the
old splendour was shaded.

Britain had ended the war apparently the strongest of all States.
Her industries were intact, her finances far from crippled, she
possessed the strongest air force and the strongest navy in the
world, and one of the strongest armies. But in the sadness of it all
she had lost the *brio* of success, and she had no grand idea to offer,
no message of hope or change, to answer the challenges of Communism from revolutionary Russia, Wilsonian liberalism from
America. At Versailles, where the peace treaty with Germany was
signed and the future of the world decided, the British did not play
the decisive role. On the one hand they failed to curb the vindictive
intentions of the French: on the other, though their chief representative was the inspired and fascinating Lloyd George, though their
delegations were attended by all the glamour of the Empire's age,
scale and experience, still they were upstaged in the Hall of Mirrors
by the presence of the Americans. The British Empire represented
tradition and continuity, but the USA represented a fresh beginning, and the idealism of the new world seemed marvellously

hopeful and exciting, set against the plumed and fatal loyalties of the old.

For though self-determination was a clumsy word, it was full of lucid suggestion. It spoke not merely of national freedoms, but of personal liberties too, of all those inalienable rights that the Americans had won for themselves, and now seemed to be claiming on behalf of everyone else. And just as the British Empire had been the enemy of the Founding Fathers, so inevitably it seemed to stand now as a vast and ancient barrier to these aspirations. The very notion of self-determination was incompatible with the Empire's survival; the whole trend of affairs, the whole conception of a world order embodied in the League of Nations, ran directly counter to British imperial positions. The British Empire delegates at Versailles, mustered by the Australians, narrowly prevented the inclusion of a clause in the League Covenant actually declaring all races to be inherently equal, a close shave indeed for the imperial comfort.

Sir William Orpen painted a conversation piece of the signing of the peace treaty in the Hall of Mirrors. It is mannered, but telling. Physically it is dominated by the British Empire, for around the compelling figure of Lloyd George, centre-stage, are assembled aides and delegates from all the great overseas possessions of the Crown— a turbanned Indian officer over Clemenceau's shoulder, a swarthy Boer at the edge of the scene, the ponderous Sir Robert Borden from Canada, the mercurial William Hughes from Australia, besides the familiar imperial figures of Balfour, Curzon, and, his tie askew, Lord Milner of Cape Town and St James.

But somehow the eye strays, away from the Maharaja of Bikaner, away from the mordant Milner, away even from Clemenceau and Lloyd George, until it alights upon the stiff ascetic person of President Wilson: for he is looking directly, deliberately at the artist, with an almost accusatory expression, as though he is staring hard into the future, and willing it his way.[1]

---

[1] Orpen (1878–1931) may have intended it so—he was an Irishman with a caustic eye for politicians. He spent nine months as an official artist at the Peace Conferences, but was to be best remembered for the compassionate pictures of life in the Great War trenches which he presented to the nation, and which are now in the Imperial War Museum in London.

4

The peace treaty was signed not only by Great Britain, but by Canada, Australia, South Africa, New Zealand and India. This seemed at first a majestic demonstration of imperial brotherhood. When the League of Nations met for the first time, in 1920, all the Dominions were again represented by their own delegates, giving the Empire six separate votes. The Americans had withdrawn from the League; the Germans and Russians were excluded from it; to foreigners it sometimes seemed that it would be dominated by a British caucus, voting imperially against the world as Rhodes and Joe Chamberlain would have wished it.

But it was really less a declaration of imperial solidarity than of Dominion independence. In 1917, when the imperial Prime Ministers assembled in London in conference, they had unanimously voted that after the war the Dominions should have an 'adequate voice in foreign policy and in foreign relations'. Smuts indeed described them frankly as 'autonomous nations', and thought they should not consider themselves an Empire any more, but a British Commonwealth of Nations. The victory had strengthened these impulses. The white colonials had gone to war trustingly, innocently almost, satisfied for the most part to be loyal assistants to the Mother Country. They had been inexperienced still, as soldiers and as statesmen, and they were as indoctrinated as the British at home in their ingenuous respect for British traditions and achievements. Though they often made fun of the British, their toffs, their drawls and their domesticity, they still looked up to the Old Country, and believed as the British did themselves in the value of its systems and the skill of its leaders.

But they had gone home with different feelings. If they had been patronized at the start of the war, at the end of it they were patronizing themselves, sometimes scornfully. They had seen the structure of British society forlornly exposed once more, and the myth of omniscience, to which they had been educated, proved a fraud. The British private soldier, so passive, so uncomplaining, they looked upon with a fraternal sympathy, often offering him cigarettes from their own more plentiful supplies, or giving him a pair of their superior boots. The British senior officer they grew to despise.[1] Their impertinence to the brass, which began as a cheerful lark, grew into an expression of resentment, as they saw all their high purposes, their journeys across half the world, the lives of their comrades, so often wasted by the incompetence of the British high command. They believed themselves to be better soldiers than the English, with some reason—most people agreed, for instance, that the Canadians were the best troops on the western front—and thought they were all too often given the bloodiest jobs and the least reward. In the early years of the war the boys at Scotch College, Melbourne, had actually cried when they heard the stanzas of 'Bugles of England': after 1918 their eyes were drier.

Their leaders too, loyally though they supported the war to the end, chafed against the leading-strings of Westminster. From the start the Canadians had demanded complete control of their own armies, and among Australians the story of Gallipoli, which began as heroic legend, degenerated into object-lesson—it had been planned without reference to the Dominions at all, but never again would Australian divisions be committed to war under absolute British command. The British were often antagonized by the colonials, too. General Sir Douglas Haig, the British commander in France during the later stages of the war, found his colonial generals 'ignorant and conceited', and described a delegation of visiting Canadian politicians, including the Minister of War, as 'well-meaning but second-rate sort of people'. As for the Australians, their soldiers were said to exert such a bad influence on the English that they were kept as far apart as possible—their desertion rate was four times higher than that of any British unit, and twelve times as many of them went absent without leave.

These rancours found political form after the war, when the colonials contemplated their own growing strength. The Canadians in particular, who had always led the way in constitutional reforms,

[1] His wife too, sometimes. It was said that Lady Godley, wife of the British commander of the New Zealanders at Gallipoli, complained when she visited the wounded in an Egyptian hospital that they were not lying properly to attention.

rebelled against the last trappings of imperial authority. It was a quarter of a century since Rudyard Kipling, in a famous poem, had commemorated Canada's decision to impose her own tariffs upon foreign goods, regardless of British policies—

> *A Nation spoke to a Nation,*
> *A Throne sent word to a Throne:*
> *Daughter am I in my mother's house,*
> *But mistress in my own.*

Now it was the Canadians again who forced into definition a new relationship between Britain and the Dominions—daughters still perhaps, in poetic or propagandist terms, but distinctly come of age.

5

In September 1922 the British found themselves looking apprehensively once more towards the Dardanelles. Under the peace treaty Turkey had been dismembered: the Greeks occupied parts of Asia Minor, the British maintained garrisons along the Dardanelles. In 1920, however, a virile new Turkish State was formed, centred upon Ankara in the heart of Asiatic Turkey, under the leadership of the visionary Mustafa Kemal, a general who had played a brilliant part in the defeat of the British at Gallipoli. Kemal repudiated the peace agreements, and resolved to rid his country of foreign troops. First he drove the Greeks out, without much trouble, then he turned his attention to the British. The main British outpost on the Asian side of the Dardanelles was Chanak, whose ramparts and minarets we glimpsed so tantalizingly that June morning on the Lone Pine Ridge, and upon this little town the Turkish armies now threateningly advanced.

In itself Chanak was not much: a shabby little Muslim town at the water's edge, with a row of foreign consulates along the waterfront, for it was the port of entry to the Strait, a clutter of high-walled crooked streets, and a fortress still badly knocked about by the guns of the Royal Navy. For Kemal to threaten it, though, was an astonishing challenge. He was the representative of a defeated lesser Power: the British not only represented all the victorious nations, but were themselves, in their imperial capacities, now the towering suzerains of all the Middle East. To the British Government under Lloyd George the situation was charged with emotional nuance.

The Colonial Secretary was Winston Churchill, who had first sent the imperial fleets and armies to the Dardanelles, and it was he who addressed an 'inquiry' to all the Dominion Governments, asking if they would send troops to the straits if fighting broke out. At the same time he told the Press what he had done. This was, in the context of 1922, a terrible gaffe. The Dominions had been left altogether in the dark about British policy towards Turkey, and they were infuriated by Churchill's presumption. Mackenzie King, the prickly Prime Minister of Canada, first heard about the 'inquiry' when he read about it in his own Sunday newspaper, and all the Dominion leaders were affronted by what seemed to be a bland British assumption of their support. 'Although the Dominions may speak with many voices for themselves as individuals,' *The Times* had imprudently declared, 'they speak as one when the time comes to speak for the Empire': but nothing could be less true, when it came to this dubious imperial summons. The Dominions were tired of European squabble, intrigue and bloodshed, and they were notably disinclined to send their young men once again to the Dardanelles, where so many of their brothers lay uselessly buried.

Only New Zealand and Newfoundland, the most thoroughly British of all the Dominions, unequivocally agreed to send troops if needed. The Australians agreed under protest. The South Africans did not answer. But the Canadian reply was the conclusive one. The Canadian Prime Minister was not competent, Mackenzie King coldly cabled, to commit troops to the Dardanelles upon a British request—such an action required the consent of the Canadian Parliament. Only eight years after George V's unilateral declaration of war on behalf of his entire Empire, this was a portentous rebuff, and Mackenzie King well realized its meaning. 'If membership in the British Empire', he wrote in his diary, 'means participation by the Dominions in any and every war in which Great Britain becomes involved, without consultation, conference or agreement of any kind in advance, I can see no hope for an enduring relationship.'

So the white settler Empire, the bedrock of the whole imperial structure, which had demonstrated its kinship so movingly in war, displayed its maturity in the new world of the peace. The Chanak affair finally dispersed any movement towards a centralist Empire, with a single foreign policy or executive, and paved the way for an altogether different imperial machinery, paradoxically at once more formal and less binding. The crisis itself came to nothing, for the British presently concluded an agreement with Kemal, and the town was never attacked after all: but the episode presented a very different imperial image from those brave assemblies of loyal statesmen which had expressed the unanimity of Empire in the flush of victory.[1]

6

Though the Empire was to expand still further, and there were still men eager to pursue the imperial mission, and propagate the imperial faith, from now on the story of the Empire would be the story of decline. Year by year the British vision would contract, and the abilities of the nation would chiefly be applied, not to projects of aggrandisement, risk or experiment, but to social reform at home. Economics rather than diplomacy would be the first preoccupation of British statesmen, and the prospect of dominion would no longer excite the nation's young men. The British were becoming a more ordinary people in their wisdom. Nothing revealed this truth more frankly to the world than the surrender by the British Empire, four years after that triumph at the Firth of Forth, of the maritime supremacy which had been its inalienable prerogative, and its surest protection, since the Battle of Trafalgar a century before.

In the heyday of Empire it had been a maxim of British policy that the Royal Navy should be equal in power to any two navies that might combine to oppose it—in practical terms, that it must be beyond challenge. Any lesser margin, it was thought, would be

---

[1] So ended Chanak's brief celebrity. Now called Cannakale, it has lost its consulates too, but on a bluff above the town gunners of the Turkish Army still keep watch on the straits, and are still suspicious of foreigners until kindly talked to.

suicidal, and in the 1890s the British were spending twice as much on their fleet as any other Power. By 1918 this tremendous criterion was untenable. Wasted by the war, Britain could no longer afford to maintain such overwhelming odds. Besides, two of her allies, Japan and the United States, were now great maritime Powers themselves, with oceanic commitments of their own. In 1922, symbolically in Washington D.C., a new ratio of sea-power was devised by international agreement, and for the first time since Nelson's day the Royal Navy was no longer the guarantor of the world's seas, nor even *primus inter pares*. In future, it was agreed, the navies of Great Britain, the United States and Japan would be limited to the ratio 5:5:3, with those of France and Italy at 1.75. The Royal Navy would no longer be able to design its ships to its own imperial requirements, for there was agreement too on what type of ships each fleet might have, and what size they might be. Even Britain's imperial fortresses, Fisher's 'keys to the lock of the world', were no longer hers to use as she pleased: under the Washington Treaty she specifically undertook not to develop Hong Kong as a base, in deference to American and Japanese opinion, and presently, under the same pressures, she would withdraw from Wei-hai-wei altogether.[1]

As a result of this treaty the British scrapped 657 ships, with a total displacement of 1,500,000 tons: they included 26 battleships and battlecruisers, among them many a proud stalwart of Beatty's Grand Fleet. Never again would a Fisher at the Admiralty be free to set the standards of the world's navies according to British requirements. No such magnificent fighting ships as the *Queen Elizabeth*, the apex of British naval assurance, were ever again constructed in British dockyards. Compromise set in. So ended Britain's absolute command of the seas, the mainstay and in some sense the *raison d'être* of her Empire.

[1] In 1930. This most absolutely forgotten of imperial outposts was forcibly leased from China in 1898, in response to Russia's acquisition of Port Arthur (now Lushun) across the Gulf of Chihli, and emulatively renamed Port Edward. The Royal Navy, which used it as a coaling station and sanatorium, grew very fond of it—the summer climate was delightful, it was a free port, and 'the inhabitants', reported a correspondent to *The Navy and Army Illustrated*, 1898, 'are a comfortable set, easy to deal with.'

The British public did not object, and even the representatives of the Dominions signed the treaty without demur. As it turned out the Washington Treaty was the only international armaments agreement that ever really worked—the Powers abided by its terms, and it provided a decade of respite from naval scares and extravagances. Hard times were coming for the British people, as the world declined into economic depression, and only sailors, imperialists and shipyard men much resented, or perhaps even noticed, the end of Rule Britannia. The glory days were gone already. Better a safe job and a home of your own than heavenly commands to splendour.

CHAPTER ELEVEN

A First and a Last Blow

FOR another generation, though, Empire would not let them be. The Afghans were troublesome again; the Persians rejected their protectorate; the Indians were restive; Sierra Leoneans went on strike; Kenyans rebelled; the Egyptians murdered their Sirdar, second in succession to Kitchener himself. Above all the British were plagued by the anxieties of the closest, oldest and most reproachful of all their possessions, Ireland: for it was in Ireland, even before the Great War ended, that the prototype of imperial revolution was launched—the precursor of all the coups, rebellions and civil wars which were to harass the British Empire from now until the end.

The English had been in Ireland for nearly 800 years. The Empire in France had gone; the Empire in America had come and gone; the vast Empire of Victoria had achieved its climax and now entered its decline; through it all they had kept their hold upon Ireland from the fortress-palace of their Viceroys, Dublin Castle. The Irish, for their part, consistently resented this occupation. Overwhelmingly Catholic in a predominantly Anglican Empire, proud of traditions as ancient as the English, they were never really reconciled to government from London: even the deliberate settling of Scottish Protestant families in Ireland, which gave some counties of the north a loyal nonconformist majority, had failed to stifle the perennial Irish instinct of rebellion. Time and again the Catholic Irish had risen, always to be subdued, and nothing it seemed could quench their spirit for long. For several centuries an Anglo-Irish gentry, the Protestant Ascendancy, had owned most of the land and governed the destinies of the country, but the Irish still honoured their own ways, preserving their national identity through every rebuff, and periodically giving their hearts to some new political Messiah.[1]

Many Englishmen refrained from reading Irish history, wrote Sir Charles Gavan Duffy, 'but it is a brand of knowledge as indispensable to the statesman or publicist as morbid anatomy to the surgeon'. He was right.[2] Ireland was the running sore of English politics. The Liberal Party had been split asunder by it, political reputations were made or broken on it. In the years before the Great War the House of Commons habitually devoted two days a year to Indian affairs, and one to colonial, but seldom a debate passed without a passionate exchange on Ireland. To most Englishmen it was a domestic problem, concerning a constituent part of the United Kingdom: but to the Irish, and their sympathizers abroad, it was a matter of Empire, and the Irish patriots habitually claimed to represent all the subject peoples in their struggle for liberty.

The Liberal solution for the problem was Home Rule—limited self-government for Ireland—but two Home Rule Bills had failed to get Parliamentary assent, one being defeated in the Commons, the other vetoed in the Lords. The Conservative-Unionists had shelved the issue during their years of power, concentrating instead on social reform in Ireland, but in 1911 the Liberals came back into office dependent upon the Irish vote, and a third Home Rule Bill was introduced. The veto of the peers was now limited to a delaying power of two years, so its passage seemed almost certain. All being well, Ireland would be self-governing within five years. The Irish Nationalist Party, Ireland's constitutional representatives at Westminster, accepted the promise and worked to implement it, and several Government committees, in London and in Dublin, began to prepare the machinery of Home Rule.

Not everyone, though, viewed the prospect sanguinely. The Conservatives remained immovably hostile—and there was always

[1] Three Irish chapters trace this melancholy story through the previous volumes of this trilogy.

[2] And he should know. As a violent Irish nationalist he was indicted for treason in 1848, but the jury disagreeing, he emigrated to Australia, where he became Prime Minister of Victoria and a Knight Commander of the Order of St Michael and St George.

the chance that they might return to power before Home Rule became law. The more extreme of the Irish Catholic patriots, including the ultra-secret Republican Brotherhood, would accept nothing but absolute independence. And most fiercely of all, Home Rule was opposed by the Protestants of Ulster in the north.[1] Tough, down-to-earth, implacably anti-Catholic, they wanted nothing of a self-governing Ireland. They vehemently distrusted the southern Irish, with whom they shared little but their island, and they believed that under Home Rule their whole manner of living, not to speak of their thriving industries and well-run commerce, would soon fall prey to Papist or Socialist bigotries. An Ireland run from Dublin would be an affront to the organic hierarchy of Empire—and it was after all Mrs Alexander, wife of the Bishop of Derry, who had most famously expressed the imperial instinct for order and decorum—

> *The rich man in his castle,*
> *The poor man at the gate,*
> *God made them, high or lowly,*
> *And ordered their estate.*

The peculiar situation of Ireland was to prove harbinger to the disintegration of Empire itself, and it was brought to a head by the arrival on the coast of Antrim, one spring night in 1914, of a small and undistinguished steamship.

2

She was a collier, the *Clydevalley*, 460 tons, twenty-eight years old, ex-*Londoner*, ex-*Balniel*, and familiar for years on the regular run between Glasgow and Belfast, the capital of the Irish north. On April 25, 1914, this unlovely vessel, its hull red with rust and blackened with decades of coal-grime, sailed quietly into the small packet-port of Larne, eighteen miles north of Belfast, with a cargo of

---

[1] I use the term 'Ulster' in shorthand. Of the nine Ulster counties, four were overwhelmingly Protestant, two were about equally Catholic and Protestant, and three had Catholic majorities. Six of them now form the province of Northern Ireland.

25,000 German rifles and 2½ million rounds of ammunition. They were to be used, if need be, to prevent by force the creation of a united self-governing Ireland.

Larne was a pleasant humdrum town, very typical of the Protestant north. Its streets ran gently down a hillside to the harbour, and there was a prominent Protestant church, and a discreet Catholic chapel, and a few shops, and a straggle of offices and warehouses along the water-front, where twice a week there docked the ferry-boat from Stranraer, forty miles away on the Scottish coast. Larne did not seem at all a passionate place. Its social order was secure—there was the Smiley family in the mansion outside town, there was the Protestant Minister in the manse, there was the usual handful of professional families, there were the dockers and seafarers in their cottages at the waterside, there was the Orange Order, the ancient society of Protestant militancy which bound all ranks and classes in sectarian loyalty. The Catholic minority kept to itself, and life in Larne was generally orthodox and uneventful—the rich man in his green demesne, the poor man at the docks.

A visitor indeed might have thought himself not in Ireland at all, that island of vitriol, but over the water in Scotland. The light was a washed, Scottish light, the sea had a clear Scottish glitter, the accents of the people sounded nearer Inverness than Dublin, and when the weather was right you could actually see, dimly beckoning across the Irish Sea, the distant outline of the Wigtown coast. The people of Larne were proud enough of being Ulstermen, but were still attached to Scotland too, one of their most popular holiday trips being a weekend visit, at excursion rates, to the home of their ancestors over the water.

All Protestant Ireland looked metaphorically towards Larne that night, for upon the supply of weapons, so its people believed, depended Ulster's future. 'Home Rule', they had been told by their leaders, 'is Rome Rule', and they were prepared to resist it even at the price of rebellion against the British Government at Westminster —of treason, in fact, against the Crown. All classes were united in this resolve, from the baronial industrialists of Belfast, who foresaw economic catastrophe in a self-governing Catholic-dominated Ireland, to the labourers' wives of the Londonderry slums, who simply hated Catholics. A Provisional Ulster Government was already in being, a shadow-regime for the north, and the public resolve was passionate, sometimes fanatic. Only guns were needed, to give it teeth.

It was a kind of resistance movement, but of the oddest kind. In Africa it would have been called a revolt against the Imperial Factor, the interference of Whitehall with the affairs of settlers: the Ulster Protestants were intensely loyal to the Crown and to the Empire, but they felt that the Liberal Government's policies were treacherously mistaken. Far from wishing to leave the Empire, they wished only to remain for ever part of the United Kingdom itself. This underlying loyalty, all the more poignant because it was to become, over the years, increasingly unrequited, gave to the Ulster Movement a character of its own. Though it was potentially revolutionary, half the British Establishment sided with it. The British Army was almost unanimously with these rebels, and so were all the imperial activists, from the cleverest Milnerites to the crudest surviving Jingos.

So as the *Clydevalley* tied up at Larne that night, the Ulster Movement stood ready to split Ireland and bewilder the Empire—for its leaders maintained that compulsory Home Rule for all Irishmen was a threat to that venerable abstraction, the imperial trust.

3

There was nothing haphazard to it. It was intricate and calculating. Its political chief was the formidable King's Counsel Sir Edward Carson, leader of the Irish Unionist Party in the House of Commons. Carson was not an Empire man: he was an Irishman, born in Dublin, a lawyer, and his concern was habitually concentrated, in the lawyers' way, on small intense issues. He was a heavy-weight with a narrow imagination, a Rhodes without an Africa, whose harsh and resonant brogue could be terrifying, and comforting, and even inspiring, but never poetic. Carson was Protestant of course, and his particular kind of rhetoric, ominously flamboyant, exactly suited the passions of the Ulster Protestants, whose dour manners masked such impetuous beliefs. Carson was a large man, but vulnerable—a worrier, a hypochondriac, a man of second thoughts and hidden doubts: but none of it showed, his imposing unsmiling figure gave confidence to any Protestant assembly, and as long before they had called the enigmatic Parnell the Uncrowned King of Ireland, so now they called this very British sort of Irishman 'King Carson'.

Beneath Carson's aegis an army of resistance awaited the outcome of the third Home Rule Bill, laboriously passing through its successive Parliamentary stages. The Ulster Volunteer Force was no raggle-taggle army of idealists. It was as professional and thorough as Carson himself. Its organizers were mostly men of the Ulster gentry, retired soldiers very often, who believed passionately in the unity of the British Empire. Its financiers were the businessmen of Belfast. Its patrons included great Ulster grandees like Lord Londonderry and Lord Dunleath. It had no uniforms and was armed with no better weapons than sporting rifles, shot-guns or even dummy rifles (supplied on demand, 1/8d in pitch pine, 1/6d in spruce): but its organization was sophisticated, and its activities were ubiquitous. Every village in Ulster had its members, and the police knew all about it from the start, loopholes in the law making it theoretically legal. When Carson visited Portadown in 1912, he was escorted through the streets in an open carriage by cavalrymen with bamboo lances, field guns made of wood, infantry with wooden rifles and pipers in neo-military dress.

By 1914 some 50,000 men, aged seventeen to sixty-five, had enlisted in this force. No military sanctions kept them there, but so proud and resolute was the Ulster spirit that the mere threat of expulsion from the ranks was enough to maintain discipline. The army was organized conventionally in divisions, regiments, platoons, and all military services were represented. There was an astute intelligence unit. There were supply and medical branches, artillery, signallers and despatch riders, three squadrons of cavalry. Half the car-owners of Ulster had pledged their vehicles to the transport branch, and there was a register of farmers willing to lend horses and wagons. The force had its own postal service and its own devoted corps of nurses. It had large secret stores of food. It had chaplains. It had a pension scheme for the wounded. It had a headquarters (the Old Town Hall, Belfast), a slogan ('For God and Ulster'), a flag (the Red Hand of Ulster).[1]

Above all it had a Manifesto. In September, 1912, Carson, consulting Scottish precedents, presented to the people of Ulster a declaration of intent, in the form of a pledge. It was called the Ulster Covenant, it was paraded at meetings throughout Ireland, and nearly half a million Ulster men and women put their signatures to it, some of them in their own blood. This is what it said:

'*Being convinced in our consciences that Home Rule would be disastrous to the material well-being of Ulster as well as the whole of Ireland, subversive to our civil and religious freedoms, destructive of our citizenship, and perilous to the unity of the Empire, we whose names are undersigned, men of Ulster, loyal subjects of His Gracious Majesty King George V, humbly relying on the God whom our fathers in days of stress and trial confidently trusted, do hereby pledge ourselves in solemn Covenant throughout this our time of threatened calamity to stand by one another in defending for ourselves and our children our cherished position of equal citizenship in the United Kingdom, and in using all means which may be found necessary to defeat the present conspiracy to set up a Home Rule Parliament in Ireland. And in the event of such a Parliament being forced upon us we further solemnly and mutually pledge ourselves to refuse to recognize its authority. In sure confidence that God will defend the right we hereto subscribe our names. . . . God save the King.*'

This declaration, with its archaic echoes, its undertones of pendantry, its Olde English script, was a recognizably imperialist document, but then Ulster had always been an imperialist place. Half the generals of the British Army, it sometimes seemed, were Ulstermen, and all over the Empire Irish Protestants flew the Union Jack above the Orange Lodge. In many ways the UVF was a truly imperial army. Some of its senior commanders were retired Indian Army men—they sometimes used Hindustani as a code, as in previous imperial crises the British had used schoolboy Greek. Many more had seen service in the Empire, and absorbed its styles and methods. The force commander, Sir George Richardson, was

[1] Derived from a legendary boat race between an O'Neill and a McDonnell. The O'Neill was losing, so he cut off his own hand and threw it ahead of both boats to the winning post.

the grandson of an East India Company soldier, the son of an Indian Army officer, and he himself, besides fighting in many a frontier skirmish, had commanded the cavalry brigade which stormed the Temple of Heaven at Peking in 1900. His chief of staff was an Indian Army colonel, his commander in Antrim was a retired general of the Royal Marines, his supply chief had been the youngest officer on the British General Staff, until he resigned in sympathy with the Ulster cause. One of the most active officers of the force was Patterson of Tsavo, whom we have already met leading his Zionist mulemen into action at Gallipoli.

Across the water, too, many men of Empire pledged their support. Bonar Law, the Canadian-born leader of the Conservative opposition, was a son of the manse, and an implacable opponent of Home Rule. He had been described as being 'as unimaginative as a ledger', but in the Ulster cause he was almost recklessly outspoken. 'I can conceive of no length of resistance to which Ulster men might go', he once said, 'in which I would not be prepared to support them.' There were things more important than Parliamentary majorities: if the Home Rule Bill went through the King himself should use the Royal Veto, never brought into action since the days of Queen Anne. Lord Milner was just as fervent. He devised a British Covenant, a kind of rider to Carson's, which was signed by nearly two million people, and he presided over a magazine called *The Covenanter*, whose motto was 'Put your trust in God and keep your powder dry': if the Government tried to coerce Ulster, he wrote in it, 'we may hope to paralyse the arm which is uplifted to strike.'[1]

Lord Roberts had nominated the commander of the Volunteer Force, and might have taken on the job himself if he were not eighty-two years old. The Chief of Military Operations at the War Office, General Sir Henry Wilson, was an Ulsterman, and had signed the Covenant. Other eminent supporters included Lord Rothschild the

---

[1] 'It is not desirable', he darkly added, 'to be too explicit. . . .' In fact he probably had in mind a plan to block in the House of Lords the passing of the annual Army Act, which had since 1689 provided Parliamentary authority for the maintenance of the standing army. It never happened, if only because the Army presently made it clear that it would not fight in Ulster anyway.

banker, Edward Elgar, Starr Jameson of the Jameson Raid, and Rudyard Kipling, who gave £30,000 to the cause, and printed a poem about it in the ultra-Conservative *Morning Post* (a pointed change from his usual outlet, *The Times*, which supported Home Rule):

> *The blood our fathers spilt,*
> *Our love, our toils, our pains,*
> *Are counted us for guilt,*
> *And only bind our chains.*
> *Before an Empire's eyes,*
> *The traitor claims his price.*
> *What need of further lies?*
> *We are the sacrifice.*

When in March 1914 officers at the Curragh, the British military base outside Dublin, were asked for an assurance that they would be ready to deal with the Ulstermen by force, fifty-eight of them, including their commanding general, threatened to resign rather than march against the Ulster Volunteers. The King himself had doubts about coercing Ulster—'Will it be wise,' he asked, 'will it be fair to the Sovereign as head of the Army, to subject the discipline and indeed the loyalty of his troops to such a strain?'—and the proposed operations were cancelled. Bonar Law was highly gratified, and assured the Ulster Unionists that they were holding the pass not just for Ulster, but for the British Empire—'You will save the Empire by your example.'

4

This was the inflammatory situation, then, into which the *Clyde-valley* sailed that April night. Her arrival at Larne had been the conclusion of a complex and hazardous operation. The guns, acquired by cloak-and-dagger, had been shipped from Hamburg first in a lighter, then in a Norwegian steamer, the *Fanny*. When the ship was inspected by Danish officials as it passed through the Kattegat, her master decided to run for it, and, leaving his papers behind, slipped moorings in the night and sailed into the North Sea.

Next day the story was in every newspaper in Europe, and everybody guessed that the *Fanny*'s arms were bound for Ulster.

They changed the appearance of the ship, they changed her name first to *Bethia*, then to *Doreen*, for days they steamed here and there, evading patrols and pilot boats—to Yarmouth on the east coast of England, to Lundy Island in the Bristol Channel, to Tuskar Rock off Rosslare in the Irish Sea: and there on the night of April 19 they met at last the innocuous old *Clydevalley*, and transhipped the guns at sea —the ships lashed together with one set of navigation lights, as the crates passed from one to the other.

At Larne all was ready for them. Every member of the Motor Car Corps had received a warning instruction: '*Sir, in accordance with your kind agreement to place a motor-car at the disposal of the Provisional Government in a case of necessity, it is absolutely necessary that your car should arrive at Larne on the night of Friday/Saturday 24th/25th instant . . . for a very secret and important duty.*' Larne was virtually commandeered. A regiment of volunteers was assembled in the demesne of the Dowager Lady Smiley, at Drumalis House; another, under Lord Massereene and Ferrard, formed a cordon on the hills above, blocking every road into the town. Telephones were cut. Food was prepared for 300 men. Down at the docks the arrangements were supervised by the chairman of the harbour company, and the local Volunteers, nearly all dockers, stood by to unload the ship. It was raining slightly.

As night fell the first of the cars and lorries approached Larne, in slow convoys down the narrow lanes, until watchers in the town could see the flashes of their headlights all over the hills, and hear the distant throbbing of their engines. All the lights in the harbour were switched on; at eleven o'clock the *Clydevalley* slipped into harbour and made fast. The arrangements went perfectly. Methodically the dockers worked there in the rain, and one by one the cars made off into the darkness with their loads of guns, and the cranes swung in the arc-lights, and the nurses in the harbour buildings kept their tea-urns on the boil. Lady Smiley looked out approvingly from her tall windows at Drumalis, young Lord Massereene inspected his check-points through the night, and the Catholic citizens of Larne, like the police, tactfully kept to their beds.

By 2.30 a.m. the last of the cars was away, and the guns were on their way across Ulster. *Clydevalley* had done her job. The army of Ulster had weapons, and Home Rule could never be imposed upon the Irish Protestants without a civil war.[1]

5

In the south a very different populace reacted to these events. The solid Protestants of the north were vehement in rejecting Home Rule: the volatile Catholics of the south awaited it more phlegmatically. The worst of the Irish miseries were over now, the famines, the evictions, the laws which condemned Catholics to permanent helotry, and the Conservative policy of 'killing Home Rule by kindness' seemed to many observers to have worked. The patriot cause had lost much of its fire since the great days of Parnell, and most Irish Catholics were not actively anti-British. The Irish Nationalist Party at Westminster was led by moderate Home Rulers, Irish volunteers still filled the ranks of the British Army, and it was generally assumed that Home Rule was on the way.

The Crown's chief representatives in Dublin were anything but bullies, and Augustine Birrell in particular, the Chief Secretary, seemed to personify the very spirit of conciliation. A charming fellow of literary tastes, the son of a Methodist minister, he loved the company of Irishmen, believed the best of them, and was a popular guest at the homes of the Dublin intelligentsia, where Gaelic art and literature were all the rage. Birrell's ambition was to be the last Chief Secretary of Ireland, and he admitted that the plays at the Abbey Theatre, where W. B. Yeats and J. M. Synge were in

[1] Some of the *Clydevalley*'s rifles, stamped with the Red Hand, were still being used in Ulster in the 1970s, and Larne's opinions have not changed. When I was there in 1974 somebody had just blown up the Catholic church. 'Not again have they?' said a man at the docks. 'One of these days they'll make a proper job of it.'

Carson remained at the head of the Ulster Unionist Party until Ulster separatism became a legal fact in 1921. He became Lord Carson of Duncairn, is buried in St Anne's Cathedral in Belfast, and stands uncompromising as ever in effigy outside the Northern Ireland parliament buildings at Stormont —unless, that is, *he* has lately been blown up.

their glory, meant far more to him than security reports from the Royal Irish Constabulary. There really seemed a chance that all the centuries of bitterness would be peaceably concluded. The days of the old rebellious mobs, singing their wild ballads, brandishing their knobkerries, cheered on by poteen whiskey and led by passionate demagogues—the high old days of Irish fury seemed long ago and half forgotten.

But as Gladstone had warned Parliament long before, Irish nationalism was not a passing mood, but an inextinguishable passion. Beneath the placid surface of things, as always the Irish revolutionaries were at work. Their motives and attitudes varied. Some were simply patriots, some social revolutionaries. Some did not believe that Home Rule would ever really come, some thought it inadequate anyway. Some hoped violence would not be necessary; some thought it inevitable; some wanted it for its own sake, believing that the shedding of blood, in sacrifice or in sacrament, was necessary for the cleansing of the Irish soul, and the fulfilment of true liberty.

The defiance of Ulster came as a shock to moderate Catholics, and foreseeing that Protestant resistance might prevent Home Rule and wreck the cherished unity of Ireland, thousands of Irishmen joined their own private armies and nationalist organizations, the Gaelic League, the League of Women, the Gaelic Athletic Association. Boys of patriotic families were recruited into Fianna na h'Eireann, 'the Fianna Boys'. Socialists and Marxists had their own Citizen Army, founded by the powerful trade unionist James Larkin and trained by Captain Jack White, ex-Gordon Highlander and son of Field Marshal Sir George White, the defender of Ladysmith in the Boer War.[1]

Much the largest organization was the Irish Volunteer Force, formed in 1913 in direct emulation of the Ulster volunteers. Its 10,000 men wore grey-green uniforms with peaked caps, and drilled openly enough in parks and squares across Ireland. It too had its agents and sympathizers everywhere, in every police station, in

---

[1] 'I feel odd', White once wrote to his father, while serving in the British Army, and asked for his advice. 'My dear boy', the Field Marshal admirably replied, 'I should be a little less odd, if I were you, and get on with your work.'

every Government office, and especially in every post office, giving
it an excellent intelligence system. It was, though, quite unlike the
tight-lipped and splendidly organized militia of the north. There
were no retired generals to command it, no traditions of Empire to
sustain it, few great demesnes to offer it parade grounds, munition
stores or refuges. It was a ramshackle, amateurish, thoroughly Irish
affair, all at odds. Its commander was a lecturer in Gaelic literature,
and it numbered in its ranks many teachers, not a few poets,
cranks, eccentrics and folk-enthusiasts of all kinds, together with a
mass of simple Irishmen who joined it out of guileless patriotism.

Authority, having turned a blind eye to the Ulster Volunteer
Force, could hardly suppress the Irish volunteers. The last thing
Birrell wanted was to antagonize the Irish public on the eve of their
emancipation, but balancing security with benevolence was a
difficult problem for him—one horn of the dilemma, it was said, was
as sharp as the other.[1] At least the movement had no arms. The
Irish patriots had many friends abroad, powerful bodies of Irish
exiles in America and Australia, enemies of Britain everywhere: the
Irish had been the most consistent of all the Empire's opponents,
and wherever there were dissidents of Empire, or critics of the
imperial philosophies, they had their supporters. But though there
was always money available, weapons there were not. The Volunteers continued to drill with broomsticks and wooden rifles, and
they looked with envy and chagrin upon the exploit of the *Clydevalley*.

6

Two remarkable members of the old Ascendancy, in particular,
believed that Ireland must be ready to fight for her independence.
The first was Sir Roger Casement, one of the saddest figures of the
whole imperial story. Like many another patriot of Catholic Ireland,
he was a Protestant, the son of a British Army officer. His devotion
to the Irish cause was not inherited, nor exactly personal, nor even
basically political: it was imaginative, aesthetic perhaps. An
instinctive and often muddled supporter of underdogs, wherever

[1] 'If not more so.'

they were, Casement identified himself with the oppressed not out of reason but out of sensuality. He was a very sensual man, tall, distinguished, rather quixotic, melancholy, whose urges were homosexual, and whose life seemed to lead him unerringly down dark and terrible paths. He had great beauty. He looked beautiful, he spoke beautifully, and there was beauty to the sense of tragedy that attended him, first to last.

Casement had become well known as a member of the British consular service in West Africa. A report he made about conditions on the rubber estates of the Belgian Congo horrified the British public with its revelations of cruelty, and later he repeated the performance after a visit to the rubber estates of Peru. His reputation stood high in England. He was knighted in 1911, retired in 1913, and went home to Ireland apparently full of honour, achievement and duty satisfied. There in his late forties he became possessed by the enchantment of the island—'bewitched', so a contemporary wrote, 'by the beauty of his own country'—and devoted himself to its causes as to a late love affair.

But he was a tortured soul. His frustrated longing for lasting affection, his considerable vanity, his half-incoherent poetic impulses, his Celtic love of the theatrical, all made him restless and dissatisfied. Immersing himself ever more distractedly into Irish history and culture (though failing to master the Irish language), Casement became one of the most extreme of patriots, and reached the conclusion that in the coming world war, which he thought inevitable, it might be better for the Irish to side with Germany, and so achieve freedom by treason. Why not? They had no quarrel with the Germans, they were no closer in race or religion to the English than they were to Prussians or Bavarians, and it was unlikely that a victorious Germany would wish to repeat the melancholy history of Ireland by occupying the island herself. When, in April 1914, the news of the *Clydevalley* coup reached him, Casement decided that his duty lay in enlisting German sympathies for the Irish cause. As the last months of peace passed, and Europe mobilized for war, he steeled himself to treason, and prepared secretly to go to Germany, via New York, to conclude an alliance with the King's enemies.

The other conspirator was Erskine Childers, a popular Dublin writer and man-about-town. Like Casement, Childers had come to Irish nationalism by way of the English Establishment. A Protestant too, the son of an eminent Anglo-Irish oriental scholar, he was educated at Haileybury and Cambridge, and had started life as a clerk at the House of Commons at Westminster. He had fought in the Boer War, and had made himself famous with an imaginary account of a German plot to attack Britain, *The Riddle of the Sands*.[1] It was as a militia soldier in 1904, during the first peacetime visit of British armed soldiers to the United States, that he had met and married Miss Mary Osgood of Boston, and he had written three books on military subjects, to one of which Lord Roberts himself, that potential Marshal of Ulster, had contributed an introduction.

But the Boer War had made a liberal of him; the passing years made him an Irish patriot; and gradually love of his country, a feel for its past and its tragic passions, a generous yearning for its liberty, turned him into a revolutionary. For him Home Rule would never be enough. By 1911 he was advocating full dominion status for Ireland, and when in 1914 the *Clydevalley* brought her guns into Larne, he volunteered to match the feat for the patriot forces of the south. So there enters our story a second fateful vessel, the 49-foot gaff ketch *Asgard*.

She had been given to the Childers as a wedding present. Built at Larvik in Norway, she was modelled upon Nansen's Arctic exploration vessel the *Fram*, and was very strong, and exceptionally seaworthy. Childers loved her dearly. Though his wife was a cripple,

---

[1] Whose hero bequeathed his name, Carruthers, first to the imperial myth and later to the imperial self-mockery:

Old Etonian braces gleam through a match-seller's rags. An Authentics' blazer shows for a moment in the noisome portals of an opium den. A beachcomber quotes Horace between hiccoughs.

'Don't look, my dear,' says the hero, thrusting his new-won bride into a taxi, or a rickshaw or a dhow.

'Why not,' she asks (girl-like) as they get under way.

'That was Carruthers,' replies the hero, in a husky voice. 'I didn't want him to know we saw.'

—*Variety*, by Peter Fleming (1933).

the two of them sailed the *Asgard* on long oceanic voyages, frequently to the Baltic, and their seamanship was expert. In April 1914 they took her to Hamburg, where Irish emissaries had already bought a consignment of arms. With them went a crew of four men and the Honourable Mary Spring-Rice, another fervent Anglo-Irish patriot. The voyage home was complex. The boat had no engine and no radio, they had to evade frequent Royal Navy patrols, they ran into a terrible storm in the Irish Sea. Off Devonport Childers brazenly sailed his yacht, deep in the water with German guns, clean through the vessels of the British Home Fleet. But all went well, and on July 4, 1914, early in the morning, the *Asgard* sailed into Howth Harbour, five miles north of Dublin itself.

The setting was lovely, and ironically rich in imperial association. Howth harbour had been built as a packet-boat station for the English crossing, but had silted up at the bar, and was now used only by small craft. On the hill above it, though, was the fort-like terminus of the Irish Sea underwater cable, and all around were symptoms of the Ascendancy—a castle, pleasant villas, a lighthouse inscribed G III R AD 1818, a coastguard station at the end of the mole. Beyond the harbour lay the island called the Eye of Ireland, to the north there stood the high Hill of Howth, open country in those days, haunted by rabbits and seabirds. The day was fine, the tide was high, sea and shore, headland and promenade glinted with the peculiar brilliance of an Irish summer day, whose dazzle is tempered always by something opaque in the atmosphere—moisture perhaps, or sadness.

The yacht tacked with some difficulty into the harbour entrance, and tied up within the mole beneath the lighthouse. The coastguards took no notice, and within an hour a detachment of Irish volunteers, soberly dressed in hats, caps and suits, marched on to the quay to collect the guns. No Simla strategists directed this operation. No 12-cylinder Lanchesters back-fired their way to Howth. The volunteers had marched out that morning from the city, and along the way they had been joined by an eager rabble of sightseers and hangers-on. With more enthusiasm than organization they unloaded the crates of weapons and ammunition, some into handcarts, some into taxis. Everyone helped. There were excited schoolchildren from the village, and Fianna Boys, and fishermen from the boats across the harbour, and one small girl was astonished to see her own teacher, a gaunt and bespectacled scholar, working manfully there among the volunteers.

It has all gone into legend, and was to be described over and over again, with growing embellishments and disagreements, and a deepening sense of veneration, as the years passed. But it was only a small episode really. There were no more than 1,000 rifles, and within an hour they were all off. Childers and his crew sank back limp but satisfied in their cockpit, the volunteers set off for Dublin with their weapons, on their shoulders or in their carts. The coast-guards, belatedly realizing what was happening, had fired a couple of distress rockets into the sky, but nobody had interfered with the unloading, and as the column marched back along the Dublin road a priest blessed it from the top of a passing tram.[1]

When the volunteers approached Clontarf, around the curve of Dublin Bay, they found a force of Scottish soldiers and policemen blocking their way. There was a fierce little engagement. Shots were fired, bayonets fixed, the police made a baton charge, and when the soldiers withdrew towards Dublin they were jeered and stoned by an angry crowd. In the confusion some of the Scots opened fire, killing three civilians and wounding thirty-two in what was to become notorious as the Bachelor's Walk incident. When the victims were buried vast crowds of Dubliners attended the funeral in a gesture as much of anger as of mourning.

But the guns had got through anyway, only nineteen rifles being seized by the British at Clontarf, and were hidden away in clandestine armouries in the shabby labyrinth of Dublin. Now the Catholics of the south, like the Protestants of the north, could fight if the need arose.

---

[1] Captain Nicholas Reid, harbour-master of Howth, most kindly introduced me to these scenes, and showed me the commemorative text from Virgil which was put on the harbour wall beneath the lighthouse: HOS SUCCESSUS ALIT: POSSUNT, QUIA POSSE VIDENTUR—*To them success was good, and the appearance of power gave power indeed*. It was Captain Reid's Aunt Mary who saw her teacher among the volunteers: the teacher was Eamon de Valera, one day to be President of the Irish Republic.

7

Few of them expected to use their guns to seize their independence. They were satisfied with the promise of Home Rule, and they needed weapons, they thought, only to prevent the Ulster Protestants from wrecking it. When war came, John Redmond, the leader of the Irish Nationalist Party at Westminster, at once declared his party's support for it. Half a million southern Irishmen fought with the British Army, in eighty Irish battalions, and Redmond's own son was killed. Childers himself went off to the Royal Navy, and was decorated for his services. The Ulster volunteers had mostly joined the army too, so that the threat from the north seemed to be suspended, and though Home Rule was postponed again, this time for the duration of the war, most Irishmen accepted the delay as reasonable. Ireland grew prosperous in the war—unemployment was down, farm products got high prices, and the vast majority of Irish people, north and south, were loyal to the Crown. 'The one bright spot in the very dreadful situation', Sir Edward Grey told the Commons, 'is Ireland. The position in Ireland . . . is not a consideration among the things we have to take into account now.'

But the fiercest of the patriots were not mollified. In August, 1914, the Supreme Council of the Republican Brotherhood, still so secret that most people in Ireland had never even heard of it, resolved that there must be an Irish insurrection before the end of the Great War. The Citizen Army was also plotting a rising, and its allied newspaper, the *Worker's Republic*, bitterly opposed Irish participation in the war—

> *Full steam ahead John Redmond said*
> *That everything was well chum;*
> *Home Rule will come when we are dead*
> *And buried out in Belgium.*

Among these irreconcilable patriots one now emerged to prominence who believed in Irish independence in fantastical terms, and saw in the preoccupations of wartime England the perfect realization of his dreams. Patrick Pearse, an IRB man, was half

Irish, half English, a poet and a kind of patriotic voluptuary. His instincts were ritual. His pudgy face, loose-lipped, round-jawed, had the melancholy intensity of a true zealot, and he really did believe that the cause of Irish liberty was more sacred than life or human love.

More than that, he believed that it depended, spiritually, upon the shedding of blood. Peaceful Home Rule would itself be a kind of of betrayal, or desertion. Those were years of blood, of course, the blood that was soaked into the soils of Flanders or Gallipoli, blood being shed for causes of varying merit and conviction in every corner of the world. 'The old heart of the earth', Pearse wrote of this terrible spectacle, 'needed to be warmed with the red wine of the battlefields. Such august homage was never before offered to God . . . .' That blood must be shed in Ireland too became an *idée fixe* to him—'from the graves of patriot men and women', he wrote, 'spring living nations.' He was not speaking figuratively. He believed in the regenerative power of death: to die was to rise again refreshed. With these philosophies Pearse was the inspiring genius of an IRB cell within the Irish volunteer movement: so secret was this group, so muffled were its activities, that even the commanders of the volunteers, mostly Gaelic enthusiasts of innocuous intent, did not know of its existence.

By now Roger Casement was in Germany, where he was trying to enlist German support for an Irish revolution, and to recruit volunteers for an Irish Brigade among the prisoners of war taken in France. He did not have much success in his recruitment, most of the Irish prisoners being old sweats of the British Army, impervious to subversion, but by 1915 he had persuaded the Germans to back a full-scale Irish rebellion. He arranged that he himself should be taken by U-boat to the west coast of Ireland, and that a shipload of weapons would run the British blockade. The plan was co-ordinated, via America, with the leaders of the IRB in Dublin. The rising would take place at Easter, 1916, throughout Ireland, and it was hoped that the Germans would help by diversionary Zeppelin raids on London, by submarine attacks on shipping in Dublin Bay, and by providing German officers to stiffen the revolt. Dublin Castle would be seized, and an Irish Republic would be declared with Pearse as its first President.

It was a desperate, almost ludicrous plot—the Knight Commander of the Order of St Michael and St George to be deposited by dinghy on the Galway coast, the half-mad poet-patriot calling to arms a people more content than it had been for generations, the deliberate shedding of blood in that placid corner of Europe, at a time when men were dying in their millions—Irishmen themselves in their thousands—in the most titanic of all battles. Nothing went right. Admiralty intelligence in London had been privy to the plot almost from the start. They knew the number of Casement's U-boat, they knew about the arms shipment, they reported that the rising was planned for April 22, 1916—Easter Sunday. Trawlers, sloops, destroyers and a light cruiser were deployed to intercept the arms ship, the Norwegian trawler *Aud*: inevitably she was caught, and scuttled herself with all her arms off Queenstown, in Cork. Casement himself tried to cancel the rising, but he was too late, and coming ashore from the submarine U20 at Banna Strand, near Tralee, was picked up almost at once by the local constabulary, and shipped away to England, where he was held in the Tower of London and presently tried for treason.

The Germans provided no officers and no diversions, and the grandiose plan for a revolution throughout Ireland fizzled out. Only about a thousand patriots rose to arms in Dublin and raised the green flag of an Irish Republic—not over the Castle indeed, where the Viceroy remained impregnable, but at least over the General Post Office.

8

The significance of the Easter Rising is debated to this day. The British in Ireland were unprepared for it, thanks to Birrell's easy-going optimism, and for intricate reasons of security the Admiralty's intelligence had not been passed on to Dublin. Many senior officers and officials had gone to the Bank Holiday races at Fairyhouse, twelve miles away. The Army commander had gone home to England for the weekend. Still, such a toy revolution had no hope of success. Casement's mission had aborted, the British Army surrounded Dublin in overwhelming force, and the ordinary Dubliner regarded the whole enterprise as treasonable madness. If there was one sure way of *preventing* Home Rule, most people thought, it was by stabbing the British in the back at their moment of greatest peril.

Yet in a genuinely mystic way it succeeded, for its power lay in the idea of it—as W. B. Yeats wrote,

> *when all is said*
> *It was the dream itself enchanted me:*
> *Character isolated by a deed*
> *To engross the present and dominate memory.*
> *Players and painted stage took all my love,*
> *And not those things that they were emblems of.*

After the Easter Rising nothing in Ireland was the same again, and perhaps nothing in the British Empire either. We shall never know whether Home Rule would in fact have come to Ireland after the war, if events had been allowed peaceably to unfold themselves— whether the Ulstermen would have fought a civil war to prevent it, whether the Irish themselves would long have been satisfied with limited sovereignty or Dominion status. The Easter Rising made these speculations worthless, and in the long run made it inevitable that Ireland would break from the British Empire altogether. It was the breath of death, thought the dramatist Sean O'Casey, that brought the seeds of new life. 'God is not an Englishman,' sang one of the Dublin balladeers, 'and truth will tell in time.'

We will not trace the course of the Rising, so petty by the terrible standards of 1916—the seizure of key points across Dublin, the inexorable massing of British troops, the agony of Pearse and his men in their besieged headquarters, the GPO in Sackville Street. It lasted only five days, and ended inevitably in the suppression of the rebels. Let us instead stand upon Carlisle Bridge, at the bottom of Sackville Street, on the evening of April 27, 1916, two years to the week since the docking of the *Clydevalley*, and look at the scene around us. The central city is half in flames, and flickers and crackles horribly. Sackville Street, one of the most splendid streets in Europe, with O'Connell's bulky statue at the southern end, Nelson's column half-way up, and the green mass of the plane trees beyond—

Sackville Street, that pride of the Dublin Wide Streets Commission, is absolutely deserted, a no-man's-land. Rubble and litter from looted shops are scattered over it, and here and there in the half-light human bodies are sprawled. Two horse carcasses lie near the northern end of the street, and up there by the Rotunda we may dimly see the line of a barricade, with sandbags and barbed wire. Sometimes there is a spatter of rifle-fire down the street. It is like a bull-ring, part in brilliant light, part in shadow, and beyond its perimeters, beyond the sandbags, down the darkened side-streets, on rooftops all around, dim shapes of men are crouched or prowling.

To our right, down-river, lies the dark silhouette of a ship, the Royal Navy's *Helga*, now and then erupting into gunfire.[1] Behind our backs from Trinity College, from Liberty Hall the trade union headquarters, from Boland's Flour Mills to the east, upstream from the Four Courts, come sporadic sounds of fighting, rifle-fire, bursts of machine-gun fire, the thump of artillery. The smell of war hangs over Dublin, compounded of dirt, death and explosive. The night sky glows with fires, and we may hear heavy lorries moving about somewhere, shouting from hidden alleys, an occasional scream, the sudden crash and rumble of masonry.

In the very heart of all this hideousness the Post Office stands scarred, scorched and blackened. Tattered above its classical portico there flies the green, white and yellow tricolour of Ireland, and perhaps we may just see, still tacked to its barricaded door, a torn and dirty scrap of paper. It is the Irish Declaration of Independence.

9

Irishmen and Irishwomen: In the name of God and of the dead generations from which she receives her old tradition of nationhood, Ireland, through us, summons her children to her flag, and strikes for her freedom . . . We declare the right of the people of Ireland to the ownership of Ireland, and to the un-fettered control of Irish destinies, to be sovereign and indefeasible . . . We

---

[1] Generally described as a gunboat, she was actually a 300-ton yacht of the Navy's Auxiliary Patrol, armed with two 12-pounders.

hearby proclaim the Irish Republic as a Sovereign Independent State, and we pledge our lives and the lives of our comrades-at-arms to the cause of its freedom, its welfare, and of its exaltation among the nations . . .

If there was something callow to this announcement, with its absurd claim to immediate authority, at least it had actually *happened*. This was the real thing, after all. From the Marconi Radio School across the street, rebel operators had already broadcast to the world the news that an Irish Republic had been declared, and everywhere lovers of Ireland were watching events in Dublin with wonder and despair. Pearse had achieved his blood sacrifice, and his fantasy had become fact.

For within the Post Office the leaders of the Irish Revolution, the first true revolutionaries of the British Empire, were trapped and doomed. They had no chance. There was Pearse himself, radiant with the prospect of martyrdom. ('Any hope?' somebody once asked him. 'None at all', he cheerfully replied.) There was the Marxist Jim Connolly, who was fighting from dual convictions, nationalist and ideological. There was Tom Clarke, a wispy, bespectacled little figure, and there was Joseph Plunkett, an Anglo-Irish aristocrat, foppish with his ringed fingers and the sabre always at his side, twenty-four years old but dying already of tuberculosis. There was Sean MacDermott twisted by polio, and Michael Collins, 'The Big Fellow', gigantic and relentless.

With them a small band of men and women fought back uncomplainingly as the cordon closed. It was blazing hot in the Post Office, the wounded lay all about, upstairs the women were always at work bandaging, typing orders or cooking food. Nobody could relax for an instant. Now and then foraging parties crept out, but almost from the start there had been a sense of entombment in the building, as the *Helga*'s guns boomed from the river, as the unseen masses of British troops waited and watched behind their barricades—and most eerily of all, perhaps, when one of the British Army's improvised armoured cars, a steel boiler on a lorry chassis, rattled slowly down Sackville Street like a messenger from the grave.

By the evening of April 28 much of central Dublin was in ruins, and the toppling walls of offices and stores, the barbed wire and the empty streets, the piles of rubble everywhere, the looted shops, the patrolling soldiers, made it look like a city enduring some much greater war. When, on April 29, Pearse and his dazed survivors emerged from the scarred Post Office to surrender, they were greeted with contempt by the British, with hostility by their fellow-countrymen. Stones and vegetables were thrown at them as they were marched away, angry Dubliners jostled them with obscenities. They were called traitors to their own country and even to their own cause.

But over the next fortnight fourteen leaders of the rising, among them Pearse, Connolly, Plunkett and MacDermott, blindfolded against a courtyard wall in Kilmainham Prison, were shot in ones and twos by the British Army. It was done in the utmost secrecy, the city being under martial law, after macabre rituals of justice. The patriots were court-martialled within the prison, and only in later years did the details of the proceedings become known. The dying Plunkett was given permission to marry before his execution: the ceremony took place at midnight, in the prison chapel, and the condemned man and his bride had about fifteen minutes together before his execution at dawn. James Connolly, who had been severely wounded in the fighting, was court-martialled in his bed and taken to his death in a chair. The men were shot in the tall narrow execution yard of the prison, only a few yards from Inchicore Road outside, in the one corner of the prison that was not over-looked by the cells of other prisoners. The men were shot, their bodies removed, the blood cleared up from the yard, before Dublin realized what was happening: they were dead before their families knew of their court-martials.

Thousands of other patriots were arrested, and 2,500 were sent to prison camps in England and Wales. The retribution of the English was swift and terrible, and when the Irish realized what was happening the Easter Rising acquired a new meaning. Trust in the British was shattered once more, and the very citizens who had thrown rotten tomatoes at the patriots a few weeks before now mourned their memory in horrified remorse. It was, wrote the Countess of Fingal, one of the great Irish chatelaines, 'as though they watched a stream of blood coming from under a closed door'. The promise of Home Rule, which might have been a reconciliation, now became a mockery, and many a loyal Dubliner wondered for the first time if the English had really intended it at all.

'We seem to have lost', Pearse had told his court-martial. 'We have not lost. . . . You cannot conquer Ireland, you cannot extinguish the Irish passion for freedom. If our deed has not been sufficient to win freedom, then our children will win it with a better deed.' This was a true prophecy. Tactically the Easter Rising was a pitiable failure, but it changed everything in the end, and if the blood shed at the Post Office and in Kilmainham Gaol led to more blood in later years, to more sacrifices, more tragic ironies, still Yeats was right to see in the story a terrible beauty too.[1]

10

During the fighting in Dublin a small English girl, Pamela Nathan, was staying for the Easter holidays in a house in Phoenix Park. From there the sound of the gunfire from the city was easily heard. The child was urged to regard the rising as an excitement, something like a Zeppelin raid for instance, but she was not persuaded. 'This is *much* worse,' she sobbed. 'This is in the *Empire*.'

[1] The scenes of these tragedies may still be visited, though Sackville Street is now O'Connell Street and Nelson's statue has been exploded from his pillar. The statue of O'Connell the Liberator, by the bridge, is still chipped from the shell-fire of 1916, and inside the reconstructed Post Office the dead are mourned by a figure of the legendary hero Cuchulain, who lashed himself to a post so that he might face his enemies even in death. Kilmainham Gaol is now a sad and shabby museum, where veterans of the republican movement direct tourists to the execution yard, or to the chapel where Plunkett and his bride were so pathetically united.

Casement was hanged in London, inevitably, but his body was returned to Ireland in 1964 and now lies at Glasnevin Cemetery in Dublin, the Valhalla of modern Ireland. Childers' end was perhaps sadder still. Returning from the Great War with a Distinguished Service Cross, he furiously opposed the creation of the Irish Free State, arguing for nothing less than full republican independence. When civil war broke out in 1922 he joined the republican insurgent army, was captured by Free State forces, and shot by a firing-squad of Irishmen in a former British barracks in Dublin. Such are the ironies of Ireland that his son Erskine Childers II, my kind mentor in Irish politics in 1960, later became President of the Republic.

It was the fact of Empire, the imperial illusion, which had, through all these events, dictated the British attitude to Ireland. 'I think it very unwise', Queen Victoria had said, 'to give up what we hold', and it was an ancient conviction of the English that Ireland must be theirs. The instinct of imperialism had inspired the Ulster Unionists; to the British soldiers who put down the Easter Rising, it was hardly more than a species of colonial riot—'All quiet but the 'ill-tribes' replied a sentry in the city, when a passer-by asked him how things were going. The man who ordered the Kilmainham executions was General Sir John Maxwell, one of Kitchener's promising young men at Khartoum all those years before—'Conkey' Maxwell, who had once put down a mutiny of Sudanese askaris, and had spent twelve years in Egypt, where he was known for his posture of 'patriarchal militarism'.

Yet perhaps even the soldiers felt some presentiment. It rankled with the English, and disturbed them, that the most persistent threat to their omnipotence should come from this sister isle, whose people were not even black, brown or yellow, nor even exactly pagan. That they should have chosen to rebel at the very moment of England's greatest peril made their attitudes more disquieting still, for it demonstrated even to the most unimaginative English mind that the deepest Irish loyalties were altogether alien to their own. 'Nobody in Ireland,' said Lord Wimborne the Viceroy, 'North or South, is, or has ever been, loyal to England in the true sense of the word.' Irish patriotism might be nonsense, was certainly treasonable, but was evidently *true*. Could it also be just? It was George Bernard Shaw who described the rising as 'a fight for independence against the British Government, which was a fair fight in everything except the enormous odds my people had to face'.

The Easter Rebellion was soon overshadowed by the vaster events of the Great War—a few weeks later the Battle of the Somme began, and 20,000 British soldiers died on its first day. Ireland relapsed into a sullen resentment. Birrell resigned, broken by 'the horrible thing';[1] nobody believed any longer in a peaceful transition

---

[1] Which ended his political career. He lived until 1933, and is perhaps best remembered for his agreeable collection of essays *Obiter Dicta*—for he was essentially, as everyone admitted, a very agreeable man.

to Home Rule when the war was over. The events in Ireland were to play a seminal role in the slow retreat of the British Empire. The militancy of the Ulstermen was copied by white settlers in Kenya and Rhodesia, when they felt the imperial government to be neglecting their interests: the example of the southerners was watched with admiration by nationalists everywhere. Violence might not win freedom in itself, but its effect could be cumulative: not only Patrick Pearse, but a host of unknown comrades across the Empire believed in blood-sacrifice as an instrument of liberty. Here for instance is the Nyasaland rebel John Chilembwe, a contemporary far away, expressing his version of the argument in a last exhortation to his men:

This very night you are to go and strike the blow and then die. . . . This is the only way to show the whiteman that the treatment they are treating our men and women was most bad and we have decided to strike a first and a last blow, and then all die by the heavy storm of the whiteman's army. The whitemen will then think, after we are dead, that the treatment they are treating our people is almost bad, and they might change. . . .

The whitemen did not as a rule consider the treatment they treated their subjects bad, but the Easter Rising gave them cause to think more deeply about it, and slowly over the years it dawned upon many of them, as they contemplated the deaths of many another Patrick Pearse, many a nameless Chilembwe—as they struggled first to subdue, then to understand, finally to make friends with subject patriots across the world—gradually it occurred to them that perhaps their opponents were right after all, and that the idea of Empire itself, the very conception of one race having the right to rule another, was unjust.

The concept of fair play was the truest ideology of the English, and it had been violated too often in Ireland. Herbert Asquith the Prime Minister soon realized that the British had gone too far, in their savage response to the Easter Rising. He sacked General Maxwell, and, hoping to undo some of the harm done, went over to Dublin himself. There he visited some of the prisoners of the Rising, held by the British Army in Richmond Barracks, and found them 'very good-looking fellows, with such lovely eyes . . .'.

11

The Ulster Volunteer Force, volunteering almost to a man when war broke out, went to France virtually *in toto* as the Ulster Division, wearing the Red Hand of Ulster as its shoulder badge, suffering fearful casualties and fighting with a stubborn gallantry to the end of the war. After the armistice, when British Governments turned their attention to Ireland once again, Ulster was rewarded for its loyalty, or its contumacy, was specifically excepted from Irish independence arrangements, and became a self-governing province of the United Kingdom.

The Catholic Irish, on the other hand, became ever less loyal to the Crown as the years went by, and mere Home Rule was never again a possibility. Kitchener, an Ascendancy man himself, refused to allow Irish Catholics to fight under their own officers, or form their own division, and almost as soon as the war was over the Connaught Rangers, 'The Devil's Own', one of the most celebrated of the Irish regiments of the British Army, mutinied in India and were disbanded. Ireland itself fell into chaos, until out of the turmoil of rebellion and civil war, recrimination and revenge, there emerged in 1923 the Irish Free State—within the Empire still, subject to the Imperial Crown, shorn of six counties of the Protestant north, but at least a nation of its own, with its own Government and its own Parliament.

So the Irish won in a way: but they lost too, for they never made friends with themselves, and *Asgard* never sailed in consort with *Clydevalley*.[1] The old enmity of sect and loyalty was to simmer on,

---

[1] Though both are still afloat as I write. The *Asgard* plays an honoured role as a training ship, and when I inspected her in 1974 was undergoing her annual overhaul by Portuguese shipwrights at Malahide, almost within sight of Howth Harbour. The *Clydevalley* has been less lucky. For thirty years she traded in Canada, between the Great Lakes and Nova Scotia, but in 1968, eighty-two years old, she was bought by a group of Ulster militants and sailed back to Larne to be a floating museum. Money ran out, and in 1974, by now the oldest registered steamship afloat in British waters, she was sold to a Lancashire scrap metal company and towed across the Irish Sea to Lancaster. There beside St George's Quay I found her in 1975, her future uncertain, her

sometimes latent, sometimes murderous, until the British across the water no longer much cared what happened to Ireland, and the British Empire itself, the cause of all these sadnesses, was hardly more than a memory.

hull rusted, her funnel and masts gone, her bulwarks scrawled with graffiti, while the Carlisle trains clanked over the bridge upstream, and the Sunday anglers fished stolidly among the mud flats of the Lune.

CHAPTER TWELVE

The Anglo-Arabs

ELSEWHERE the Empire, like an old father of young sons, was finding a temporary new lease of life, for the conquered possessions of the Ottoman Empire offered fresh fields of enterprise. There was almost a new empire in itself out there, extending with territories old and new from the frontiers of Libya in the west to the hills of Kurdestan in the far north-east. Much of it was mandated territory, but in one form or another by the 1920s the British Empire among the Arabs comprised Egypt, Palestine, Transjordan, Iraq, much of the western shore of the Persian Gulf, most of the southern Arabian coast and the haggard port of Aden, considered by many imperialists the least desirable of all the imperial properties. It was a vast and vital region. A new class of imperialists, the Anglo-Arabs, came into being to administer it, and some people hoped it might one day mature into a new brown Dominion, standing loyal, grateful and useful in oil between India and the Mediterranean.

2

For centuries the British had dealt with the Arabs. The Levant Company had begun operations in Syria in 1581, twenty years before foundation of the East India Company, and for generations Britons had been familiar figures in most of the Arab countries. The British had governed Aden since 1839, had effectively ruled Egypt since 1882, and had long been *de facto* suzerains of the Persian Gulf. In the Levant they maintained a special consular service, demi-imperial in character, and Cooks the travel agents so dominated the Middle Eastern travel market that even the Kaiser Wilhelm II availed himself of their services, when he took his white horse and eagle-helmet to the Holy Land. British merchants, explorers, surveyors, spies, had wandered everywhere in these countries: they ran the steamers on the Nile and on the Tigris, they controlled the Suez Canal and the Persian oilfield, and they had maintained their own Anglican Bishop in Jerusalem since 1841.

Inevitably, when war with Turkey seemed likely, they coveted this half-familiar territory for themselves. The implanting of British power there was one of the earliest and most consistent of their war aims—they might be past their expansionist prime, but every instinct told them that no other Power must dominate the land-mass between Europe and India. Anglo-Indians wanted to create an Arab province of British India, governed absolutely in the Indian manner; Anglo-Egyptians wanted to clinch their control of the Suez Canal; strategists and financiers eyed the Mosul oil deposits in Kurdistan; War Office planners dreamed of a Haifa to Basra railway line; men of God wanted to restore the Holy Land to Christian guardianship. Milner envisaged an enormous imperial protectorate to include the whole of Arabia and Persia—from India to the left bank of the Don, he argued, 'is our interest and our preserve'. Fisher wanted a new canal cut from Alexandretta to the Persian Gulf, giving the Royal Navy an alternative route to India. Asquith was attracted by the idea of a new Viceroyalty of the Middle East, with its imperial capital at Baghdad. Kitchener wanted to be its first Viceroy.

Before the war the British had dealt warily with the Arabs of the Arabian peninsula, theoretically subjects of Turkey, in practice largely autonomous. The sheikhs of the Persian Gulf, as we have seen, they made more or less their vassals, but the chieftains of the interior they generally left well alone. 'The cardinal factor of British policy', said a Foreign Office minute in 1913, 'is to uphold the integrity of the Turkish dominions in Asia,' and even when the redoubtable Ibn Saud of Nejd openly revolted against the Turks that year, he was told that there was no chance of concluding a treaty with the British Empire instead. Still, they kept in touch with him, from their bases on the Persian Gulf: and at the same time, on the other side of the peninsula, they cautiously contacted the Grand Sharif Hussein of Mecca, head of the Hashemite clan, descended directly from the Prophet Mohammed, which held the hereditary guardianship of the Holy Places.

They liked the desert Arabs. The Bedouin struck a responsive chord in them. With his patrician style and his picturesque appearance, his great flocks of goats and camels, his taste for coffee and beautiful boys, his blend of arrogance and hospitality, his love of pedigree, his fighting ability and what would later be called his *machismo*, the Bedouin was every Englishman's idea of nature's gentleman. He seemed almost a kind of Englishman himself, translated into another idiom. It was upon this romantic fixation, this idealization of a type or a legend, that the British were precariously to build their new positions in the Middle East.

3

Their chosen vessels were the Hashemites, the most high-flown and ornate of the Bedouin clans, whose desert rawness had been smoothed by long acquaintance with Turkish habits, and who had pretensions to some kind of primacy, religious and temporal, over all the Arabs. Hussein was a picaresque but highly ambitious figure, coveting the position of Caliph, spiritual leader of all the Muslims, which was held *ex officio* by the Sultan of Turkey. He had no love for the Turks, having spent some years as a political prisoner in Constantinople, and after the outbreak of war he had conspired with the British to rise against the Ottoman Empire, in return for British arms, money and expertise, and promises of favours to come.

The Empire's compact with the Hashemites was deliberately vague and opportunist. It was wartime, and the British were concerned first to win the war. The Arab Revolution, to be led by Hussein and his sons, was thus seen differently by its several participants. The Hashemites represented it as a national movement, to unite all the Arabs in an independent united kingdom. Their rivals in the peninsula, notably Ibn Saud, saw it as an unprincipled attempt by one Arab clan to impose its hegemony on the others. Their tribal levies saw it as a kind of prolonged *ghazu*, a war raid in pursuit of booty. The sophisticated Arabs of Syria and Lebanon, who had their own revolutionary movements, saw it as an archaic threat to the Arab future, mounted by bucolics out of the wilderness.

And the British saw it as a tool of their own intent. The conspiracy with Hussein was hatched from Cairo and Khartoum, and fostered by an intelligence agency called the Arab Bureau, working under the British High Commissioner in Egypt, Sir Henry

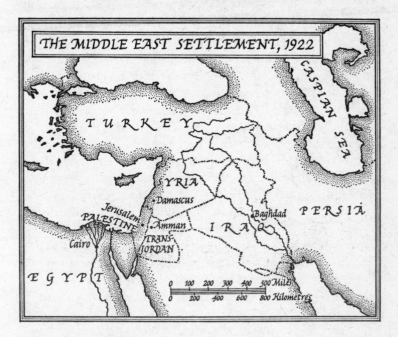

McMahon. The men in Cairo did not really know much about the desert Arabs: they had none of the long tradition of expertise which gave the Indian Political Service its insight into Arabian affairs. McMahon himself was a cautious freemason who spoke neither Arabic nor French, and was ignorant of Arabian matters. His advisers were mostly men of extreme intelligence but somewhat amateur enthusiasm, fired by the exigency of war and intrigued by the allure of Arabness from across the Arabian Sea. Everything about the liaison was at once vague and disingenuous. McMahon's promises to Hussein were deliberately vague; Hussein's claims to universal Arab leadership were necessarily vague; the geographical

terms employed were unavoidably vague, for nobody had really
defined, for example, the limits of Syria or the extent of Palestine.
All was veiled in a courteous opacity, and the messages which
McMahon exchanged with Hussein, encouraging him to rebellion,
were to become, as 'The McMahon Letters', synonymous with
diplomatic ambivalence.

They were not a treaty, had no legal force, and carefully left all
British options open, but their implication seemed to be that in
return for his help Hussein would be recognized after the war as the
titular head of an independent Arab kingdom, embracing much of
the old Turkish Empire. The British treated Hussein with a
skimble-skamble deference, couching their letters in sickly honorifics, and addressing Hussein as 'The excellent and well-born
Sayid, the descendant of Sharifs, the Crown of the Proud, Scion of
Mohammed's Tree and Branch of the Koreishite Trunk, him of the
Exalted Presence and of the Lofty Rank, Sayid son of Sayid, Sherif
son of Sherif, the Venerable, Honoured Sayid, His Excellency the
Sherif Hussein, Lord of the Many, Emir of Mecca the Blessed, the
Lodestar of the Faithful, and the cynosure of all devout Believers,
may his Blessing descend upon the people in their multitudes!' They
designed a flag for his new kingdom, symbolizing Arab unity—
black for the Abbasids of Baghdad, white for the Ommayads of
Damascus, green for the Alids of Karbala, red for the Mudhars.
Careful though they were to avoid commitments, still their
assurances to Hussein came to be regarded, at least by the Grand
Sharif himself, as a pledge: if he rebelled against the Turks, they
would make him King of the Arabs.

In fact they did not take him very seriously, or perhaps the Arabs
either. 'What we have to arrive at now', McMahon wrote, 'is to
tempt the Arab people into the right path, detach them from the
enemy, and bring them on to our side. This, on our part, is at
present largely a matter of words, and to succeed we must use
persuasive terms and abstain from haggling over conditions.'
Hussein they regarded generally as a tiresome and faintly comic old
rogue, knowing very well that there were moments when he
seriously considered joining the other side after all, and the idea of a
true Arab State, taking its place in the comity of the nations,

probably seemed so remote to them that their assurances of Arab independence were given lightly and heedlessly.

Far from frivolous, however, were the assurances they gave elsewhere about the future of the Middle East, for in fact the British were working to contradictory plans. With the Grand Sharif they had apparently agreed that the Hashemites should rule over the whole of Syria, Transjordan and Palestine, the northern provinces of Iraq, and most of the Arabian peninsula, though they had ambivalently excluded from this promise 'portions of Syria lying to the west of the districts of Damascus, Homs, Hama and Aleppo', whatever that might mean. With their French allies they had concluded a quite separate pact, 'greatly confusing', as Winston Churchill observed, 'the issue of principles'. Under this, the Sykes-Picot agreement, Syria, Lebanon, Transjordan and Iraq would be divided into British and French spheres of influence or exploitation, with Palestine under some kind of international control, and the Arabs only truly independent within the Arabian peninsula. Finally, in another *quid pro quo*, they made a fateful pledge to the leaders of the Zionist Movement, the powerful international organization dedicated to the establishment of a Jewish National Home in Palestine. Jewish money, talent and sympathy were very important to a Britain at war, and the British Government promised the Zionist leaders that they would encourage the development of a Jewish National Home in Palestine—a country whose population was, in 1916, 93 per cent Arab. The British saw advantages to themselves in a Jewish State there, British-sponsored and perhaps British-protected: 'A Jewish Commonwealth', Lloyd George envisaged, while Chaim Weizmann, the most eminent of the Zionists, said that they aimed to make Palestine 'as Jewish as England is English'.

All these cross-purposes, equivocations and contradictions entramelled imperial policy towards the Arabs. The British in the field never lost their affection for the desert Arab, but they never lost either a nagging feeling of dishonesty or betrayal, a guilt-sensation that would never have troubled their forebears in the robuster imperial adventures of old.[1]

[1] Britain's wartime promises to the Arabs are debated to this day, especially

4

For it was to prove a febrile relationship, and the man who set the tone of it, the showiest figure in the acquisition of these new provinces, was perhaps the most introspective of all the varied activists of Empire. We have glimpsed him once already, when we followed Allenby through the Jaffa Gate to accept the submission of Jerusalem in 1917: he was the untidy young staff officer in what appeared to be borrowed uniform, with a long sensitive face, a slight but sinewy body, and a donnishly distracted air.[1] This was T. E. Lawrence, archaeologist, scholar of Jesus College, Oxford, the confused and enigmatic exhibitionist who was to be known to the world as Lawrence of Arabia.

He entered the arena modestly enough. He had worked as an archaeologist in northern Syria, spoke Arabic and had taken part in a clandestine intelligence survey in Sinai. When the war came he was recruited into the Geographical Section of the General Staff and posted to Cairo; and from there he was sent to the Hejaz as one of the British officers lent to Hussein to stiffen his revolt. Lawrence was an amateur soldier of a kind more familiar in the Empire's later

---

as to whether or not Palestine was included in the area of Arab independence. This is how it seems to me a reasonably unprejudiced observer might interpret them:

Region	To Hussein	To France	To the Zionists
Syria	*Independent*	*French-Arab*	—
Lebanon and			
coastal Syria	*Reserved*	*French*	—
Northern Iraq	*Independent*	*French-Arab*	—
Central Iraq	*Independent*	*British-Arab*	—
Southern Iraq	*British*	*British*	—
Palestine	*Independent*	*International*	*National Home*
Transjordan	*Independent*	*British-Arab*	*National Home*

[1] The uniform *was* borrowed, in bits and pieces from members of the general staff. Like Allenby, Lawrence thought the entry into Jerusalem 'the supreme moment of the war'.

campaigns, an Oxford intellectual who remained obdurately and often infuriatingly civilian beneath his uniforms. His pose was shy but superior, in the maddening Oxford way, but for every man he antagonized with his self-conscious theatricals, he entranced another with the riddle of his style and sexuality.[1]

His was a strange genius. Confused in his own spirit by doubts and anxieties of the profoundest kind, sexually ambivalent and perhaps despairing, he exerted an astonishing power over the most unlikely subjects—statesmen, common soldiers, society women, Arab tribesmen, even regular soldiers of the British Army. He was a very good man, kind, generous, and perhaps this, the deepest trait of his nature, was apparent to people of perception beneath the flummery and the deceit (for he was a gifted and enthusiastic liar). Nobody remotely like Lawrence had ever played a part in the extension of the British Empire, but though it was only the chance of war that made him an Empire-builder, and though he later persuaded himself that he was doing it all for the Arabs, still he was one of those who saw the Arab countries as a potential Dominion of the British Empire—'our first brown dominion, not our last brown colony.'

Lawrence was an inspired guerilla leader, and soon became the effective commander of Hussein's revolt. He himself led its first foray out of the Peninsula, into the country at the head of the Red Sea, and he presently came to see himself as a King-maker, escorting the Hashemite family to the thrones of the Arabs. He became not simply a colleague of the Arab leaders, but actually a friend, so restoring to imperial affairs a relationship between imperialist and client that had scarcely existed since eighteenth-century India.[2]

[1] 'As I was working in my tent last night . . . in walked an Arab boy dressed in spotless white, white headdress with golden circlet; for the moment I thought the boy was somebody's pleasure-boy but it soon dawned on me that he must be Lawrence whom I knew to be in camp. I just stared in silence at the very beautiful apparition. . . . He then said in a soft voice, "I am Lawrence, Dalmeny sent me over to see you." I said "Boy or girl?" He smiled and blushed, saying "Boy". . . .'
　　　—from Richard Meinertzhagen's *Middle East Diary* (London 1959).
[2] Among several Bedouin children named after him was Lawrence al Shaalan of the Rualla tribe, born in 1918, who survived both the rise and the

It might have come to nothing, though, and the Hashemites might have faded from the imperial scene, if Lawrence had not persuaded Allenby, then organizing his invasion of Syria, of the potential importance of the Arab revolt. We have a picture, from Lawrence's own writings, of the first meeting between the two men, when Lawrence implanted in Allenby's mind the idea that the Arabs might have great meaning for the British future. It took place at the general headquarters in Cairo. Lawrence had ridden overland from Aqaba, at the head of the Red Sea, and was dressed self-indulgently *à l'Arabe*, flowing white robes, gilded dagger at the waist, sandals flip-flopping incongruously along the military corridors. He was the local equivalent in fact of the Anglo-Indian irregulars who, in turban and sashed tunics, had ridden with their cavalrymen through so many Victorian adventures, but in the ramrod world of Allenby's command, he was a rarity indeed. He looked faintly absurd, suggestively feminine, highly unmilitary.

Across the table sat The Bull, tremendous with command. One can almost feel the tension, even now, as the temperaments faced each other. They looked at one another with suspicion. Allenby thought Lawrence rather a fraud, with too high an opinion of himself as a soldier, and an altogether disproportionate view of the importance of his Arabs. Lawrence feared Allenby to be just another brass-hat. But even as Lawrence offered his plans, we may feel the atmosphere changing. Lawrence recognized the hidden spark in Allenby. Allenby dimly saw in Lawrence some hint of greatness. The idea of the Arabs as allies, not merely mercenaries, gained a new meaning when Lawrence talked of it, so fresh from the peninsula. He wanted to spread the Arab revolt northwards, into Syria, and he wanted more weapons and ammunition for the Arabs, more gold, more air support.

The general, so Lawrence himself thought, could not make out how much was genuine performer and how much charlatan—'the problem was working behind his eyes'. But he was convinced. 'At the end he put up his chin and said quite directly, "Well, I will do

---

fall of the British supremacy in the Middle East to die in 1976 as the last of the great independent chieftains—smuggler, gun-runner, tribal raider and friend of Britons to the end.

for you what I can"'"—an improbable cadence, but then Lawrence preferred a mannered prose.

5

When Allenby resumed his campaign through Syria, his right flank consisted of the Hashemite army, commanded by the Emir Feisal, third son of the Grand Sharif, and directed by Lawrence; and when Damascus fell Arabs and British rode into the city together, something new in imperial victories. Feisal set up an Arab administration in Syria; Abdullah, his brother, was promised by his father the throne of Iraq. Everywhere the flag of the Hashemites flew, and the wide kingdom of the Arabs seemed to be at hand. Lawrence had given McMahon's promises meaning, it seemed, and Allenby had sealed them.

From Cairo and London, though, the prospects looked different, for the British were already preparing to divide the Arab lands with their European allies in a muffled version of the African Scramble. They had admitted to nobody that their instinctive purpose had been to gain an imperial supremacy over the Middle East, but it was so. Their Allies assumed it, and sometimes one could read it between the lines even of their own most Wilsonian announcements. At the end of the war the British published a declaration, jointly with the French, assuring the Arabs they would be absolutely free to choose what Government they wished, but there was a sting to the rider. The only concern of the Allies, it said, was 'to offer such support and efficacious help as will ensure the smooth working of the governments and administrations'.

Cynics knew what 'efficacious help' meant, in the imperial vocabulary, and so it was to prove. At the peace conference Feisal represented his father, actually as King of the Hejaz, in his own eyes as King of the Arabs: the only other Asians present were the Indians and the Chinese. He argued for the complete independence of the Arabs—'as representing my father, who, by request of Britain and France, led the Arab rebellion against the Turks, I have come to ask that the Arabic-speaking peoples of Asia . . . be recognized as independent sovereign peoples, under the guarantee of the League of

Nations.' But he had no chance. Nobody bore himself with more dignity at Versailles than this descendant of the Prophet, immaculate in his Sharifian robes, guarded at the Hotel Metropole by two huge Nubians with drawn swords, and attended often by the now celebrated Lawrence. But dignity availed him nothing in the end. The United Arab Kingdom collapsed in disillusionment ('a madman's notion', Lawrence himself called it, 'for this century or the next probably'), and Feisal was presently ejected by the French from his throne in Damascus. When the final arrangements for the Middle East were agreed by the Supreme Council of the Peace Conference, the Hashemites were not represented at all.

This is what was decreed. In the Arabian peninsula the *status quo* would be maintained, with King Hussein confirmed in his sovereignty of the Hejaz. France was given a mandate over Syria and Lebanon, traditionally French spheres of activity since the days of the Crusades. The Zionists got their National Home. The rest would be British, embodied in mandatory Governments in Palestine (including Transjordan) and Iraq. From the half-truths and evasions of their wartime policies, from the rivalries and intrigues of Versailles, from the discomfiture of the Arabs, from the pressure of the Zionists, from the power of Allenby and the complex imaginings of T. E. Lawrence there really had come into being a new province of the Empire.

Among many British Arabists a profound sense of shame set in, to dog their attitudes and even affect their policies until the end of the Empire. They felt they had betrayed their friends, and believed the imperial policies to have been dishonourable. This is how one of them, Walter Smart of the Egyptian service, summed up the wartime exchanges in hindsight: 'The Anglo-French bargaining about other peoples' property, the deliberate bribing of international Jewry at the expense of the Arabs who were already our allies in the field, the immature political juggleries of amateur Oriental experts, the stultification of Arab independence and unity . . . all the immorality and incompetence inevitable in the stress of a great war.' The Arabs were no less bitter in their disillusionment. Feisal retired sadly to the Hejaz; the Iraqis burst into rebellion against their British overlords; Transjordan subsided into squabbling groups of tribes and petty States, precariously held in check by a handful of Englishmen.

6

Among those most deeply affected by this denouement was Lawrence, whose private mortifications were thus sublimated into a public emotion. Shame was to be the *leitmotif* of his epic memoir, *Seven Pillars of Wisdom*, and he was outspoken in his view that Britain had ill-treated the Hashemite family, now reduced once more to their guardianship of the Holy Places. Fortunately there presently came into office one of those public men improbably held in thrall by the Lawrentian enigma, Winston Churchill. Setting up a Middle East Department at the Colonial Office, and appointing Lawrence as his particular adviser, he set out to straighten accounts, if not with the Arabs in general, at least with the Hashemites. In 1921 he summoned a conference at Cairo to interpret in imperial terms the decisions of the Peace of Versailles.

'Practically all the experts and authorities on the Middle East', he later wrote, 'were summoned', which meant in the context of the times that of the thirty-eight participants, thirty-six were British and two Arab. There is a famous photograph of this conference, taken at the Mena House Hotel—where Churchill, who spent much of his time painting pictures, guarded by an armoured car, held a one-man exhibition of Pyramidical portraits. There we may see, frozen for ever in their official poses, the true progenitors of the Anglo-Arab empire. The chubby-faced, balding man in the centre is of course Churchill himself, twenty years older than he was at Spion Kop, and by now an experienced imperialist. The young man in the three-piece suit, papers untidily protruding from his jacket pocket, is T. E. Lawrence. The only woman in the group, wearing a wide flowered hat and a fox fur, is Gertrude Bell, writer, orientalist, explorer, and forceful protagonist of the British presence in Iraq. On Churchill's left is Sir Percy Cox, 'Kokkus' to the Arabs, who so fatefully advocated a British advance to Baghdad in 1915, on his right is Sir Herbert Samuel, Britain's first High Commissioner in Palestine, charged with the fostering of the Zionist National Home. Tactfully among the paladins stand two Iraqis, the soldier Jaafar al Askari, in a spiked helmet and Sam Browne, the politician Sasun Effendi Haskail, in a tarboosh and wing collar, and in front of them all, guarded by a kneeling Sudanese, two young lions gambol emblematically on the gravel.

It was scarcely a conference really. Churchill and Lawrence had already made its decisions—'over dinner', Lawrence said, 'at the Ship Restaurant in Whitehall'. But the grandees at Mena House ordered the new arrangements, and gave to all the British officials in the Middle East some sense of unified purpose. Churchill's solution was to create two new kingdoms in the Arab world, the Kingdom of Iraq, the Emirate of Transjordan. Both would have Hashemite monarchs, Feisal in Baghdad, Abdullah in Amman, but both would be unmistakably protégés of the British.

So the new empire was established—'knitting together the old', thought the historian Arnold Toynbee, then a Foreign Office official. By the middle 1920s Britain was overwhelmingly paramount in the Middle East, and her control of the Arab world was absolute, if not in principle, at least in fact. The routes to India were safe as never before, the oil wells of Iraq and the Persian Gulf, the Abadan refinery, all were securely in British possession. 'I must put on record my conviction', Lawrence wrote after surveying this consummation, 'that England is out of the Arab affair with clean hands' —or if not with clean hands, he might have added, at least with full pockets.

None of the new territories became colonies, and Aden was to be, first to last, the only true British possession among the Arabs. Elsewhere the new suzerainty was veiled in euphemism. Egypt was proclaimed independent, in 1922, but remained a British fief just the same. The Persian Gulf emirates were officially Protected States, or States in Treaty Relationship, but did what they were told. Iraq and Transjordan had their own monarchs and Governments, but were effectively run by British advisers, and policed by British forces. Palestine was a Mandated Territory, but was governed by the familiar Colonial Office hierarchy of Chief Secretary and District Commissioner, Director of Public Works and Conservator of Forests. The vassal dynasty of the Hashemites, the front of British control among the Arabs, adopted under the tutelage of the Empire all the trappings of western kingship, and there were Baghdadi tailors By Appointment to the Royal House, and two new royal palaces for the penniless Emir of Transjordan, one for the summer in the Moab mountains, one for the winter by the blue Dead Sea.

7

This was a little forlorn. It was the example and policy of the British that turned old Hussein's sons and grandsons into Arab parodies of Windsors or Hanoverians. They were more easily controlled, of course, as members of the royal brotherhood: like the subservient rulers of Indian princely States, or the more docile chiefs of black Africa, they were absorbed into the purposes of their patrons. When the Emir Abdullah of Transjordan remarked one day that the only lady whose hand he would ever kiss was Queen Mary of England, he was expressing as much a political as a domestic condition.

The Iraqi dynasty, whose first king was the fighting prince Feisal, soon degenerated into pastiche. Though in 1930 Iraq theoretically became an independent Kingdom, it really remained a British puppet, and the royal family willingly connived in the sham. Only a generation removed from the arcane chieftaincies of the Hejaz, now they seemed almost as British as the British themselves. English tutors and governesses, English nannies, English coachmen, English grooms and mechanics were installed in Kasr al-Rihab, the Palace of Welcome, on the outskirts of Baghdad. An English architect designed a Royal Mausoleum. The royal sons were sent to Harrow, and though the Queens of Iraq remained generally in purdah, the Kings made frequent public appearances, in immaculate lounge suits, or feathered topees and gauntlets, driven in crested landaus by white-gloved grooms. Feisal died in 1933; his headstrong successor Ghazi was killed driving his sports car too fast across Baghdad; the climactic years of the monarchy were dominated by the regency of the Crown Prince Abdulillah, who looked after the throne during the boyhood of his cousin, King Feisal II.

This interregnum was the true allegory of the British presence among the Arabs, for there was never a client prince more subtly anglicized than Abdulillah, or a regime which seemed, except to its struts and sycophants, more inevitably fated. Abdulillah was your true Anglo-Arab, the most transient of all imperial types, for it had scarcely a generation in which to appear, reach its fulfilment and be obliterated. He was educated at Victoria College, Alexandria, a transplanted English public school specifically designed, like the Princes' Colleges in India, to produce surrogate Britons. A slim and elegant man, with a small dark moustache, sloping eyebrows and a Mongolian puffiness about the cheeks, he had adopted all the externals of the British style. Reticent, formal, courtly, shy, strict about protocol, devoted to pedigree, he was made to be a well-heeled monarch in exile, growing old charmingly in house parties of the English shires, or on hospitable Italian terraces. Instead he was the all too proper figurehead of the Hashemite Kingdom of Iraq.

See him there now, in the garden of his palace. He is sitting in a deck-chair beneath a fronded tree, wearing a dark suit in the brilliant sunshine, smoking a Turkish cigarette. Before him, in the shade, the young king is playing chess with his brother, using big wooden chess men on a carved table, but Abdulillah is not watching them. His legs elegantly crossed, he is looking thoughtfully across the lawn towards the distant roofs of the capital, spiked by minarets and hazed in dust. It is a highly suggestive scene, but in a charged way: the two boys intent upon their game, the enamelled guardian aloof and preoccupied at his cigarette. It seems timeless, changeless, as though the three figures have always been there, and are condemned to remain for ever motionless in that Baghdadi garden: but at the same time it seems a fictional or contrived tableau, arranged there by some producer of plangent gifts. Will they ever grow up, those young innocents? Will that urbane regent ever be old? Are they real princes, or only tokens, placed there for a purpose, like pieces on a wider board themselves?

8

Among and around the Hashemites a group of Anglo-Arab satraps rose to power. Some were Arabs themselves, like the Iraqi politician Nuri-es-Said, who became a living emblem of the British connection.

Many more were Britons, discovering for themselves fine opportunities for adventure and advance. The ruling establishments of the British, swiftly erected through the Arab countries, became at once late outposts of the imperial idea.

In Jerusalem the British High Commissioner first occupied the great Augusta Victoria Hospital, originally built to commemorate the Kaiser's visit, but supposedly designed as the Government House of a German-conquered Palestine: later he was given a palace of finely dressed Jerusalem stone, golden as the city itself, encouched in pine trees, and there, looking splendidly out across the Holy City, he was constantly reminded that he stood in direct descent from Pontius Pilate in the imperial satrapy. In Amman, the capital of Transjordan, the British Resident was so much more powerful than the King that the affairs of the country revolved naturally not around the royal palace on its hilltop but around the Residency in the valley. And beside the Tigris in Baghdad the British High Commissioner's headquarters very soon assumed a somewhat baleful dominance, with its wide gravelled courtyard and its Persian gardeners, its lions and unicorns and punctilious aides, its dogs, its hunting talk, its flowering verbena, its familiar dinner lists of visiting grandees and fawning locals, its languid Union Jack above the river, its varnished motor launches, its statue of General Maude the conqueror—a structure, born by the Great War out of the Indian Raj, which became the truest monument to the British presence among the Arabs.

There was no shortage of Britons to man these frontiers. Not many were needed anyway, and bright young men who had soldiered in those parts welcomed the chance to make careers there. Some were veterans of the Revolt itself. Here is Fred Peake, for instance, 'Peake Pasha', the founder of the Transjordan Arab Legion, wearing an Arab headdress with his khaki uniform, standing with a hand on his stomach among a troop of his Bedouin swashbucklers: with his piercing blue eyes and his beautifully cut Saxon features, he looks just like one of the fighting Anglo-Indian patriarchs of a century before, and indeed he had come to the Arabs by way of India and the Egyptian Army. Here is Alec Kirkbride, Abdullah's chief British adviser, a huge kindly Scot whose pungent friendship with the Emir gave extra subtleties to the imperial theme. ('Why don't you like that idea?' Abdullah asked him once of a new development scheme. 'Who said I didn't like it, Sidi?' 'Nobody, but I know that when you flick your head away like that, it means you disapprove.') Here is Charles Belgrave, Adviser to the Ruler of Bahrein, tall, worldly, witty, almost theatrically handsome, who got his job by answering an advertisement in the personal column of *The Times,* whose tastes ran to roulette, pantomime, watercolour painting and *The Fairchild Family,* and whose ugly house above the water-front bulged in every cranny, corridor and bathroom with the books of his eclectic library. Or here is John Bagot Glubb, the most famous of them all, nicknamed by the Arabs Abu Hunaik, Father of the Little Chin, because of a disfigurement of his jaw: succeeding Peake as commander of the Arab Legion, he went much further, and with his merry Irish smile, his quiet manners and his quick emotions, became in the end the grey eminence of Anglo-Arabia.

Few of these men went to the Arab countries with any imperial mission. The basis of this enterprise was pure opportunism: its protagonists seldom pleaded moral obligation or White Man's Burden. At least towards the desert Arabs, the British felt no prejudice or even condescension, and most of this new imperial class went to the Middle East partly because they relished the adventure of it all, but partly because they genuinely admired the Bedouin ethos. On a personal level, it was a meeting of equals. Most of the Britons were men of the rural gentry, and they felt at ease and at home with Arab gentlemen, affectionate and paternal, like squires or conscientious subalterns, towards the Arab rank and file. They were happy men, probably the happiest of all the imperialists. More than any colonial servants since the great days of British India, they felt themselves fulfilled: they believed the British presence to be good for the Arabs and for the world in general, they felt, some of them, that they were expiating a betrayal, and they genuinely wished their subjects well.

9

For a couple of decades it worked. What Britain needed in the Middle East, Balfour once said, was 'supreme economic and political control, to be exercised . . . in friendly and unostentatious co-operation with the Arabs, but nevertheless, in the last resort to be exercised.' This is exactly what she got. For thirty years the British were able to safeguard their oil supplies and their strategic interests at minimum cost to themselves, and more than any other of their suzerainties, more even perhaps than their Empire in India, it was their position in the Middle East that kept them among the ranks of the Great Powers into the middle years of the century.

Arab nationalist opinion, of course, soon turned against them. They did little enough to conciliate the younger patriots. Their puppet regimes were essentially law-and-order governments, and their affinities were with the traditional ruling classes of Islam. The kings, emirs, sheikhs, princes, and sultans of the Middle East saw in Britain their best protector against the dimly perceived but all too clearly apprehended dangers of radicalism: the British believed that by bolstering the conservative forces of Islam, the ruling families, the desert heritage, they could best maintain the stability of Arab society. They often quoted Curzon's dictum about the most unselfish page in history, and they continued to claim, until the end of the Empire, a particular kinship between Englishman and Bedouin Arab—'we understand each other you see, we use the same language so to speak. . . .'

But it was to go sour in the end, all the spirited pleasure of the British presence, all the comradeship of Briton and Hashemite, even the devotion of so many British administrators to the ideal of Arabness. In the Middle East the British never had time to acquire the profound expertise they had gathered over so many generations in India, while the Arabs, especially the urban intellectuals, proved to be nationalists of a sophistication and intensity unknown to the imperialists elsewhere. The British, so late in the imperial day, were nagged by a sense of incongruity: the Arab patriots believed the whole imperial structure in their midst, disguised as it was in mandate, protectorate or formal independence, to be false.

So it was. Iraq and Transjordan, the bulwarks of the British position, were only semi-nations. Their kings were creatures, their diplomatic missions were mere sops to their self-esteem, their trade, commerce and industry were ancillary to imperial needs, their armies were trained, equipped and often commanded by Britons. The British Empire was a true ally of reaction in the Middle East, depending as it did upon the alliance of sheikhs and princes, distrustful of urban values and intellectual tastes. The progressives were bound to rebel against its presence, sooner or later, and their antipathies were given an extra focus by the problem of Palestine, where the Zionists were busily building their National Home under British auspices—a permanent imperialist bridgehead, as every Arab agreed, upon the shore of Islam.

The Arab Awakening, as it came to be called, was sporadic and scattered—an assassination in Egypt, a mutiny in Iraq, a riot in Palestine—but it was never altogether quiescent. As the decades passed the British stance among the Arabs became more and more defensive, and the High Commissioners, the Residents, the Advisers, the General Officers Commanding, the Conservators of Forests, fortified themselves with their sheikhly partners against the assaults of change.

10

We will take a journey now, through the British Middle East some time in the 1930s, when the imperial suzerainty was complete still, but precarious—powerful, but challenged nearly everywhere by subversion and dissent.

We will start from Cairo, then as always the power-base of the British presence—from the Embassy in fact, formerly the Residency, before that the Agency, in whose offices the Anglo-Arab conspiracy had first been hatched. This is a building of sombre dominance, set among wide lawns on the edge of the Nile, with its back to the mediaeval city and its front to the pyramids. British sentries stand guard at its gates, and when the Ambassador drives out, ostentatiously from the mudguards of his Rolls fly the twin flags of his plenipotence.[1] Through the gates we pass ourselves, accepting to the manner born the crash of the sentry's salute, and drive away through the tumultuous streets of the capital. Everywhere, though this is a sovereign kingdom now, the British Empire shows. There stand the hideous brown barracks of Kasr-el-Nil, with the soldiery's khaki shirts drying from its windows, and there to the right is the Turf Club, awnings down against the sun[2]—and we wave to Reggie and Lorna taking breakfast on the terrace of Shepheard's, and catch a glimpse of the 9th Lancers polo team limbering up beyond the hedges of the Gezira Sporting Club. My goodness, are they *never* going to get that Anglican Cathedral finished?[3] Great God, isn't that Andrew Holden there, in the tarboosh, just getting off the tram? Doing things the hard way, isn't he?[4]

Away we sweep beside the Sweet Water Canal, a thin thread of green through the eastern desert, and in an hour or two, as we approach Ismailia, we see the masts and upperworks of ships eerily gliding above the banks of the Suez Canal. British of course, we discover to our gratification when we scramble up the levee to watch them go by, and very British too is the luncheon they give us at the officers' mess up the road ('What a bloody awful place,' says our host as he waves us away. 'Do we *have* to have an Empire?') Over the canal next, and at Kantara, on the east bank, a troop train stands in the siding waiting to move into Palestine ('Keep your fingers off the window-ledge, lads, or the wogs'll get 'em for wedding-rings'):[5] the soldier who waves us through the check-point into

[1] Even franker were the sun-awnings of the house—which, though Egypt had officially been independent since 1922, were still decorated with the Arabic advertisement *Dar-el-Himaya*—'House of the Protectorate'.

[2] Which had been Lord Cromer's house, by the way, and survived to be burnt down by the mob in 1952.

[3] Not until the Second World War, when it came splendidly into its own for church parades, and General Montgomery used to read the lessons, to the despair of his security advisers, during weekend leave from the western desert.

[4] Yes, and he continued to do things the hard way, living on a super-annuated Nile steamboat, taking the tram to his office at the Egyptian Ministry of Finance, wearing an old tarboosh, until his retirement in 1954, when I succeeded him as tenant of the SS *Saphir*. May he rest in peace.

[5] 'Wogs' were a new invention then. The word was applied originally by

Sinai is an Egyptian in a tarboosh, but the officer you may see behind him in his office, chatting over a cup of coffee with a fat Arab in a green sports coat and *khuffiya*, looks remarkably like that fellow Jarvis, you know the man, used to write those frightfully funny skits in the ship's paper coming out. . . .[1]

By the evening we are in Palestine. Here the British seem to be embattled. There are ancient armoured cars on patrol, and road-blocks, and when the District Commissioner of Gaza, in his straw hat and suede shoes, meets us at the garden gate of his bungalow above the town, we find it guarded by watchful infantrymen. 'Welcome to the Holy Land!' he says. 'They shot poor Andrews up in Galilee last night. You will keep your eyes skinned, won't you, when you get up towards Bethlehem tomorrow?—always a tricky spot, Bethlehem.'

But Bethlehem is quiet in the sun, and before long we are in the streets of Jerusalem. The Holy City is full of police and soldiers, but retains something of its old serenity nonetheless. The Arabs meditate in the courtyard of the Dome of the Rock, the Jews touch the Wailing Wall reverently with their foreheads, the British laugh over their gins-and-tonics in the bar of the King David Hotel or are to be heard blasphemously banging balls about on the tennis courts of the YMCA. The sun-bleached streets are orderly, but watchful. Rolls-Royce armoured cars trundle through the shadows of the Jaffa Gate. At dinner an Arab economist tells us Islamic unity is inevitable, a man from the Jewish Agency expounds the nature of *kibbutzim*, and Mrs T explains how to get to a dear little shop near the Citadel that sells divine embroideries.

Next morning Wing-Commander W kindly flies us to Baghdad, in the back seat of his Wapiti biplane. We resolve to press our suits and get our hair done, as we check into the Regent Palace Hotel

---

the RAF to any native, allegedly from 'Wily Oriental Gentleman', more probably, says Eric Partridge in his *Dictionary of Slang*, just from 'golliwog'.

[1] Possibly C. S. Jarvis (1879–1953), who was Governor of Sinai from 1922 to 1936, and was indeed a prolific comic writer. He ran the desert province almost single-handed, and might well have been sitting in that customs post, for he was a specialist in drug smuggling.

(*cuisine anglaise*, and conveniently up the road from the Eastern Bank), for Anglo-Arab life on the Tigris is very formal. You never know whom you may meet, in the shady old Turkish houses, dusty in the palm groves along the river, in which the English community prefers to live. Sniffy will be the conversation after the soirée, if you neglect your curtsey to the Regent, or speak to him too frankly, man to man.[1] With luck, it is true, we may bump into the delightful Freya Stark, home in the flower of her enthusiasm from some unimaginable journey, and we are sure to meet Lady Drower, whose speciality is the language of the Mandaeans, or even Agatha Christie the popular novelist, whose husband is excavating up at Nineveh. Look out for Sir X Y, though, we have been warned, mind your Arab sympathies in front of Mrs Z, and whatever you do, don't go spouting any Fabian nonsense to Nuri Pasha.

And before we leave these neo-imperial scenes, pleasure bound to Isfahan perhaps, or back to duty in Baluchistan, we must be sure to attend a parade of the Assyrian Levies, who guard the British airfields in Iraq. They will leave a proper after-taste, for if any of the Middle Eastern peoples are protégés of the Empire, these are they. The Empire indeed brought them here, snatching them from Turkish persecution in the confused aftermath of the war, and now they are the Empire's devoted wards. Their commanding officer is very proud of them. Splendid little chaps. Almost like Gurkhas. One would never know, would one, that they had been bullied, massacred, exiled and humiliated so constantly down the centuries?

They're safe now, anyway, many of them with English names, many more with English manners, every one of them loyal to the Crown. Whatever the Iraqis do, we can always depend upon the Assyrians! They march by in their stocky hundreds, brown and solemn in puttees and forage caps, with no trade but their soldiering, no loyalty but to the British Empire, no pride except pride of service

[1] As I did during an interview in 1955, when we were discussing the Arab dress Abdulillah occasionally wore. 'You would be surprised how comfortable it is', he said. 'Yes,' I foolishly replied, 'and it does make one look so fearfully dignified.' This was not a success, and our conversation soon ended. Within a few years the poor man was to be dismembered by the mob, and a hawker walked through the streets of Bagdad offering one of his fingers for sale.

—heads very high, arms swinging to the regulation height, marching and counter-marching in the Mesopotamian heat to the strains of a florid Colonel Bogey from the Levy band.[1]

'I say, your car's waiting. Don't want to rush you, but you've got a long way to go, haven't you? Do remember me to Molly, if she's still in Karachi. . . .'

11

They are sad scenes really, ironic scenes: an Empire in menopause, ruling with a benevolent obscurantism a people just realizing its potential. Soon it will all be swept away, leaving very little behind: but though it is a transient and infertile association, still there is much charm to it, and much goodwill. Let us cheat now, and return just for an evening to Amman, that modest country capital in the hills of Moab: for there we may pay a call upon the most beguiling of all the Anglo-Arab chieftains, Abdullah ibn Hussein, the second son of the old Sharif, and now a greybeard in his own right.

No need to be nervous or diplomatic, as he was himself when, half a lifetime ago, he first encountered the British Empire in the person of Lord Kitchener of Khartoum, eyes ablaze upon the steps of the Cairo Residency. Abdullah is a hospitable prince, and hardly have we signed the book at the British Residency than we are invited up to the palace for dinner that evening. Here is Empire at its happiest, an ambiance without grand principles or purposes perhaps, but without arrogance or sycophancy either. The palace stands on a ridge outside the town, guarded by long-haired Arab Legionaries in the street outside, by stalwart Circassian bodyguards, in astrakhan hats and polished jack-boots, inside its high gates. The evening is fresh; the food smells excellent; an evening with the Emir Abdullah is likely to be fun.

The style of it is cosmopolitan. Though neither of the Emir's

[1] They were to serve the British until 1955, when the Royal Air Force bases in Iraq were handed over to the Iraqis, and the Assyrians were left to look after themselves. I went to their farewell parade, at Habbaniyah, and thought it touching that they marched away to oblivion to a martial arrangement of 'The Old Folks at Home'.

two wives are present, nor his black concubine either, there are several Englishwomen at the table: and though there is no alcohol there is excellent Turkish food, stuffed aubergines, sizzling kebabs, baklava or savoury pastries. Abdullah loves his victuals and the black Berberine servants silently keep his plate well-stocked, or replenish the table with steaming silver platters. The conversation is variously in Arabic, Turkish and English. The English guests are in evening clothes; the Emir wears a black Turkish frock-coat, with a white waistcoat buttoning up to his neck, and a spotless white turban on his head.

He is in his merriest form, his face all a-twinkle, and as he worries his kebab he chaffs the imperturbable Kirkbride, making outrageous comments in Arabic or Turkish, or telling him comic stories. Glubb is there, rubicund and amiable, and Lancaster Harding, the Director of Archaeology, describes his new dig at Jerash, and all the wives laugh a good deal, and all the men wish they had a drink. It is a homely evening. Arab and Briton, they have been brought into each other's lives by the processes of history, and share the same experiences of war and peace. Artificial though it is, still there is to the occasion a wistful breath of might-have-been; and here, authentically to capture its fugitive beguilement, is the Emir's own after-dinner story, a favourite of his for many years, and fondly familiar to all his guests. It is called the Tale of the Upstairs Donkey.

On the outskirts of my native Mecca (the Emir Abdullah says) there used to be a haunted house, empty for many a year. One day the merchant Abdel Kader, passing it on his way home, heard sounds of revelry from its upper floors. He was a sceptical man, and thought some young bloods were having a party up there. When he stopped beside the gate to listen the noise stopped and a voice called to him out of the darkness: 'Come up, Abdel Kader, come up!' Determined to teach the young rascals a lesson, he tethered his donkey and entered the house. At once the noise started again, but the higher he climbed the stairs, the higher the revelry seemed to be, until when at last he reached the flat roof of the house, there was nobody there.

As he stood there cursing, though, he heard a peal of raucous laughter from the garden below, and set off angrily down the stairs again. Again there was nobody in the house, and nobody in the garden, and nobody at the gate: but when he reached the street once more, he found that his donkey had vanished into the dark. Poor Abdel Kader, how he cursed now! but as he stood there in the half-light, vowing retribution and reviling the younger generation, he heard another peal of laughter, high above his head. He looked up at the roof above him, determined to catch a glimpse of his tormentors, and this time saw a head silhouetted against the parapet of the house, clear against a starlit sky.

It was (concludes the Emir Abdullah, spitting out a datestone and half-winking at Mrs Glubb), his donkey.[1]

---

[1] Abdullah was assassinated in 1951, in the Haram esh Sherif in Jerusalem, but his grandson King Hussein, long since emancipated from the Empire, is still guarded by Circassians and served by Berberines in the palace. Sir Alec Kirkbride was my kind eye-witness of this imaginary dinner-party, and to him also I owe the story of Abdel Kader and his donkey.

Sir Henry McMahon, the true progenitor of this chapter, survived until 1949, achieving his greatest eminence as Sovereign Grand Commander of the Masonic Supreme Council 33°. He won immortality twice over, as the author of the McMahon Letters and as the negotiator of the McMahon Line, which defines to this day the frontier between Assam and Tibet.

As for his correspondent the Grand Sharif, he was at least recognized by the British as King of the Hejaz: but in 1924 Ibn Saud seized his kingdom from him, incorporating it into the new State of Saudi Arabia, and poor Hussein spent his last years in exile on the British island of Cyprus, where his greatest joy was his Arab mare Zahra—'Draw nigh!' he would gently call to her in his lonely old age—'Come nearer, Cooling of the Eyelids!'

CHAPTER THIRTEEN

A Muddled Progress

IN the meantime, while this frail new empire came fitfully into being, a terrible event occurred in the greatest of the old possessions, India, providing one of those markers in time, like the Indian Mutiny or the Boer War, by which imperial patterns can best be traced. It was the massacre at Amritsar, which was recognized even then as the worst of all stains upon the imperial record.

It happened in a public enclosure, something between a square, a rubbish dump and a garden, called the Jallianwalla Bagh, in the very heart of Amritsar. This was a venerable city of the Punjab, and it contained the holiest shrine of the Sikhs, the Golden Temple, where the splendid major-domos of the faith, sashed and staved, all day long marshalled the pilgrims towards their holy places beside the pool. The city was a brown maze of narrow streets, mud-paved, always crowded, with open-fronted bazaar shops, and toppling merchants' houses, and foetid lanes with open drains. It was a highly-strung and volatile place, like most such sacred sites, and in April 1919 it was in a state of uproar. A wave of nationalist protest had been sweeping India, yet another epilogue to the Great War, and in Amritsar there had been riots and demonstrations, culminating in the deaths of five Englishmen and an assault upon an English woman missionary, riding her bicycle innocently through the town.

Nobody doubted that the British would retaliate, but by now the city was too inflamed to count the risk; though public assemblies had been forbidden, on April 13 some hundreds of people deliberately and defiantly crowded into the Jallianwalla Bagh for a political meeting. It was an ominous scene. The Bagh, surrounded by high walls, was sunk below the level of the surrounding streets, and overlooked on all sides by towering houses, and this gave it rather the feeling of a prison exercise yard, or a place of execution. There were only three entrances: two gates at the southern end, and a narrow passage, hardly wide enough for two men to pass, at the north-west corner.

Near this passage a patriotic orator, clambering on to a pile of rubble, began to read to the crowd a passionate poem of liberty. With every word the tension rose. The people swayed, stirred, sighed, and sometimes shouted responses: and the heat was so great, the place so jammed, the emotion of the occasion so high, that one could almost feel the heart of the crowd thumping there, and hear its excited breathing. This was elemental India—'life chained to an imperfect mind', as the poet Rabindranath Tagore described it—uncouth but innocent, too big for itself, furious when excited, docile and endearing when calm. It was a crowd in short, like most Indian crowds, assembled there partly out of conviction, partly in the hope of profit, partly for something to do. It was a dangerous crowd.

Suddenly there was a rumble of heavy automobile engines outside the walls, and people near the entrance passage could see, gleaming in the street outside, the brown steel shape of an armoured car. In a moment the corridor was full of armed men, pushing their way fiercely into the garden, and on to the higher ground behind the speaker. One or two were English officers, some were Baluchis, some were Gurkha riflemen, taut purposeful little mercenaries like tamed wild beasts, with rifles in their hands and kukris at their belts. In a matter of moments they were briskly deployed along the top of the garden, and were kneeling with loaded rifles facing the crowd.

The speaker did not at first notice the commotion behind him, but some of the crowd at once made for the gates, or took cover behind a small stone shrine that stood half-way down the garden. When the orator looked over his shoulder, and saw the soldiers kneeling there, he shouted to the crowd not to be alarmed—they would never shoot—they only had blank cartridges. Hardly had he spoken than a command rang out, and the impassive Gurkhas, obedient as machines, began to shoot at point-blank range into the crowd. The panic was frightful. People fought each other to get to the gates. They scrabbled at high walls, they trampled one another down, they rushed this way and that, they tried to hide, to take shelter behind each other, to lie flat on the ground. The Gurkhas were unmoved. Loading in their own time, they aimed especially at the two exits at the bottom of the garden, until the gates were jammed with dead and wounded Indians, and nobody else could escape.

The shooting went on for about six minutes. When the soldiers withdrew again down the alley, and the armoured cars returned to their barracks with a reverberation of gear changes among the city lanes, 379 people had been killed, and another 1,500 wounded. The Bagh was littered with corpses and wounded men, blood trickled through the dust, and when dusk fell upon the carnage, and the last survivors had left the square terrified and aghast, women and children crept around the garden searching for their menfolk—turning bodies over, inspecting shattered faces, or simply squatting helplessly among the horror. Later the jackals and pi-dogs arrived: and since a curfew was in force, by ten o'clock there was scarcely a sound in the Jallianwalla Bagh, but the rustling and gnawing of these animals, the moaning of the wounded, and the echoing voices of the British patrols outside, tramping with cocked rifles through the shuttered streets.[1]

---

[1] Amritsar has never been forgotten by the Indians, and the name of Jallianwalla has become, according to one Indian historian, 'a word of concentrated import—as intense as "Bastille" in the history of modern Europe'. The Bagh has been turned into a memorial garden, but seemed to me nevertheless, when I spent an hour sketching there in 1975, to have no ghosts. A food pedlar had set up his stall by the entrance passage, where the armoured cars parked that day, children played on the terrace where the Gurkhas knelt to fire, the domes of the Golden Temple shone brilliantly over the houses, and the little group of Indians who presently joined me did not once refer to the massacre, but were more concerned with practical philosophy ('What place in the world is there', one veteran earnestly asked me, 'for a man of my age?') Nobody really knows how many died in the massacre. The British official figure was 379, in Amritsar they now speak of 'over 800'.

2

This was the tragedy of Amritsar, when Brigadier-General Reginald Dyer, CB, felt it his duty to make an object lesson of the demonstrators in the Jallianwalla Bagh. He was obliged to resign his commission, but he had many supporters, for he believed he was forestalling another Indian Mutiny. His superior officers condoned his action; the guardians of the Golden Temple enrolled him into the Brotherhood of Sikhs; the House of Lords passed a motion in his support; the readers of the *Morning Post* subscribed a £25,000 testimonial; in his attitude as in his orders he was only reflecting an innate sense of inadequacy that would presently debilitate the British in India. Still, as Winston Churchill said, it was an episode 'without precedent or parallel in the modern history of the British Empire . . . an extraordinary event, an event which stands in singular and sinister isolation'.

This was true: at least since 1858, the British had seldom behaved murderously towards their subjects, except in battle. But then the situation in India in the years after the Great War was itself without precedent or parallel. The truth was dawning on the British rulers of the sub-continent that their dominion, still absolute after so many generations, was foreseeably coming to an end. India was not like any other British possession, not a casual acquisition, or a hazily conceived name in a statistical chart, or a colony settled in by cousins, or even a fashionably recommended investment. India was part of the British truth, the other half of a mirror, 'the stern step-mother of our race'. So long as anyone remembered Britons had been coming and going between the two countries, noblemen to be Viceroys, vicars' sons to be civil servants, Army officers, missionaries, poor cousin Ethel off to find a husband, young Tom from the Queen's Arms going to sweat it out in Poona or Bangalore, and good riddance to 'un.

India had long since filtered back to Britain, too. Maharajahs were common figures of London society. Indian cricketers and polo players were popular performers. There had been Indian Members of Parliament, Indian university professors, even an Indian peer—

Lord Sinha of Raipur, Privy Councillor and Freeman of the City of London.[1] Monuments of the Indian connection were scattered across Britain, from war memorials to the country houses of nabobs, from village wells donated by benevolent Maharajahs to the cemetery, high on the Sussex hills, of the Indian soldiers who had died at their hospital in Brighton during the Great War. India was more than Empire. In an illogical way it was, to the British and to many Indians too, a part of England, a distant part that only a minority knew, but so interwoven with English destinies that the association seemed indivisible.

There was profit in it, even now. India was one of the most valuable fields of British investment—in 1914 some £800 million of capital was invested there. Rubber, coffee, indigo, tea, coal, jute, railways—all these Indian industries were very profitable to British financiers: jute mills in the 1920s were said to be making an annual profit of 90 per cent. India provided a sizeable army for the imperial defence (paid for out of Indian revenues), not to speak of the prestige and authority which accrued from the mastery of this vast possession in the east, and generations of Englishmen had benefited directly from the Indian link.

But the relationship went much deeper. Most men entering the Indian services now had some family connection with the country, and families like the Rivett-Carnacs, the Maynes, the Ogilvies, the Birdwoods, the Lawrences or the Cottons felt themselves to be almost of dual nationality, so old were their links with India. The Anglo-Indians had long since idealized their purposes in the east, and evolved folk-myths of their own.[2] They looked at the affairs of the sub-continent with a vision peculiar to themselves, part paternal, part loving, part contemptuous. They were selective in their affections. The simpler and more martial the Indian the better, and the more rural the countryside. Many of them went through life in a profound condition of love-hate, detesting the filth, disease and corruption of India as a whole, passionately devoted to their own

---

[1] Whose grandson, the third baron, now lives at 7, Lord Sinha Road, Calcutta.

[2] 'Because the great Harcourt Butler Sahib is taking care of us,' Indian mothers were alleged to croon, 'my baby can sleep peacefully.'

chosen aspects of it. Nobody pleased an Anglo-Indian more than an upright Rajasthani soldier of the old school: nobody repelled him more absolutely than some young law graduate of Bengal, with his progressive ideas and his never quite perfect English. In fact the British were still strangers in the land, for their knowledge of Indian culture was seldom profound, even after generations in the country, and they saw the great sub-continent only through their own experiences. The Indian totality was as baffling to them as to anyone else—'I felt', wrote one more than usually frank British District Officer, describing his relationship with his subjects, 'like a man wandering about with a dim lantern in the dark.'

By the 1920s the ancient association was changing. The British were losing interest in their Empire, and there was a falling-off of recruitment for the Indian services. The Indians were restive and disillusioned, and joined the nationalist movement in their hundreds of thousands. A million Indians had served the British during the Great War, and everybody expected concessions of independence in return. The moderates wanted Dominion status, like the white settler colonies, the extremists wanted to be quit of the Empire altogether. They were angry and disappointed when all the British conceded was the system called dyarchy, which certainly gave Indians a far greater share in government, but was a long way from liberty. The British genuinely thought dyarchy a great step forward—'the war had compelled England', wrote Lionel Curtis of it, 'to recognize that the principles for which she was fighting . . . must be extended to Asia and Africa.' The Indians thought it a miserable reward for their loyalty, or alternatively an inadequate concession to their demands.

So a sense of impending change came over India, in the years after the Great War, not unlike the rumours and superstitions that had swept through the country before the Indian Mutiny. 'The people are restless', reported a percipient Deputy Commissioner to his superiors in 1918, 'and discontented and ripe for the revolution': and in that very year the revolution had begun. The Rowlatt Acts, giving the British Government almost unlimited powers against Indian subversives, were its spark. Mahatma Gandhi was its prophet.

3

We last saw Gandhi toiling down the slopes of Spion Kop as a humble and loyal stretcher-bearer, but since then he had become a celebrity. The son of a palace official in the minuscule Gujarati princedom of Porbandar, on the shore of the Arabian Sea, he had been trained as a lawyer in London, and had briefly Anglicized himself, dressing in high white collar and dark suit, cultivating the art of small talk, even learning the violin, before gravitating to South Africa as legal adviser to an Indian firm in Durban. There he had taken to politics, by way of the grievances of the Indian community, and had become well known as a champion of Indian rights and self-respect. Returning to India in 1914, he plunged at once into the furious world of nationalism, arguing first for Indian Home Rule, later for complete independence, and reverted to his Indian origins. Now he dressed altogether Indian style, professed a frugal vegetarianism, and year by year prepared himself for the vatic role he was to play in the struggle for Indian liberty.

Nobody then, nobody later, knew quite what to make of Gandhi —Mahatma Gandhi, Gandhi the Pure Soul, Gandhi of the round disarming spectacles and the toothless smile. He shared with T. E. Lawrence the quality of enigma, so that he seemed to one man a saint, to another a hypocrite, and sometimes seemed to exchange the roles from one day to the next. A very small man, 5 foot 4 inches, and slight to the point of emaciation, he had vivid black eyes, spoke very pure English with a vestigial South African accent, and enthralled nearly everyone with his suggestion of almost unearthly wisdom. 'The unknown looked out at us through his eyes', said the most worldly of his disciples, Jawaharlal Nehru, and Lloyd George reported that when Gandhi visited him at his home in Surrey, an unknown black cat bounded through the window to settle on Gandhi's lap—it left when the Mahatma left, and was never seen again.

Gandhi was sweet-natured, but sly. He was truly innocent in some ways, calculating and self-conscious in others. Like Lawrence again, he well understood the value of publicity. Like many another

Indian *guru*, he veiled his shrewdness in platitudes and truisms, and sometimes cheapened it with opportunism. There was something sterile about him. Not only did he forswear, in middle life, all sexual activities, but he had no eye for nature, and his equipoise was essentially uncreative. His repeated political fasts to the death never *were* to the death, as the British wryly noted, and there was to him an irritating element of crankiness and faddism, not to mention sanctimony—'I cannot free myself from that subtlest of temptations,' he once wrote in all solemnity, 'the desire to serve.' 'The saint has left our shores', General Smuts wrote when Gandhi left South Africa for the last time, 'I sincerely hope for ever': yet Smuts fell beneath his charm too, for even at his most contumacious Gandhi remained a being of consummate serenity and appeal—'like a good night's sleep', is how one contemporary described the effect of his presence.

Though Gandhi came to command a universal audience, he was specifically a man of his time, place and opportunity. His time was the first half of the twentieth century, the century of disillusionment; his place was an India still primitive and illiterate, but given new cohesion by modern communications and political ideas; his opportunity was the waning confidence of a generally kindly, certainly not sadistic Empire. 'We all feel', Lord Minto the Viceroy had written in 1907, 'that we are mere sojourners in the land, that we are only camping, and on the march.' Gandhi's time had come.

He was an Anglophile. 'Hardly ever have I known anybody', he wrote of himself in his youth, 'to cherish such loyalty as I did to the British Constitution. . . . I vied with Englishmen in loyalty to the throne.' He was decorated for his courage in the Boer War, and for years continued to declare his devotion, if not to the British Empire, at least to the British: 'The Emperorship must go, but I should love to be an equal partner with Britain, sharing her joys and sorrows.' In many ways indeed his ideas were those, orientally tempered, of an English gentleman, and he often Anglicized Indian philosophical techniques. One of his favourite poems was Kipling's 'If', one of his favourite hymns was Newman's 'Lead, Kindly Light'. Even politically, he was often attuned to the public school spirit. 'True democracy', he once wrote, 'is not inconsistent with a few persons representing the spirit, the hope and the aspirations of those whom they claim to represent'—Lord Milner's views exactly.

Over the years his ideas of national independence became fused with thoughts about human dignity, raising the struggle for Indian emancipation to a level beyond the vision of the Easter Rising heroes, or the groping patriots of Nyasaland. For Gandhi Indian nationhood was only an aspect of human fulfilment, applicable to all, and it was lucky for him that the British in India were, by and large, tolerant and sympathetic rulers, so that he was able to work out his moralities on a political stage, without being shot on the spot.

Gandhi's revolutionary formula was a conglomeration of foibles, dogmas and contradictions—political ambition, social theory, religious precept, racial pride, personal intuition. He rationalized it all into a single metaphysic, and called it *satyagraha*—truth-force. If imperialism was essentially a glorification of force, *satyagraha* was just the opposite—it postulated, Gandhi said, 'the conquest of an adversary by suffering in one's own person'. It was indiscriminate in application. By its means Gandhi pulled together all the separate threads of Indian discontent, social, economic, political, historical, and wove them into a radical movement of incalculable power. *Satyagraha* was the means: the end was *swaraj*, independence, which was as much a personal as a national condition. India could not be herself while aliens ruled her, and Indians could not be altogether themselves.

To the Indian masses Gandhi became semi-divine. They believed him capable of miraculous feats, flying or vanishing, and at the peak of his powers his hold over their emotions was absolute. Women of all ranks became passionate nationalists; an army of children, the Monkey Army, became the couriers and scavengers of the move-ment, like the Fianna Boys in Ireland. The people did not under-stand him, though, and he did not always understand them. He was repeatedly warned, by friends as by enemies, that *satyagraha*, hazily grasped by a vast illiterate populace, would inevitably lead to violence: nevertheless, when the Rowlatt Acts were announced, he launched a nation-wide protest against them. The result was savagery all over India. Everywhere the mob came out in Gandhi's name, breaking windows in Calcutta, destroying offices in Bombay, molesting the missionary in Amritsar.

the existing system of government has become almost a passion with me. . . . I am here therefore to invite and submit to the highest penalty that can be inflicted upon me, for what in law is a deliberate crime, and what appears to me to be the highest duty of a citizen.'

These were confusing submissions, to a conventionally educated official of the Indian Civil Service. Broomfield was thirty-nine then, a slight man but of some presence, the son of a London barrister. He had been in India for fourteen years, but nothing in his training and experience could have prepared him for the peculiar accused who now stood before him, surrounded by his friends and supporters, looking spindly, saintly and almost demure. Gandhi made a very long statement to the court, and in the course of it suggested, not very seriously perhaps, that only two courses were open to the judge—to resign his post on the grounds that the law was bad, or to hand down the severest possible sentence. Gandhi spoke gently, and everyone was impressed, even the policemen and the sceptical Strangman, who thought the courtesies rather overdone. Broomfield himself was clearly touched by the occasion, and he produced a judgement that would be quoted always, when the manners and values of the British Raj were later to be debated.

'Mr Gandhi,' he said, 'you have made my task easy in one way by pleading guilty to the charge. Nevertheless what remains, namely the determination of a just sentence, is perhaps as difficult a proposition as a judge in this country could have to face. The law is no respecter of persons. Nevertheless it will be impossible to ignore the fact that you are in a different category from any person I have ever tried or am likely to have to try. It would be impossible to ignore the fact that, in the eyes of millions of your countrymen, you are a great patriot and a great leader. Even those who differ from you in politics look upon you as a man of high ideals and of noble and even saintly life.'

There were few people in India, said the judge (optimistically perhaps if he counted the British) who would not sincerely regret that Gandhi could not be left at liberty. But it was so. He was going to sentence the Mahatma, said Judge Broomfield, to six years' imprisonment, as a balance between what was due to the prisoner and what seemed to be needed in the public interest: 'and I should like to say in doing so that if the course of events in India should make it possible for the Government to reduce the period and release you, no one will be better pleased than I.'[1]

5

The course of events soon did, for Gandhi spent less than two years in his prison at Poona, during which time he not only read Kipling, Gibbon, Jules Verne and the entire text of the Mahabharata, but learnt Tamil and was operated on for appendicitis by the local English surgeon. Broomfield's careful and courteous judgement, though, typified the English attitude to this slippery and inexplicable opponent—by turns fascinated and repelled, trusting and suspicious, hostile and appeasing.

The British were groping in India. Once the proudest trophy of the Crown, it was becoming an awkward anachronism, a captive nation that fitted no category. Nobody really knew what best to do about it, because few really believed, in their heart of hearts, that Indians were capable of governing themselves. 'To me it is perfectly inconceivable', said Lord Birkenhead, Secretary of State for India in 1925, 'that India will ever be fit for Dominion self-government', while it was Ramsay MacDonald the Socialist who once observed that parliamentary democracy could no more be transferred to India than ice in an Englishman's luggage. H. G. Wells thought the Englishman in India was like a man who had fallen off a ladder on the back of an elephant, 'and doesn't know what to do or how to get down'.

Since the beginning of the Indian Empire there had been a conflict of views, between those who believed India must remain for ever British, and those who thought the highest purpose of the imperialists must be to prepare the sub-continent for modern

[1] Broomfield rose to be a High Court judge at Bombay, and was knighted: in 1942 he retired to Bournemouth, and there, fifteen years later, he died:
> *They look for nothing from the West but Death*
> *Or Bath or Bournemouth. Here's their ground. They fight.*
> *Until the Middle Classes take them back,*
> *One of ten millions plus a C.S.I. . . .*

nationhood—the trustee conception of imperialism, against the law-and-order school. Even in the 1920s both views were common. King George V, for example, subscribed distinctly to the permanency theory. 'I suppose the real difficulty', he wrote, 'is the utter lack of courage, moral and political, among the natives. . . .' T. E. Lawrence, on the other hand, thought the time already overdue for a British withdrawal—the Indian Empire had lasted too long, and was evidently failing. In the field there were undoubtedly many British members of the ICS who thought the Empire was only now approaching its *raison d'être*—the careful transfer of responsibility to Indian hands that had been the declared purpose of its presence for more than a century.

At home successive British Governments wavered in their policies, offering a nibble of freedom one year, apparently discouraging all progress the next. The Coalition Government of 1918 gave India dyarchy, the Conservative Government of 1927 set up an Indian Commission of Inquiry, under Lord Simon, which included no Indians at all. The Prime Minister of the Labour Government of 1929 was Ramsay MacDonald, who promptly ate his words about parliamentary democracy for India, and announced that within months rather than years there was likely to be an Indian Dominion within the British Commonwealth. Whatever their party, they were all at sixes and sevens. They had lost the touch of Empire, and far from commanding events, bemusedly responded to them. All they could offer the Indian rebels was gradual constitutional advance, embodied in conferences, legal proposals and commissions of inquiry, and having as its only aim an Indian Dominion modelled as far as possible on the Westminster model.

6

It was not enough for Gandhi, who had long lost his faith in the imperial system, and was committed now to absolute independence, like the Irish rebels of 1916. Inner voices, he said, had persuaded him that a new revolutionary campaign was necessary, and to launch it he declared an Independence Day. On January 26, 1930, at meetings all over India, citizens were invited to make a pledge of independence—MAHATMA EXPECTS EVERYONE TO DO HIS DUTY, said the lead headline in the *Bombay Chronicle*—and soon afterwards Gandhi set off upon an allegorical mission of defiance, to mark the moment when the British Raj no longer had meaning for Indian patriots. He wanted to do something that would be instantly understood by the Indian masses, whose grasp of political issues was hazy and whose conception of Empire was decidedly fluid—*most* Indians, even then, had probably never set eyes upon an Englishman. Gandhi decided deliberately to challenge an official ruling that everyone knew: the Government monopoly on the production and sale of salt.

The salt monopoly was an old staple of Anglo-Indian affairs. It affected the lives of every citizen, and it affected the state of Government—half the retail price of salt went in taxes. Gandhi conceived the idea of publicly producing a quantity of salt himself, out of sea water, and inviting the Indian people to do the same in a universal gesture of self-respect. It was a masterly application of *satyagraha*. It was easy, it was harmless, it had moral content, it would appeal to everyone and it had a fine historical quality, like throwing tea into Boston Harbor. On March 12, 1930, Gandhi set off from his head-quarters in Ahmedabad on his Salt March to the Indian Ocean. He was making for a village called Dandi, 250 miles away on the Gujarati coast, and he turned the twenty-four days of the march into a triumphant pilgrimage.

There is a film of the walk, and in it we see the prophet and his followers striding cheerfully, vigorously and jerkily towards their goal. Seventy-eight people started off with Gandhi, all dressed in white homespun; but they grew into thousands along the way, and were followed by rich but less robust supporters in motor cars, and often escorted by boys on bicycles, or straggling crowds of locals, or tooting companies of minstrels, as they marched along the dusty tracks, between the rich pasture-lands, under the heavy hanging trees of Gujarat. It is an almost Italian landscape there, like some hotter and poorer extension of the Po valley, and there was an almost Virgilian composure to Gandhi's progress.

Farmers knelt beside the road when he passed; women came out of their houses to offer him food, rest or comfort, as though he were on his way not simply to a political demonstration, but to his own

Calvary. When they came to a village they stopped, and Gandhi, accepting with a princely calm the obeisance of the elders, climbed to a platform or a mound and addressed the people: when he moved on the elders generally went with him—nearly 400 village headmen resigned their posts to join the march. Gandhi bore himself like a great leader, temporal as well as spiritual, and fragile though his figure was, in the old newsreel pictures he seems to stride through those scenes powerfully, head bowed, staff in hand, and to stand head and shoulders above the crowds of his admirers, disciples, sightseers and welcoming committees—who, tacitly keeping at a distance from him, accentuated this air of supremacy, and made it look as though some physical aura surrounded the Mahatma, like an electric field.

Nobody interfered with the Salt March. The Government stayed prudently aloof, and on April 5 the great company of patriots approached the isolated fishing and boating village of Dandi. It was only a hamlet really, a little cluster of houses, a mosque and a Hindu temple, pleasantly encouched in green trees and reachable only by crossing an expanse of sedgy and treeless flatland. It was an amphibious kind of village, rather Venetian, lagoon-like. The track there was half under water when Gandhi came—as he put his foot in the mud he exclaimed with delight 'It feels like velvet!'—and this curious environment, with the fresh sea air, and the western breezes, put him in high spirits.

Dandi was the simplest possible place. There was not even a radio in the village. The men lived by fishing or transporting goods up and down the coast in their sailing-vessels. Life was hard, and Dandi men travelled widely in search of jobs, so that the villagers had relatives in most parts of India, and knew all about the phenomenon of Gandhi. There were scenes of wild excitement as he entered the village, dancing, music, prayers beneath the great banyan tree, and he was taken along the single street to a two-storey villa, the property of rich Muslim sympathizers, which was to be his lodging. Dandi was not exactly on the sea. An expanse of tidal marsh, crossed by a wooden causeway, lay between it and the coastal reef, where the village fishing boats were drawn up on the sand: and here each day, when the tide was out, patches of salt were

Technique: on the way to Beauvais, 1930

left for the picking. The plan was that Gandhi would go to a spot immediately below the villa, early next morning when the light was good, and pluck a lump of salt from the ground in frank and exuberant contempt of the law.

The world watched Dandi that day. Interest was enormous everywhere, especially in America, and scores of reporters and cameramen had arrived from Bombay to record the occasion. The little village was crammed with sightseers, and there were crowds camping on the hard salty ground around it, and a constant flow of traffic from the town of Navsari, ten miles away across the flats. In the morning, when Gandhi's aides went to the appointed spot to make things ready, the tide having ebbed, they found that unfortunately there was no salt there—police agents, it was suggested, had brushed it into the mud during the night. It did not matter, though. They found some elsewhere, and by the time Gandhi emerged from the villa after his morning prayers, surrounded by his disciples and followed by photographers, all was ready.

It was a pity in a way that the Mahatma did not in fact, as legend was persistently to suggest, make his way into the Indian Ocean to pluck the salt from the sea itself. The brilliant light, the endless sandy beach, the boats lined up picturesquely on the foreshore, the slow surf off the Arabian Sea, the solitary saintly figure stepping into the water—it would have made an unforgettable vignette. Never mind, the truth was remarkable enough, and in a matter of hours newspapers across the world carried photographs of the Mahatma stooping like a bony seabird to pick up his illicit mineral from the mud. The crowd burst into song and slogan, and the news agency men sprinted back to their hired cars to get their pictures away.

Some nights later, when Gandhi and his followers were camping beside a river in the village of Karadi, five miles away, a police posse arrived from Navsari. Its British commander entered the hut where Gandhi was sleeping, beside the silent river, and shone a torch in his face. His Majesty's magistrate had arrived with a warrant for Gandhi's arrest, in the small hours of the morning to avoid a riot. The Mahatma was to be held under a law of 1827, allowing Authority to detain a suspect without trial indefinitely:

and when he had cleaned his teeth and said his prayers, they took him away to prison once more.[1]

7

Again he was not there long, for the events he had started with the Salt March rolled on all around him and made his imprisonment a dangerous embarrassment for the British. He had launched a gigantic new wave of patriotic protest—like releasing a spring, Nehru thought—and all over India people responded once more. Suddenly salt was everything. People ostentatiously made it, or gave it away to crowds, or dug it out of the earth, or auctioned it (Gandhi's original spoonful was sold for 1,600 rupees). People rioted over salt, newspapers were banned because of it, soldiers mutinied, professors led their entire classes to collect it from the seashore. In the cotton town of Sholapur revolutionary workers formed their own administration. In Wadila 1,500 people raided the Government salt depot. The British responded fiercely, with violent police action and mass arrests. Most of the Congress leaders were gaoled and by the end of May, 1931, nearly 100,000 Indian patriots were in prison. 'May I congratulate you', wrote Nehru without irony, from his prison to Gandhi's, 'on the new India you have created by your magic touch?'

Certainly the Salt March publicized India's discontent as nothing had before. World opinion everywhere turned against the Raj, and the British were forced more obviously upon the defensive. They were not really in a coercive mood. They were resigned rather than resentful now, and there were no massacres this time. On most

[1] The Dandi landscape has changed since then, because the tidal expanse where Gandhi got his salt has been dammed, and is now paddy-field: the village itself is much the same, and still suggests to me Torcello or Sant' Erasmo. Gandhi's visit is ecstatically remembered. The house he used is still owned by the same family; the banyan tree thrives; there are dhows still in a creek by the Navsari road; and in an India where the Gandhian message is largely discredited, the people of Dandi still live scrupulously by the Mahatma's tenets. I am greatly indebted to the village schoolmaster, Mr Dhirubhai Patel, and members of his family, for their sweet kindness to me in 1975.

levels relations between the races were cordial enough.[1] Besides, the King's Viceroy in India was above all a man of God and of peace. Lord Irwin indeed toyed with the idea of entering upon a fast himself, in the Gandhian manner, to bring peace to India by self-redemption, and though he was dissuaded from this unprecedented gesture, still his approach to Indian politics was tinged with an abnegatory mysticism.

Tall, high-crowned, grave, imposing, Lord Irwin was one of the more remarkable of the Viceroys. His presence was given a mediaeval piquancy by an atrophied and handless left arm, as though he had been maimed by some stroke of sorcery, but in many ways he was an imperial modernist. The son of a great Yorkshire land-owner, a devout High Anglican, a Fellow of All Souls and the biographer of John Keble the divine, he was by no means a reactionary. His instincts were conciliatory, and he had already made many enemies at home by publicly announcing his view that Dominion status was the natural end of Indian constitutional progress (though privately he had added, for he was not a strong man, that he thought it 'wholly improbable whether now or in the near future'). He was not at ease with the arrests, baton charges, declarations of martial law, by which the British responded to the Salt March rising, and in fact felt a strong fellow-feeling for Gandhi as a colleague in metaphysics. He had met the Mahatma several times, and felt he understood him. After all, he said, when people remarked upon Gandhi's infuriating ways, 'some people found Our Lord very tiresome.'

In January 1931, accordingly, Irwin decided to release Gandhi from his purposeless imprisonment, and invited him to come to Delhi to negotiate a settlement of India's problems. This was drastic action indeed, and the consequent meetings between the Viceroy and the Revolutionary were to become legendary and notorious. It happened that in 1929 Lord Irwin had moved into the immense new palace that had been completed for the Viceroys at

---

[1] And many Englishmen in India, of course, were sympathetic to Indian nationalism. In Bombay a group of businessmen called the Young Europeans liked to *épater la bourgeoisie* by driving around town with Congress flags fluttering from their radiators.

New Delhi—a stone epitome of British authority in India. Here it was, in the most magnificent of all the imperial palaces, that the two men met, the Viceroy aristocratically at ease in that stateliest of homes, the Mahatma, fresh out of prison, barefoot, with his staff and *dhoti*. Gandhi had first suggested the meetings, asking to see Irwin 'not as a Viceroy, but as a man', and it was as equals, almost as comrades, that the incongruous principals met. The spectacle enthralled the world. Foreign correspondents rushed to Delhi, and wondering crowds watched through the wrought-iron gates on the afternoon of February 17, 1931, as the Mahatma, huddled in a shawl against the winter chill, clambered up the monumental steps of the palace and disappeared inside. This was not an Imperial Power summoning its subjects for instruction: this was an exchange between sovereignties.

They met eight times, talking in Irwin's chintzy and comfortable study in the west wing of the palace. Gandhi's small figure, padding along the interminable corridors, became almost a familiar of the house, and innumerable anecdotes of his relationship with the Viceroy went the Delhi rounds. They laughed a good deal, it seemed. They exchanged badinage sometimes. Gandhi was quite ready to be amused at himself, and when they toasted each other in non-alcoholic liquors, and Gandhi chose water, lemon and a pinch of salt, Irwin wryly regretted that it must be *excise* salt. Once they talked late into the night, and then Gandhi, wrapping his shawl around his shoulders and grasping his willow staff, set off down the long viceregal drive to walk the five miles home to the house where he was staying. 'Goodnight, Mr Gandhi,' called the tired Viceroy into the night, 'goodnight, and my prayers go with you!'

The agreement they reached meant little. For the moment it ended the disturbances, but it displeased Indians by its moderation, and British conservatives by its concessions. Though Gandhi said the talks had been conducted 'with much sweetness', the Viceroy, looking back on their meetings in later life, admitted that the Mahatma was 'not a very practical person to deal with'. The meeting represented something altogether new, nevertheless, in the old dialogue between Britons and Indians. From the Indian point of view it was a significant advance, and Gandhi and Irwin appeared together in innumerable cartoons and posters, and even gave their names jointly to a match factory. To old-school imperialists it was a surrender—'taking tea with treason'. No longer did the Raj decree, it seemed: it bargained over non-alcoholic refreshments in the Viceroy's study. Winston Churchill, a seer among imperialists, instantly recognized the truth of it. He was revolted, he said in a famous Parliamentary anathema, by 'the nauseating and humiliating spectacle of this one-time Inner Temple lawyer, now turned seditious fakir, striding half-naked up the steps of the Viceroy's palace . . . to negotiate and parley on equal terms with the representative of the King-Emperor'.[1]

By his own lights he was right to be sickened, for he had truly seen in Irwin's hospitality the pattern of the imperial end. It was a glimpse of the inevitable. As Nehru later wrote, 'the best of individuals seem to me to play a relatively unimportant role when vast elemental forces are to play against each other': no splendour of palace, no hauteur of satrap, neither coercion nor conciliation, could now stay the decline of Empire.[2]

8

Later in the year Gandhi went to England, to a Round Table conference convened by MacDonald's Labour Government. The conference did not achieve much. Gandhi proved particularly difficult, irritating the Indian Muslim delegates as much as he did the British, and the British went ahead with their own ponderous constitutional plans for India. The Mahatma, though, made a profound impression upon the kindly British public, who never forgot him. The publicity was enormous. One day he was toothlessly smiling as the women of a Lancashire textile factory gave him

---

[1] Though he did not really mean it when he added privately that Gandhi ought instead to be bound hand and foot, laid in the dust outside the gates of Delhi, and ceremonially trampled upon by the Viceroy, riding an elephant. Or did he?

[2] Irwin, becoming Lord Halifax, went on to be Neville Chamberlain's Foreign Secretary, and later wartime Ambassador in Washington. He died in 1959, and is described in the *Dictionary of National Biography* as possessing a character of 'baffling opaqueness'.

three cheers in their flowered pinafores. Next day he was bartering quips with London street-urchins—'Hey, Gandhi, where's your trousers?'—and boldly drinking cider at a luncheon meeting of the London Vegetarian Society. He exchanged ideas with Charlie Chaplin and Bernard Shaw, and was given two woolly dogs and three pink candles by a group of children on his birthday, October 2.[1] Nearly everyone was delighted by him, but Churchill refused to see him, and a group of Oxford dons found his dialectic unpersuasive —'now I understand', exclaimed one exasperated scholar, 'why they made Socrates drink the hemlock.'

He took tea with King George V and Queen Mary, at Buckingham Palace. This was a pungent engagement. Though the King had accepted the suggestion reluctantly—'What! Have this rebel fakir in the Palace!'—still His Majesty was, reported his private secretary, 'as is his custom, very nice to Mr Gandhi'. The King took the occasion to 'warn' the Mahatma, nevertheless, that civil disobedience was a 'hopeless and stupid policy', and Gandhi must put a stop to it. 'Remember, Mr Gandhi, I can't have any attacks upon my Indian Empire.' It probably did not occur to the King, still less to his private secretary, that if he was the not very distinguished monarch of an island kingdom off the coast of Europe, Gandhi was the almost deified prophet of a sub-continent. The irony did not escape the Mahatma, though. When he left the palace reporters asked him if he had felt properly dressed for the occasion, in his loin-cloth and sandals. 'It was quite all right', he replied. 'The King had enough on for both of us.'

9

The longer the British stalled, the more impossible their task became—between 1921 and 1931 alone the population of India had increased by 34 million. The Victorian imperialists had not only felt themselves to be masters of physical nature, so that almost nothing seemed impossible to them: they had also been men of commanding character, confidence and interest. Now the truly

[1] Mine too—we are Librans. Gandhi was sixty-two, I was five, and I can dimly remember the commotion caused by the Mahatma's visit to Britain.

imaginative men of Empire, the really striking characters, the people who could catch the imagination and fire the spirit, were more often the leaders of their subject peoples. The passion and the poetry of conviction were on the other side—and two or three men with a new song, as the poet Arthur O'Shaughnessy had written, could 'trample an empire down'.

By now Gandhi was a great world figure, equivocal but inspiring. No imperialist, not even Kipling, had been so universally celebrated, analysed in a thousand essays, visited by a constant stream of pilgrims, ranked with national monuments and rhymed with Napoleon Brandy by Cole Porter himself.[1] Letters reached him from everywhere, sometimes addressed 'Gandhi, India', or 'His Excellency President Gandhi', or even simply with his picture pasted to the envelope. Cartoonists all over the world had merely to sketch a loin-cloth and a pair of spectacles, and every reader would understand.

Gandhi's opponents, imperialism's rearguard, were men of a different calibre and carriage—decent British gentlemen, diligent to a man, bravely grappling with the bickerings, sectarian rivalries and snaky intrigues of Indian politics, but lacking now the magic and mystery of success—the dreaming of dreams (O'Shaughnessy again) that could 'go forth and conquer a crown'. Obsessed as they were with the day-to-day conduct of affairs, they did not often seem to grasp the grandeur of the historical processes in which they were playing a part—only Churchill sensed the scale of it all, and George V could see nothing more to the Gandhi-Irwin meetings than a somewhat disagreeable social ordeal. 'His Majesty was only troubled', wrote his private secretary to the Viceroy, 'by the comical situation of the religious fanatic with his very restricted covering being admitted to your beautiful new house for what, His Majesty feels, must be rather interminable and irksome conversations.'

Irksome conversations, concerning the fate of millions and the destiny of the world! The British were working to the wrong time scale. They did not realize how imminent and how irresistible was the crisis of nationalism, as the subject peoples broke out of their docility into the consciousness of strength. They thought of the

[1] In 'Anything Goes' (1934).

Asians and Africans still as children, or more pertinently perhaps as social inferiors, and talked to them habitually *de haut en bas*. The class system of England was reflected everywhere in the Empire, and nowhere more revealingly than in India: if the mandarins of the ICS looked down upon the box-wallahs of Calcutta, and the business-man looked down upon the railway employee or the British soldiery, one and all looked down, with varying degrees of benevolence, condescension or dislike, upon the indigenous Indian. Even now ordinary Anglo-Indians considered themselves innately superior to Mahatma Gandhi himself. 'Just a bit of a nuisance', is how one ICS man described him, 'and slightly absurd.' 'I never allowed anyone to shout "Mahatma Gandhi" ', said a member of one of the oldest Anglo-Indian families, 'without giving him six on the bottom with a stick.'

More profoundly the British, whatever their public views or political convictions, instinctively hated the idea of leaving India. LET CURZON HOLDE WHAT CURZON HELDE. '*I think it very unwise to give up what we hold.*' As Thucydides said about the Athenian Empire, in an observation popular among the imperial classicists, 'it may seem wickedness to have won it, but it is certainly folly to let it go.' To abdicate India would be a terrible psychological wrench for the British governing classes, let alone for the old Anglo-Indian families who had made it their lives. It was trauma for them.

On one level of thought, since they were honourable men, they were genuinely concerned to create, before they left, a self-govern-ing India that would be peaceful, prosperous and likely to last. On another their interminable hesitations, procrastinations, proposals and inquiries were delaying actions, conscious or unconscious, postponing the time, year by year, conference after conference, when the old adventure must end at last. Throughout the 1920s, well into the 1930s, they argued and conferred, seeking always, if they did not always recognize it, a few more years of grandeur.

Each year the Indian nationalists became more uncompromising. The British offered safeguards, vetoes, reservations, balances, divided responsibilities, executive interventions. The Indians simply wanted freedom—harder than it sounded, because they were bitterly divided among themselves, Hindus, Muslims and

Princely States all seeking their own ends or even their own sovereignties. By 1935, under the pressure of all these separate interests, practical and emotional, obvious or innate, the British had got as far as a new Indian Constitution, giving the Indian people for the first time a substantial share in the running of their country.

This was the Government of India Act, 1935, the longest single act of legislation ever passed by the British Parliament. It was algebraic in conception, its purpose being to unite within a federalist India not only the Muslims and the Hindus, but also all the hundreds of Princely States, each one bound by a separate treaty to the Empire. Inevitably it smacked of pedantry. Churchill called it, with his preference for the bold and simple, 'a gigantic quilt of jumbled crochet work'. It was certainly elaborate, but it was a true step towards Indian self-government. If the federal Government was to be essentially British still, the provincial Governments would be autonomous, and Indians would be getting their first experience of Parliamentary responsibility. The British applied it diligently, and thought it generous. Elections were held, provincial Governments came to office, some of the forms of democracy were for the first time introduced to the Indian people.

But it was not a solution, only a starting-point. Churchill called it a starting-point for 'some downward slurge', while its progenitors believed it to be only a starting-point for a very protracted progress towards self-government. The British were generally thinking in terms of generations, before India became a Dominion: the Congress leaders were counting the years before she became a Republic. Nehru indeed had already appointed a kind of shadow-Government at the centre, appointing its own Ministers to the new provincial Governments, and issuing its own directives from its Allahabad headquarters.

By 1935 only some 500 Britons were left in the Indian Civil Service, and it was becoming more difficult each year to attract recruits from home—what ambitious man wanted to outlive his own career? The outline of an Indian State was already becoming visible, like a pattern discerned in the fire. Gandhi took a year off from politics, devoting himself to spinning, meditation and social work; Lord Irwin became Lord Halifax, and went home to be

Foreign Secretary in Neville Chamberlain's Conservative Government of 1938. General Dyer, paralysed by a stroke at his home near Bristol, died uncomforted by the expressions of support he still received from eminent Anglo-Indians, the comradely letters from his Indian soldiers, the fact of his honorary Sikhship or the unfailing confidence of his family. 'I don't want to get better', he told his daughter-in-law. 'I only want to die, and to know from my Maker whether I did right or wrong.'

As for Winston Churchill, when Lord Irwin urged him to bring his views on India up to date by talking to some Indians, he remained immovable. 'I am quite satisfied with my views on India,' he said, 'and I don't want them disturbed by any bloody Indians.'

CHAPTER FOURTEEN

Sweet, Just, Boyish Masters

A T home the Empire held little interest for the masses now, and H. G. Wells estimated that nineteen Englishmen out of twenty knew no more about the British Empire than they did about the Italian Renaissance. The British had lost their taste for the far-flung. Emigration declined so rapidly after the war that soon more migrants were entering the British Isles than were leaving them, and being better educated than they used to be, and more experienced in the world, the people were less ready to accept what their betters told them. 'Look as though you mean it!' was an old Lancashire injunction, popular among ex-servicemen: but as an imperial nation, the British no longer looked as though they meant it very fervently.

The ruling establishment, however, even when Government was in the hands of the Labour Party, had by no means abandoned the imperial purpose. Its members knew that the world influence of Great Britain, its wealth, its self-esteem perhaps, depended upon the possession of the Empire, and few politicians, not even Socialists, believed that it should be allowed to wither. Besides, the 'sacred trust' which the League of Nations mandate had bestowed upon civilization as a whole was still best administered, they thought, by the British Empire. The effects of the New Imperialism having long worn off, then, repeated attempts were made to boost the image of Empire, make it exciting to a new and more cynical generation, and contrive fresh realities for it.

2

The most lavish exercise in indoctrination was the British Empire

Exhibition at Wembley in 1924. The biggest fair Britain had ever known, this was a vast advertisement, inspiring to some, laughable to others, innocently entertaining to the vast majority, for the principles, practice and above all continuation of Empire.

Wembley was a suburb on the edge of the countryside, at the end of the London underground railway. The Exhibition made it famous, for to its 220 acres of pavilion and display ground there came, during its 150 days' opening, 27 million people—more than half the population of Great Britain, and more than four times as many as went to the Great Exhibition of 1851. The exhibition was mounted with the maximum publicity throughout the Empire—the very first broadcast in Johannesburg was an organ recital to raise funds for it—but it was chiefly aimed, of course, at the domestic public, a public becoming more urbanized and domesticated every year, more degenerate, readers of the *Morning Post* would say, and apparently less ready to respond to the calls of patriotism and derring-do.

Everything that was imperial was packed into Wembley's fifteen miles of streets. There was a great stadium, to signify sportingness. There was a Palace of Engineering, the largest concrete building in the world, to illustrate invention. There was a statue of the Prince of Wales made of Canadian butter. There was a reproduction of the Niagara Falls, which, being on the Canadian-American frontier, were at least half-imperial, and a reconstruction of the tomb of Tutankhamun, who, since he had been disinterred by Lord Carnarvon, was at least posthumously British, and a posse of Tibetan trumpeters whose origins somewhere east of the Karakoram surely made them more or less Indian. The whole arena was fluttered over by a perpetual cloud of Union Jacks, and at night it was 'floodlit', as a recent Americanism had it: while round and round the entire exhibition, night and day, there trundled on elevated tracks, like a figure of the imperial momentum, its private railroad line, the Never-Stop Railway.

The show was opened imperially, too, by the King himself, his speech being 'broadcasted' into a million homes by the transmitters of the British Broadcasting Company, and turned into a gramophone record that same afternoon by His Master's Voice. 'This great achievement', said His Majesty slowly, 'reveals to us the whole Empire in little. . . . We believe that this Exhibition will bring the peoples of the Empire to a better knowledge of how to meet their reciprocal wants and aspirations; and we hope further that the success of the exhibition may bring lasting benefits not to the Empire only, but to mankind in general.' Lord Milner hoped Wembley would provide a 'powerful bulwark' against the decay of the Empire, and so clearly did the organizing committee of generals, retired pro-consuls and imperial tycoons—the exhibition cost £4 million, but much of the cash was provided by commerce and industry (and it lost £1½ million in the end).

Many of the surviving imperialists helped. Kipling named the Exhibition streets—Drake's Way, Dominion Way, and running grandly down a slight hill through the centre of the great display, Empire Way itself, with the huge pavilions of Canada and Australia, the Palaces of Industry and Engineering ('six and a half times the size of Trafalgar Square'), and the Empire Stadium standing Romanly at the end of it. Edward Elgar, now Master of the King's Musick, conducted the massed choirs at the opening ceremony on St George's Day, and they sang his great hymn, almost an alternative national anthem, 'Land of Hope and Glory'. Edwin Lutyens, the architect of the Viceroy's palace at New Delhi, designed the Queen's Doll's House, which was probably the most popular exhibit in the entire show. The architecture was distinctly gubernatorial, neo-classical with oriental interludes, the gardens were landscaped in the best Jamaican or Ceylonese manner around a cool English stream. Imperial scholars published a twelve-volume Survey of the British Empire to commemorate the event, and as for the British Government's own pavilion, it frankly described itself as 'British civilization in microcosm, illuminating and epitomizing the imperial ideal'.

By the standards of the day the publicity arrangements were very expert. There can hardly have been a soul in the kingdom who did not know about the British Empire Exhibition, and the lion rampant which was its symbol became part of the nation's visual currency. For a whole generation, Wembley came to typify a moment of life, as the Crystal Palace had frozen time for their grandparents seventy years before. Even so, many people saw something flaccid, even ludicrous, to this self-conscious projection of the imperial theme. *Punch* thought it downright silly, and published a cartoon by H. M. Bateman captioned 'Do you Wemble?', typifying the whole exhibition in terms of its roller-coaster. The intellectuals of Hampstead, by now almost unanimously anti-imperialist, disapproved of it on principle, and some of them formed a society called WGTW—The Won't-Go-To-Wembleys.

The millions who did go often went for the wrong reasons, paying altogether too little attention to the New Zealand dairy products, and altogether too much to the amusement park, the dance hall and Joe Lyon's gigantic grill room. The Bright Young Things of Mayfair treated it as a spree, misbehaving themselves in the Nigerian Handicrafts Exhibition, or strumming the ukulele as they paddled about in symbolic pools. As P. G. Wodehouse's Bertie Wooster put it, 'I mean to say, millions of people, no doubt, are so constituted that they scream with joy and excitement at the spectacle of a stuffed porcupine fish or a glass jar of seeds from Western Australia—but not Bertram. . . . By the time we had tottered out of the Gold Coast village and were looking towards the Palace of Machinery, everything pointed to my shortly executing a quiet sneak in the direction of that rather jolly Planters' Bar in the West Indies section. . . .'

'I've brought you here to see the wonders of the Empire,' says Father in Noel Coward's *This Happy Breed*, 'and all you want to do is go to the Dodgems.'[1]

[1] Another brand-new Americanism, recognized by the *Oxford Dictionary* only in 1972. Many traces of the exhibition remain at Wembley, besides the stadium: the Empire Way was tactfully renamed Olympic Way for the Olympic Games of 1948, and looks rather down at heel now, but around it still stand the remains of the great pavilions, metamorphosed into warehouses and offices, while the Never-Stop Railway terminal is now an electrical repair shop, patches of ornamental garden survive, and many a crumbled concrete arch, half-decipherable name or touch of arabesque speaks still of the Bright Young Things, Bertie Wooster and the butter prince. On Sunday mornings an open-air market thrives around the stadium, being especially popular with the Indians, Pakistanis and Jamaicans of the district.

3

It was difficult to remember, as one watched the fashionable young clowning at the British Empire Exhibition, that only twenty-five years before their parents had seen Queen Victoria's Diamond Jubilee as an apex of pride and excitement. A generation that had survived the Somme, Kut and Gallipoli had different standards of indulgence. Few ordinary Britons were much moved, for instance, by the prospect of new imperial responsibilities in the Middle East, or the news of riot and disturbance that arrived week after week from India. The middle-aged and the governing classes might care about the British Empire: the young and the working people did not.

Its symbols had gone threadbare. The opulence of the Victorian heyday had been suited to an age of hope and progress, presided over by an almost mythical mother-figure. Since then the British had lost their grand optimism, and all the paraphernalia of imperial power, the banners and the elephants, the guards of honour and the peregrinating battleships, seemed to many citizens irrelevant to the times. The time for posturing was over. So was the age of Carlylean hero-worship which had been essential to the old romance. Hardly a new national hero had emerged from the Great War: the generals were discredited, the admirals were no Nelsons, Lloyd George was a Welshman and even the genuine epic of T. E. Lawrence, when dressed up for the public by the popular Press, came out dated and unconvincing, like a late yarn from Henty.[1]

But there were other sources of excitement. There was Hollywood, for instance. Hollywood easily out-dazzled the fading splendours of the pro-consuls and the Empire-builders, and through its illusory window people glimpsed new worlds more compelling, more voluptuous, than ever the Raj had been. Once the mill-girls and the bank-clerks had dreamed of Indian adventures, African romances, even death or glory upon the imperial battlefields—

[1] 'Somewhere in the wild hills of Afghanistan', reported the Sunday newspaper *Empire News* quite untruly in 1929, 'a gaunt holyman wearing the symbols of the pilgrim and a man of prayer proceeds along his lonely pilgrimage. He is Col. Lawrence, the most mysterious man in the Empire. . . .'

Yet ever 'twixt the books and his bright eyes,
The gleaming eagles of the legions came,
And horsemen, charging under phantom skies,
Went thundering past beneath the oriflamme.[1]

Now they were transported instead to the glamorous new world of the film stars, where all life seemed to be played to incidental music, and where even stories of the Empire itself, a not infrequent subject, were given the gloss of fairy-tale. How could a Viceroy compete with a Fred Astaire, or a Governor's lady, however gracious, match up to a Norma Shearer? Wembley was a poor competitor for the Sunset Strip, and the chairman of the Empire Day Movement was bound to admit that in the Britain of the 1920s there were 'many dark corners where the rays of our Empire sun have not been able to penetrate'.

Thick and fast came rival historical movements, too, which seemed to make the existence of Empire more than ever peripheral to English life. The rise of the Labour Movement to power had nothing whatever to do with Empire. Communism was passionately opposed to it. Empire did not save the British people from the Great Depression or the General Strike. Empire contributed nothing to the new functional architecture, the new abstract art, the social experiments that were gradually changing the form of English society. So immediate and so vivid in the 1890s, Empire now seemed dimmer and more distant, and was becoming, like the British institutional architecture of the day, pompously retrospective. The message of the New Imperialism, wrote Philip Guedalla in 1924, so thrilling to the generation of *fin de siècle*, now held no more than 'a dim interest for research students'.

With lack of interest went doubt. They doubted everything, and many now doubted the justice of the imperial idea. Gandhi's influence was everywhere. Opposition to Empire was not yet a

---

[1] From *The Volunteer*, by Herbert Asquith (1881–1947)—

And now those waiting dreams are satisfied;
From twilight to the halls of dawn he went;
His lance is broken; but he lies content
With that high hour in which he lived and died.

political platform, not even among the Socialists, but it was a temper of thought growing ever more fashionable among the British middle classes, and was allied with pacifism, internationalism and a woolly sort of Marxism. Among students it was the contemporary orthodoxy—'This House', as the Oxford Union blandly voted in 1933, 'will on no account fight for King and Country.' Among readers of the progressive papers it was almost taken for granted. Kipling found himself reviled by the trendier critics; 'Land of Hope and Glory' no longer seemed to sing the majesty and beneficence of Empire, only its vainglory; Rupert Brooke was out, and only the lines of more bitter war poets, Wilfred Owen, Siegfried Sassoon, were quoted now in the common-rooms.

There was a reaction then against all things Victorian, oratorios to antimacassars, and this inevitably rebounded against that ultimate Victorianism, the British Empire. The social structure of the nation was slowly changing, the landed classes were giving ground to the up-and-coming bourgeoisie, and so the hold of Empire weakened too, for it took loyalty, consistency, discipline to rule a quarter of the world—as the young American observer Walter Lippmann perceptively observed, no Empire in the history of the world had long survived without a governing class at the centre.

Most damagingly of all, the imperialness of Great Britain was now more often treated frivolously. It became rather funny. The New Zealander David Low was tilting at the Empire when he created his preposterous cartoon character Colonel Blimp, forever declaiming the glories of the past to his Turkish-bath cronies, and here is the humourist 'Beachcomber', J. B. Morton, in the *Daily Express*:

ADVERTISEMENT CORNER

Will the gentleman who threw an onion at the Union Jack and repeatedly and noisily tore cloth during the singing of 'Land of Hope and Glory' at the Orphans' Outing on Thursday, write to Colonel Sir George Jarvis Dela-maine Spooner, late of Poona, telling him what right he has to the Old Carthusian braces which burst when he was arrested?[1]

[1] Nobody laughed at the Empire more persuasively than Morton, who was born in 1893, who was educated at Harrow and Oxford, who fought in

4

In the field some of the imperialists, too, especially the younger men, now had their niggling doubts. 'It is the virtue of the Englishman', wrote Goldsworthy Lowes Dickinson in 1913, 'that he never doubts. That is what the system does for him.' But it was no longer true. The men who ran the Empire, many of them ex-servicemen themselves, could hardly help sharing the new national attitudes. To some of them it was now apparent that the British Empire was not eternal after all, that parts of it indeed might not out-last themselves. They understandably resented this prospect, not merely because it might cost them their jobs, but because they believed still in the British mission, and thought they knew more about the true state of things in Burma or Somaliland than did President Wilson in Washington or the radical intellectuals of NW3. The villains of the Victorian Empire-builders were the villains of their successors too: meddling MPs, Americans, leftist agitators, lady do-gooders, self-righteous newspapers and their ignorant reporters. Their fathers, though, had argued from positions of power and certainty: by the 1920s the young imperialists were playing, as they might say, upon a losing wicket.

One senses consequently a new modesty in their approach, a new frankness with their literate subjects. Public education had never been a forte of the British Empire, which had frequently left the task to missionaries and private enterprise, but even so an educated subject class had emerged in several parts of the Empire. In India the western-style universities, established in the previous century, had by now produced thousands of men with western manners and ideas, and Indians vastly out-numbered Britons in all but the highest

---

France in the Great War, and whose cast of imperial characters included Big White Carstairs, Mrs Elspeth Nurgett MBE, the Resident of Jaboola, the Gogo Light Infantry, the Harbour-master of Grustiwowo Bay and the M'Babwa of M'Gonkawiwi, M'Gibbonuki and the Wishiwashi hinterland ('Is the M'Hoho mentioned in your report,' asked the Colonial Office once in reference to this particular vassal, 'the M'Hoho near Zumzum?' 'No,' Big White Carstairs replied, 'the M'Hoho near Wodgi.')

ranks of the administration. In Africa there was a sizeable class of educated blacks, mostly mission-trained, and many young men were now going for further education in England. In the new provinces of the Middle East there existed already an urbane and sophisticated intelligentsia, rooted in the Islamic culture, and as the years went by many more young men there, too, absorbed western ways and values, until the Anglicized subject of Empire, with his upper-class English accent, his freedom with English literary quotations, his acquaintance with the Wars of the Roses, became almost a generic figure, whatever his own language, origin or religion, from Sierra Leone to Calcutta.

All this meant a modification of the ritual aloofness of the imperialists, maintained as a matter of policy, as of taste, at least since the Indian Mutiny. It was impossible now for the British to live altogether separately from their subjects, as they had with such success for nearly a century: and though the new proximity often led to irritation, and sometimes to more rather than less misunderstanding, still it meant that the British in the field were less arrogant and disdainful than they used to be. We need not doubt that for most of them, in 1930 as in 1860, the white race was inherently superior to the black, the brown, the yellow or the half-caste. A note of condescension, at best, coloured every instinct of the Briton in his Empire. Now, though, he was more likely to be self-conscious about his attitudes. It no longer came quite so spontaneously, that clap of the hands for the bearer, that stick on the bottom for the Gandhian demonstrator. Even the forms of Empire came to be questioned sometimes. Was it really to impress the Orientals, younger men sometimes wondered, that the Empire maintained its pomp and pageantry, or was it to sustain the self-esteem of imperialists?

Racial bigotry, one of Empire's ugliest aspects, was past its worst, and the imperialists in the field were matured, softened perhaps, weakened almost certainly, by the changing order of things. 'All over India', wrote George Orwell, 'there are Englishmen who secretly loathe the system of which they are part': and he told the story of a journey he made by train with a man of the Indian Educational Service, in whom he gradually discovered a common antipathy to the imperial values. 'We parted as guiltily as any adulterous couple'—and with reason, for these half-hidden self-questionings were to prove as effectively treasonable, in the long run, as any armed rising against the Crown, or conspiracy of Bolsheviks.

5

India was lost, anyway, and the imperialists were more concerned now with the Crown Colonies, once the poor relations of Empire, now its chief hope. None of them enjoyed any real responsibility, and most of them seemed likely to remain within the Empire for ever and ever. The dependent colonies were expected to pay their own way—the total British expenditure on them in 1930 was only £3 million, and the tropical possessions were mostly in an appalling state of dereliction: but they seemed to represent the imperial structure of the future. 'The Empire is Still in Building', said the Empire Marketing Board in one of its neo-Biblical slogans, and the allegorical figures likely to appear now in the imperial propaganda were smiling Negroes of Jamaica or West Africa, garlanded Fijians, resolute Malays or diligent junk-men of Hong Kong.

Since it seemed likely to last longer, the colonial administrative service now offered more coveted careers than India. There was in fact no Colonial Service as such. Some colonies chose their men by competitive examination, but most recruits were selected by patronage. Officially the patron was the Colonial Secretary: unofficially, throughout the 1920s, it was one of his private secretaries, Major Ralph Furse, and it was Furse more than any other man who set the tone of the imperial services in the post-war years.

He was a conservative of a complicated sort. The son of a crippled agnostic—'he taught me to ride a horse, to tell the truth, to love my country and to honour soldiers'—Furse was a member of Pop, the ruling society of Eton, and he remained a very responsible schoolboy all his life. He liked to call his seniors 'Sir', and had a sensible weakness for the great and famous: 'I bowed as we shook hands,' he recorded of his first meeting with Milner, 'then, on an instinctive impulse, I drew myself up to my full height and looked him straight in the eye. He gave a perceptible start. . . .' Though he had an unexpected passion for ballet, he stood for manly values, straight, prefectorial values: during his service on the western front he took a cold bath every morning, often in the open air, and there was a seven-year engagement before he married the daughter of Sir Henry Newbolt. Furse was not a brilliant man, but he had many of the traditional qualities of the Englishman: courage, patience, fitness, sympathy, good humour.[1]

For thirty-eight years this man chose the rulers of the colonial Empire. He liked to call his method 'one of the arcana imperii', for it was altogether unwritten, instinctive and customary. He worked like a mole, he said, burrowing, tunnelling, establishing private contacts with headmasters and university tutors, so that likely men were sometimes unwittingly shunted, by one means or another, along the corridors of the establishment to his office in Westminster. A new genre of imperial service had come into being during the past half-century, since the acquisition of Britain's vast African empire. Those ragbag black territories, it was thought, strewn across a continent without culture, without history—those bold and earthy possessions did not require intellectuals, but all-round men of practical skills. The men they needed, said Frederick Lugard, Governor of Nigeria, were plain English gentlemen, 'with an almost passionate conception of fair play, of protection of the weak, and of playing the game'.

These were Furse's men, not especially clever, not particularly ambitious, but healthy, and brave, and cheerful. In the 1920s and 1930s, as the Indian Empire faltered, they gave to the colonial empire a new cohesion. They were not zealots. They had principles but not beliefs, says a character in one of the novels of Elspeth Huxley, herself an Anglo-Kenyan, and if they were seldom gifted men, and perhaps unlikely to rise to great office at home in England, still they were seldom prigs or bigots either. The African empire did not require ideologies in the field. A recruit for the Nigerian service in 1930 spent a year at Cambridge learning the rudiments of law, tropical medicine and Nigerian languages, but learnt no local

[1] Though not, to judge from his memoirs, *Aucuparius* (London 1962), modesty.

history at all, not even imperial history, and indeed went out to the colony without ever having heard of its founder, the Rhodes of West Africa, George Goldie.

Furse had got a third at Oxford, and it was the game man with the third-class degree that he favoured for the Empire. He recruited thousands, for after the war there was a great expansion in the service. Most of them were ex-servicemen, most of them public school boys—'the public school spirit', it was said, 'is greatly valued in the colonial service, and it is a matter of conscious policy to ensure that the supplies of it shall be constantly replenished.' For the most part the new recruits had no lofty sense of mission. They generally assumed the colonial empire would last indefinitely, and took the job because it offered them honourable responsibilities, excused them the drab British grindstone, sounded fun and promised a pension. They were very decent men. They were Sanders of the River. They were Great White Carstairs. 'Never since the heroic days of Greece has the world had such a sweet, just, boyish master', wrote the American philosopher George Santayana, in one of the most widely quoted of imperial compliments, and he was thinking of Furse's men.

They were often very close to their subjects, closer by far than the administrators of India, for the colonial officials were less hamstrung by tradition or convention, and were also, not infrequently, very fond of their charges. Relations with chiefs and potentates were often easy and friendly, and the concept of Indirect Rule —allowing the native peoples to run their own affairs, by their own cultures—meant that racial prejudice was never extreme. Settlers might talk of damned niggers, or mock the customs of the indigenes: Furse's men would think it, by and large, hardly cricket. Here is part of a minute circulated by a Governor of the Gold Coast among his staff:

'I wish all officers to remember that a very high standard of work and conduct is expected from members of the service. We must always remember that we are Civil Servants—servants of the public. We are in this country to help the African and to serve him. We derive our salaries from the Colony and it is our duty to give full value for what it pays us. I

attach considerable importance to good manners, especially towards the African. Those people who consider themselves so superior to the Africans that they feel justified in despising them and insulting them are quite unfitted for responsible positions in the colony. They are, in my opinion, inferior to those whom they affect to despise, and often betray, by their arrogance and bad manners, the inferiority of which they are secretly ashamed. . . .'[1]

It was a resurgence of the trusteeship ideal, but it was weakness too. Furse's colonial service was perfect for the imperial decline—not too aggressive, not too dogmatic, not even too sure of itself. These post-war imperialists were, without doubt, the *nicest* rulers the Empire ever sent abroad, but they were not the strongest. They saw the other side too generously, and if ever it came to My Empire Right or Wrong, one did not need to be a medicine-man to prophesy their resignation. 'In such dangerous things as war,' Clausewitz had said—and Empire was essentially a risky business—'the worst errors are caused by a spirit of benevolence.'

6

The imperialists were undismayed. 'The British Empire stands firm', announced Stanley Baldwin, Kipling's cousin and Prime Minister for most of the 1920s, 'as a great force for good. It stands in the sweep of every wind, by the wash of every sea. . . .' The Never-Stop Railway trundled round and round; every few years the Dominion Premiers met in London; the Council of the Federation of Chambers of Commerce of the British Empire recommended a uniform monetary system for the whole Empire; George V combined imperial duty with kingly pleasure by going several times to *Rose Marie*, Rudolf Friml's smash musical about the Royal Canadian Mounted Police, and by planting an Empire Plantation of

[1] I was sent this quotation by the late J. A. de C. Hamilton, formerly of the Sudan Political Service, who spent the last years of his life pondering 'the swing of the pendulum between liberty, which seems to lead to chaos and anarchy, and the desire of most people for law and order'. I am grateful to his memory for many insights and anecdotes.

trees in Windsor Great Park, each tree representing one of his colonies. When the Viceroy of India entered his dining-room at New Delhi for an official dinner, preceded by two elegant aides-de-camp, the band played the National Anthem, and all the Indian servants, poised behind their chairs in their gold and scarlet liveries, buried their heads in their hands. The fantasy of Empire survived, and to many of its cultists remained truer than the reality.

Consider Lady B's face, as she drops her curtsey to His Excellency after dinner, and sweeps away into the ballroom beyond. She is only the wife of a provincial Governor, one of eight in India, but how sure she is of her status still, how determined to maintain the imperial proprieties. Her husband has risen laboriously through the ranks of the ICS; her father, I dare say, was an Indian Army general; she herself grew up in the long-established private world of Anglo-India, moving from cantonment to hill station, from Meerut to Bangalore, home for school or holidays perhaps, back again for the rites of courtship and marriage in Lahore Cathedral to poor dear Edward.

Poor Edward? Well, Lady B is no fool. She has watched India changing since the war, she went to the pictures last night and saw Rudolph Valentino in *The Sheikh*, she senses the shifting attitudes at home—even among her own people, my dear, even among people who should know better!—and she doubtless sees in her husband's provincial grandeur some element of pathos or even despair. Poor Edward! He will go no further: how far has he come?

But she will never admit the question, even to herself, and so as she passes through the gilded doors, between the flunkeys, her face is set in its usual expression of resolute hauteur, tinged with the querulous. It is not a happy face. It looks disdainful, but defensive too. It is creased by many emotions, none of them very becoming, and it mirrors the narrow conventions of imperial society. Does Captain C call his napkin a serviette? She always knew his origins were suspect, even if he was a Bengal Lancer. Does Mrs T speak rather too familiarly? She should remember her husband is only in Public Works. Surely that wretched man from Reuters has not been invited here? Oh dear, here comes Dr Chatterjee, one must put on a cheerful face one supposes, but really *what* a bore Indians can be—

'Oh Dr Chatterjee, how nice to see you! Leila is here too? And how are those sweet children of yours?'

Yet she is a nice woman really, kind at heart, not arrogant by nature. The Empire has made her what she is: and so she is loyal to the system, sticks to its guide-lines and conforms to its preferences. If she ever has doubts about it all, she stifles them for Edward's sake. She curtsies to the Viceroy as to her own household god, for her lares and penates are truly the deities of Empire itself: she knows no other, and if there is something unnatural to these attitudes and beliefs, if she is not really being herself at all, as she accepts without a word or a smile the footman's offer of champagne—if Lady B is a totally different person from the Penelope Arabella Honoria hidden away behind that satin corsage and velvet sash, well, that is because by the 1920s the Empire itself is not altogether organic, and it is why those sour little lines of discontent show at the corners of her mouth, even when dear Edward, clearing his throat and bending creakily from the waist, introduces her to his old friend and valued colleague, that little bounder Arwal Mukkerjee of Justice.

7

They never gave up! Milner died in 1925, but his disciples lived on, and *Round Table*, the magazine of his cult, still diligently pursued his several chimerae. A steady stream of imperial propaganda still emerged from the London publishers: books not so flamboyant as their prototypes of the New Imperialism, but still decorated with the crests of Dominions and Colonies, with chapter headings like 'The Thread That Binds Our Race', with maps to illustrate the variety of minerals within the British Commonwealth, with fanciful colour plates of Rhodes choosing his burial place in the Matopos, or Gordon serene above the black spearmen on his palace steps. Schoolchildren from Eton to Hackney Primary got a half-holiday on Empire Day. Boy Scouts still wore wide-brimmed hats derived from the Boer War, crouched around camp-fires murmuring incantations from Kipling, and shared a motto, Be Prepared, with the South Africa Police.[1] At Oxford the Rhodes Scholarships, financed by

[1] Itself supposed to have been evolved from Baden-Powell's initials.

Cecil Rhodes' will, supposedly indoctrinated a constant stream of hefty young colonials in proper modes and values.[1]

There survived, too, many of the institutions around which the New Imperialism had assembled before the turn of the century. The Royal Colonial Society had become the Royal Empire Society, but still had its great library of imperial books, its club rooms of Uganda *mvuli* or Australian jarrah, its enormous paintings of colonial occasions, its bedrooms decorated by the generosity of Colonel Wakefield of Montreal, or the Honourable T. Biggs of Adelaide. The British Empire League was active still, and the Victoria League, and the British Empire Union, and the Patriotic League of Britons Overseas, and the Empire Day Movement. There were active federationalists still about, like Lord Willingdon, Irwin's successor as Viceroy of India, who hoped for 'a great Imperial Federation, when we can snap our fingers at the rest of the world'. When, in 1933, the devastating 'body-line' bowling of Harold Larwood threatened to sour relations with Australia, the Dominions Secretary himself intervened to get him dropped from the England cricket team.[2]

Most insistently of all, there was the Conservative popular Press, which had built its original fortunes upon jingoism, and remained shrilly faithful to the theme. The *Daily Express* in particular, owned by the Canadian Presbyterian Lord Beaverbrook, *né* Max Aitken of New Brunswick, made an Empire Crusade the basis of its editorial policies, and very nearly split the Conservative Party itself. Beaverbrook's panacea was fiscal. He wanted to turn the British Empire into a self-contained Free Trade area, surrounding it with tariff walls against the world outside, but allowing absolute freedom of commerce among all its constituent parts. The British were, with progressively less conviction, still wedded to the idea of universal free trade which had made them rich a century before: even in 1930,

[1] Germans and Americans were eligible too, as suitable partners in the governance of the world. President Kennedy's administration of the 1960s included sixteen Rhodes Scholars, headed by the Secretary of State and the Secretary of the Army, and among the German scholars was Adam von Trott zu Solz, who was executed for his part in the bomb plot against Hitler in 1944.

[2] He emigrated to Australia.

83 per cent of imports paid no duty at all. Beaverbrook's plan, which after all honoured *half* the old dogma, was accordingly very persuasive. It would not only restore the prosperity of the British, it was argued, it would give the Empire new meaning, and perhaps make of it at last the economic super-Power Joe Chamberlain had imagined. Immensely vigorous, infectious, with powerful friends and heaps of money, Beaverbrook founded his own splinter-party, the United Empire Party, and plugged his theme incessantly, until the phrase Empire Free Trade, if not its meaning, was familiar in every British household, and the impish face of the millionaire, remarkably like some whiskered rodent of the lumber camps, became the face of contemporary imperialism.

It never happened. Economically the Empire did help to cushion Britain's decline, and trade with the Dominions and colonies increased in the post-war decades. Many of the great British corporations were now investing more money than ever in the tropical colonies: 70 per cent of New Zealand's exports, more than half Australia's and South Africa's, came to Britain. The Dominions welcomed the idea of limited trade preferences—in particular they wanted Britain to put higher duties on foreign food and raw materials, lower ones on Empire produce, and in 1932 this came about. Anything so grandiose as the Empire Crusade, though, which really went far beyond economics, smacked to them of imperial centralism, relegating them to perpetually second-class status—how could their own infant industries, they reasoned, compete with Britain's long-established factories?

Only the British themselves stood to gain by such imperial devices; and they were certainly no Crusaders, in the years of the Great Depression.[1]

8

Still the Empire proceeded, by the force of old momentum. Most Britons still considered it, all in all, a force for good in the world,

---

[1] Though Beaverbrook's own imperial knight-at-arms survived his creator, who died in 1964, to appear to this day cap-à-pie upon the masthead of the *Daily Express*.

and only a minority could conceive of its actually coming to an end. The diligent District Commissioners still went their dusty rounds, presiding earnestly as ever over their courts, accepting family appeals and settling tribal quarrels, attending to irrigation rights, school-book requirements, the state of the Kasungu road or the latest damned questionnaire from the Colonial Office. The fleets and armies were still disposed around the world: in 1934, of Britain's 1,008 military aircraft, 580 were at home, 175 at sea with the Navy, 96 in India, 60 in Egypt, 51 in Iraq, 28 at Singapore, 12 at Aden and 6 at Malta.[1] In Singapore they had started work on a new base for the Royal Navy, to be the main centre of defence for the eastern and southern Empire, with dry docks for the greatest battleships, protecting batteries of 15-inch guns and an ancillary air base. The colonial careerists still looked forward, eagerly as ever, to their just rewards of CMG, KCMG or GCMG.[2] There were still colonial wars—against Arabs and Jews in Palestine, against Yemenis in Arabia, against Afghans, against Iraqis, against the Mad Mullah of Somaliland and a Burmese monk who claimed to be able to fly.[3] 'I am so glad that your son Kenneth is happy as Tetrarch of Galilee,' wrote the King's Private Secretary to Mrs Harry Blackburne in 1936, 'and I only hope that he will not have such a difficult time as the only one of his predecessors in office with whose name I am familiar.'[4]

The Empire even expanded still, and it was not until 1933 that it

---

[1] Oh, and one Hawker Audax biplane was on loan to the Royal Canadian Air Force, whose entire flying equipment it constituted.

[2] 'Call Me God', the irreverent said, 'Kindly Call Me God', or '*God* Calls Me God'. The Order of the British Empire, which was instituted in 1917 to meet a wartime demand for honours, was even more ribaldly regarded, being soon nicknamed the Ought to Be Ended, but its title has survived as the last official usage of the word 'Empire'.

[3] The monk, Saya San, could *not* fly, and his rebellion was suppressed in 1932: the Mullah, Mohammed bin Abdulla Hassan, was far from mad, and skilfully defied the British for nearly twenty years.

[4] Not quite so demanding perhaps, but still Sir Kenneth Blackburne, who prints this letter in his *Lasting Legacy* (London 1976), spent three dangerous years in his house above Nazareth during one of the worst periods of Palestinian violence. He went on to be Governor of Jamaica.

reached its ultimate limit, the greatest expanse of territory ever presided over by one ruler in the history of mankind. The first acquisition of Victoria's Empire had been the port of Aden, at the south-western corner of the Arabian peninsula: now, almost a century later, it was from Aden that the last imperial frontier was reached. Even in the 1930s southern Arabia was almost insulated from the rest of mankind—the world, suggested a local schoolboy in an examination, 'consists of four parts, Aden, Ma'ala, Hedjuff and Tawahi'. Among foreigners, only the British knew much about it, for in the course of the past century they had steadily extended their influence outwards from Aden. In Spheres of Influence, in Treaty Relationships, in Protectorates, they had gradually become masters of the south Arabian littoral, and established their suzerainty over its astonishing gallimaufry of petty sheikhdoms and sultanates— who until then, obsessed as they habitually were with feud and pedigree, had been generally convinced, like the schoolboy, that to all intents and purposes the world was them.

Perhaps the last true expression of High Empire was the system of treaties by which the British, in the 1930s, pacified the last refractory chieftains of the Hadramaut. It imposed at last a universal peace upon a part of the world where incessant warfare had been for centuries part of everyday life, where tribes fought tribe, village fought village, and even neighbours, animated by obscure hereditary vendettas, battened their houses against each other. The truce was conceived and concluded by a single British political officer and his wife. The British imperial manner was not greatly admired among the proud and wilful chieftains of South Arabia: they called it *kibr Ingles*, English condescension, and they strongly resented its assumptions of superiority. Harold Ingrams and his wife Doreen, though, were of the new imperialist *genre*. He had fought in the war, she was an adventurous woman with a genuine empathy for other peoples and cultures. Posted by the Colonial Office to Arabia in 1934, when they were in the field the Ingrams lived as the Arabs lived. They wore Arab clothes, they ate Arab food, they laughed at Arab jokes, they responded to Arab poetry. Nothing could be more excruciating to your conventional Empire-builder, to whom all such conduct smacked of Soapy Sam or Going Native, especially when a memsahib was involved: but the Ingrams really meant it, and by these means, in 1937, they brought peace to the Hadramaut.

Some 1,300 leaders signed the truce, from the rulers of powerful tribes to the brawl-leaders of village factions, and life in the Hadramaut was transformed by it. In one particularly argumentative hamlet a man emerged from his own house for the first time in eighteen years, and another, walking a few yards down the street, met his own sister for the first time since 1916. 'Shut up,' quarrelsome children were now alleged to tell each other in the street, 'or I'll tell Ingrams', and so impressed were the chiefs by the potency of the peace-maker that several of them pressed their daughters upon him in marriage. ('He doesn't want your dirty daughters', was the official interpreter's habitual frank response.)

The origins of Ingrams' Peace were soon to be forgotten, but still this was, in the afternoon of Empire, an achievement worthy of its noonday—worthier, perhaps, for it was done without bluster, with only a minimum of punitive bombardment, by a single servant of Empire and his wife. This is how the chiefs of the Al 'Alawi greeted this late demonstration of the Pax Britannica:

In the Name of God the Supporter. . . The interest taken and the attention directed by His Majesty's Government towards our sacred home and the present assistance for the establishment of peace and security within our province, and the safeguarding of the nation from disturbances and troubles which destroy the country and subject it to despair and worse, that such interest taken by His Majesty's Government for the removal of all these things will make the future of the Hadramaut bright and prosperous, and it is expedient on us to express our great appreciation and thankfulness in our hearts which direct us to submit our hearty thanks to His Majesty's Government. . . . We entertain the hope that our best compliments, thanks and gratitude will be conveyed, on our behalf, to the Great Government of London which we hope may continue to be the source of peaceful arrangements and good actions. Please accept our high regards.

Signed by all the Seiyids of Al 'Alawi

CHAPTER FIFTEEN

Britishness

THE Great Government might still be great in the fastness of south Arabia: elsewhere its command of events was less than absolute, and within the white Empire its position was now distinctly equivocal, nobody quite understanding how it worked. Britishness itself had become a debatable condition. In Victoria's day it had been embodied above all in the Monarchy, the distant, unfailing source of power and justice. The Crown was the gauge by which a man could claim himself to be British. It was the one abstraction that could unite the loyalties of disrespectful Australians, half-American Canadians and distinctly un-English South Africans. It was very, very grand, surrounded by a mystic sheen of tradition: even the Viceroys, Governors, Captains-General and Commanders-in-Chief who represented it in the field were but suggestive reflections of its splendour. When Queen Victoria went to Ireland in 1900, a photographer took a picture of Her Majesty boarding the royal yacht at Holyhead in Wales. It is as though some sacred relic is being conveyed to the other island, so hushed is the scene, so gravely respectful are the admirals, noblemen and ladies-in-waiting, so ethereal does the old lady appear to be, all in black, attended by her turbaned Hindu, as she is conveyed in her chair across the gangplank to the ship.

Now all had changed. The Monarchy was still immensely popular in most parts of the Empire, even in India, even in Ireland, but its mystique had faded. Britishness needed more than a George V to keep it whole, and by now even the most British of the overseas Britons were acquiring identities of their own.

2

To most Britons at home the colonials were still hardly more than transplanted fellow-countrymen, as English really as the Welsh, say, and to be treated with the same avuncular or schoolmasterly familiarity.[1] To most foreigners, too, colonials still seemed to be Englishmen of a sort, linked by obscure constitutional duties to London, and subject ultimately, it was generally assumed, to the authority of Whitehall. The colonials themselves were confused in their sensations: some felt themselves truly British still, some considered themselves citizens of altogether new countries, and most, perhaps, felt an awkward ambiguity of loyalties, like the Welsh.

In hindsight one can see that they were, in fact, becoming more and more un-British. In physique, in character, in perceptions, they were becoming something else. The Britishness was wearing off. Even their languages were diverging fast. The Canadians now sounded much more American than English, betraying their imperial origins only by a peculiarly rounded way of pronouncing the dipthong 'ou', and adopting all the latest New York slang only a year or two after it had hit Manhattan. South African English was only just English at all, having been heavily influenced by Afrikaans: its singsong delivery, its distorted vowels, its squashed up syllables, made it peculiarly repellent to English purists, and South Africans who came to London very soon learnt to disguise it, until the more practised of them could easily be taken for Liverpudlians. The Australian language, on the other hand, had acquired a truly noble robustness, its original Cockney long since matured into something altogether antipodean. If it sounded harsh in women's voices, it could sound glorious in men's, and it had a historical meaning: for its resolutely working-class tang, like its mordant syntax, were deliberate expressions of the Australian fresh start, with nothing posh or namby-pamby to it.[2]

[1] *Exasperated Englishman*: 'Tell me, is there a word in your language, equivalent to the Spanish *"mañana"*?'
Welshman: 'There is, but not with the same degree of urgency.'

[2] It has often been phoneticized, from Marcus Clarke's convict conversa-

Smoke of defeat: the day of surrender, Singapore

Their national characters, too, had developed in different ways, pulled by different magnetisms. The British Canadians, though enterprising, hard-headed and successful, were generally agreed to be a less than stimulating people: living as they did in the shadow of the United States, they were torn between reflex and emulation, and emerged oddly pallid, their history generally anti-climactic, their very achievements, like their interminable landscape, tedious to experience.[1] The British South Africans also suffered from the proximity of a more vital people: though they occupied the most beautiful parts of South Africa, they never did achieve the intense comradeship of the Boers, found their own convictions flaccid before the zeal of the Volk, and lived through the years in a state of ingratiating compromise, seeming always like settlers in the land, never indigenes. The New Zealanders, by contrast, seemed absolutely of their soil: in some ways the most British of all the colonials, and certainly the most loyal to Crown and Flag, in other ways they were the most truly exotic, living as they did socialistically in their remote and beautiful islands, with high Alps to climb like Swiss, and geysers to bathe in like Icelanders, and the Antarctic itself hardly out of sight beyond the southern tip of their territories.

And finally there were the Australians again. They were always last, and generally loudest. Constitutionally the Canadians, since Chanak, had been more than ever the pace-setters of the Commonwealth, but in a subtler way it was the Australians who convinced the world that colonials were no longer so many English alter egos.

---

tion, 1874—'*Stow yer gaff, and let's have no more chaff. If we're for bizness, let's come to bizness*'—to C. J. Dennis's larrikin poet, 1915—*The world 'as got me snouted jist a treat;/Crool Forchin's dirty left 'as smote me soul*—or Professor Afferbeck Lauder's Strine nursery rhymes, 1965—Lilma Smarfit, George E. Porchy, Girldie Larks, *Mary Header little lamb* or *Harsh, harsh, Wisperoo Des*.

[1] 'If you don't know yourself in this country', a Turkish immigrant to Canada told me in 1974, 'you die of boredom. Mind you, if you *do* know yourself you die of boredom anyway.' When they had a competition to name Canada's first space satellite, the poet Leonard Cohen suggested 'Ralph'. If you introduced yourself to a Canadian as Alice in Wonderland, I was told once, he would say either 'Oh, I thought you'd passed on', or 'Are you published in Canada?'

Fires of victory: London in the blitz

The Australian relationship with England had been sweet-sour from the start. Born out of convict transportation, at once envious and resentful of England, sentimental about blood-ties, self-consciously egalitarian, Australia by the 1920s had evolved a national character more pungent and aggressive than that of many much older States. The Australians were a belligerent people, intensely proud of their record in the Great War, and perhaps expansionist too—in 1933 they claimed a third of Antarctica as Australian territory. At the same time they were profoundly anti-heroic and iconoclastic. Suffering as they did from an ancient sense of inferiority, they coped with their own self-doubts partly by virile postures, but partly by a wonderfully dry and self-deprecatory philosophy of life.

Here is an example of it, expressed in a favourite Australian anecdote. There's this bloke, see, sitting on the cliff up South Head ready to jump. 'What's up, digger?' says the policeman. 'Everything', says the man. 'The wife's run off with a cobber of mine, the shop's gone to the pack, the kid's crook and I've lost my false teeth.' 'Gow-orn', says the cop. 'So I'm going over', says the man, pulling his coat off. 'Maybe you are,' says the cop soothingly, 'but let's have a little talk about it first, shall we?' So they have a little talk about it, and then they both go over.[1]

3

By now the greatest cities of the white Dominions were of international stature, richer and more powerful than any in Britain except London. Look for example at Toronto, in 1926 a city of 500,000, growing fast and very pleased with itself on the shore of Lake Ontario. One could not call it a handsome city, but it had a bristly character of its own, partly in assertion against the French-Canadian metropolis of Montreal to the north, partly in self-defence against the dynamic American cities to the south, where the *Chicago Tribune* regarded the British Empire as Anti-Christ, and the Mayor of Chicago promised that if ever King George V came to his city, he would personally punch him on the nose.

[1] 'This world, the next world, or Australia': Oscar Wilde, *The Importance of Being Earnest.*

The citizens of Toronto were mostly of Scottish extraction, and on the face of it the city was still very British—Union Jacks all over the place, the *Toronto Globe* reverberatingly imperialist, the Lieutenant-Governor's mansion rigid with formality. The cathedral was intensely Gothic. The University was ineffably Oxbridge. The policemen wore bobbies' helmets. The Flag flew bravely from the Armoury, the headquarters of the Ontario Regiment. The presidents of the banks all seemed to be knights, and the shops were full of imperial familiars, Mazawattee Tea, Hovis, HP Sauce, Cash's Name Tapes for going back to school, Andrews' Liver Salts for the morning after.

It was 'a nest', the local historian Jesse Edgar Middleton suggested, 'of British-thinking, British-acting people', and Lord Bessborough, later to be Governor-General of Canada, once said that if there were two things Toronto understood perfectly, they were the British Empire and a good horse. When Royalty came to Toronto, as it not infrequently did, the city burst with loyal excitement: Canadian Scottish soldiers lined the streets in kilts, and the grand ladies of the town competed in the tastefulness of their entertainments. The most loyal citizen of all was one of the richest, Sir Henry Mill Pellatt, as fervently imperialist a financier as Rhodes himself. Casa Loma, Sir Henry's vast Balmoralesque mansion on the outskirts of the city, was big enough for his entire militia regiment to parade in its cellars, and was intended specifically for the entertainment of visiting kings, queens and princes of the blood.

But there was a falseness to it all. Toronto was not, as visiting sophisticates might suppose, simply a transplanted industrial city of the British provinces, animated by the same parochial rivalries and conceits. It was a city of tremendous drive, made to rival those pulsating giants south of the border. It was much richer than any comparable English city. It had far more cars, it had many more telephones, and its commerce was run with a gusto that was authentically of the New World. The Historical Pageant at the Massey Hall that year was billed as 'The Most Pretentious Pageant Ever Presented Under Canvas', while the Royal York Hotel, about to be built on the Front Street, would be the Largest Hotel in the British Empire, with 3,000 bedrooms. The daily wireless programme of the *Globe* listed sixty-four available radio stations, mostly in the United States: on an average day the paper contained two and a half columns of used-car advertisements, and four of Teachers Wanted.

Innovation, not tradition, was the true meaning of Toronto. It was here that Frederick Banting, working in the meticulously ivied halls of the University, discovered insulin. It was here that five-pin bowling was invented. It was here that Sammy Taft, the haberdasher of Spadina Avenue, designed the wide-brimmed fedora worn by all the best Detroit gangsters.[1] The ferries that chugged to and fro across the lake, run by Mr Lol Solomon, were like pioneer steamboats of the West, with their homely names—*Bluebell, Primrose, Mayflower*—and their nonchalant skippers chewing at the wheel. The vast new station going up by the water-front was resplendent with the names of the western wilderness, Saskatoon, Thunder Bay, Moose Jaw, while Yonge Street claimed to be the longest street in the world, extending as it did from the centre of the city far out into the farmlands.

In short, Toronto was a city of the Americas. It was a self-made city, built on grab and enterprise. Its stock market dealt in mining shares, the pulp industry, wheat and other investments of the forest and prairie. Its first families were mostly local tradespeople made good. When Sir Henry Pellatt, who made his money out of electric light, was building his great castle on the hill, its progress was watched with a not altogether disinterested concern by Lady Eaton, wife of the dry-goods king of Toronto, from her own steepled mansion down the road—and stifled though her chagrin undoubtedly was, for she was thoroughly well-bred, still it must have been galling to have seen those baronial battlements rising ever more largely above the ridge.

Behind the façade of imperial loyalty there was a strongly nationalist impulse. Toronto people liked to boast of their Canadianness—Managed and Operated Exclusively by Canadians, a Canadian Concern, Canadian Conceived—and the more adventurous of the newspaper editorialists looked ahead to a Canada that was a Power in its own right, neither British nor American. To the truest Toronto

[1] He was still in business when I walked down the avenue in 1975, and still had one in his window.

Britons this smacked of disloyalty. It was the thin edge of the wedge, and sometimes the traditionalists watched the progress of the city with dismay. Tradition was all-important, Jesse Edgar Middleton thought. 'When a young man's great-grandmother helped to embroider the colours of the Third York Militia in 1812, when his grandmother danced with the Prince of Wales in 1860, or when his grandfather was huddled in the square at Ridgeway or stormed the trenches at Batoche, his inner feelings are likely to be governed by his spiritual inheritance. . . .'

The trenches *where*?[1] The constant looking back to older forms was in fact the inner weakness of Toronto, as of Canada itself—its half-wayness, its hybrid kind, which flattened the impact of its energies, and blunted its self-confidence. By the 1920s Canada was respected everywhere, but fired no ecstasy. The price of goodness was ennui. It was a country without glamour, wrote John Buchan, presently to become its Governor-General: alive but not kicking, Rupert Brooke once said. 'A community of moderationists', is how one Toronto publicist described his city, and that sad to say was half the trouble, for the determined moderateness of the Canadians took the edge off their adventure, and made Toronto seem a more ordinary city than it was.

More insidious still was the feeling that Toronto's Britishness possessed an air of parody, a Beachcomber touch. There was something comic to a civic aristocracy, rich, titled and consequential, so inescapably bourgeois as Toronto's. There was something forlorn to the pageantry of the Toronto Scottish, panoplied in all the paraphernalia of their ancestry, but bereft of the true Highland cragginess, too pale, too pudgy. Sir Henry and Lady Pellatt were photographed at a Toronto garden party in the 1920s, and they really do look rather absurd. Behind them is a background of period elitism, a top hat here, a flowered cloche there, the glimpse of a satin shoe, somebody leaning lightly on his cane beside the lawn. Largely in

[1] The trenches at Batoche, Saskatchewan, where during a three-day battle in 1885 the Canadians suppressed the last rebellion of the Metis half-castes under their tragic leader Louis Riel. For an account of the Metis risings which Jesse Edgar Middleton might not altogether endorse, may I suggest the first volume of this trilogy, *Heaven's Command*?

front loom the First Citizens of Toronto. They are dressed very imperially, she in silk, feathers, furs and twisted pearls, he in the truest St James's uniform, from the bowler hat and the wing collar to the silver-bound walking stick.

Alas, despite it all they look like cruel caricatures of all they wish to be, so heavy-jowled, so clumsy-looking, so defensive of expression, that they might be figures in some Wodehousian farce of English manners, or cartoon images of Imperialism.[1]

4

Britishness had mutated very differently in Johannesburg, the *casus belli* of the Boer War and still the fulcrum of South Africa. In 1928 Jo'burg was only forty-two years old, just elevated to the technical status of a city, but it was already a great metropolis. Financially it was still tied closely to the City of London, and the companies that mined and marketed its gold were registered in Britain: but the style of the place was now altogether *sui generis*, and in its brief life it had developed from a seamy mining-camp in the veld into one of the richest, boldest, most striking and some would say nastiest cities in the world.

Everybody called it Jo'burg, and this somehow raffish nickname, with its suggestion of smoking-room camaraderie, or even confidence trickster, was just right for it. It was an unlovely town, but fascinating, being physically dominated by the yellowish tailings of the gold mines, visible like billious hillocks at the end of almost every street. Jo'burg was literally paved with gold, for its paving stones really were impregnated with gold dust, and since the days of the great Rand rush the mines had grown so large, so overwhelmingly the *raison d'être* of Johannesburg, that it was impossible

---

[1] Toronto is almost unrecognizable now, having changed during the twenty-five years I have known it from a recognizably imperial city into the most intense of all cosmopolitan melting-pots (page 756 of the 1973 telephone directory began with Mr A. Jentile and ended with Mr Yim Jew). Much of its money is American now, and Casa Loma is a tourist spectacle run by Kiwanis, but the Eaton family is still influential, and the Timothy Eaton Memorial Church is familiarly known as Timothy Eaton and All Saints.

to escape their presence. This was the greatest of all company towns. Monolithically in the centre of it stood the offices of the Anglo-American Corporation, the true power-house of South Africa and the greatest mining corporation on earth: no need to put a sign on those portentous offices—40 Main Street was quite enough, and money-men everywhere called it Corner House.

Around the gold, then, almost as important to the Empire as it was to South Africa, revolved the life of Jo'burg: its shrewd stock exchange, its multitudinous banks, all the offices of its brokers, speculators, usurers, investment advisers, its host of white clerks and overseers, its anonymous army of black serfs, troughed like cattle in the barrack-sheds of the mining companies, or forlornly queuing in the evening light for the trains back to their distant townships. Johannesburg stood nearly 6,000 feet up on the veld, and its climate was brilliant in summer, raw in winter. It was a merciless town, and if President Kruger could have seen it in the 1920s, he would have grunted 'Man! I told you so!'

For on the surface not much remained of the old Boer tradition. Few Afrikaner names appeared on the roster of the Stock Exchange, or in the membership lists of the Rand Club, brooding behind its brass-ringed pillars in Loveday Street. Still fewer hunted with the Rand Hunt, or watched the Test matches at the Wanderers' Club. Uitlanders of many kinds, well outside the pale of the Volk, predominated in the pages of the *Rand Daily Mail*—Mr Justice Stratford at the High Court, Miss Daphne Kincaid-Smith in beige georgette at the Chambers of Mine Staff Dance, or Mr Sydney Rosenbloom, composer of the foxtrot *Shanghai Butterfly*, 'the first successful effort of the kind harmonized by a South African composer'. Edwin Lutyens had come from England to design the new Art Gallery, the Lancs and Yorks Society met on Wednesdays, Owen Nares was playing in *The Last of Mrs Cheyney* at the Empire.

Physically the city was unlike any other in Africa, and remote indeed from the peaceful market dorp of the Afrikaner preference. It was the second largest on the continent, smaller only than Cairo, and it was built not in the Afrikaner, nor even in any imperial, but in a brash neo-American style. Size and sizzle were its characteristics—big rich buildings, broad streets in an inflexible grid, a railway passing through the heart of it, crossed by many bridges. Jo'burg was spared the heavy orthodoxies of a capital city—the South African executive buildings were at Pretoria, the High Court was at Bloemfontein, Parliament at Cape Town. This was simply a city of money, and looked like one: already, such was the momentum of its profits, three or four buildings had succeeded each other on some downtown sites.

Like most gold-rush towns, it abounded in anecdote and eccentricity. The sudden bend in Bree Street, so unusual in a city of rectangles, was said to have been caused by an assegai hitting the surveyor's theodolite: alternatively it was suggested that two men had first surveyed it, a German working in metres, an Englishman in feet. Many of the original pioneers were still about. Mr Pritchard still lived on Pritchard Street. Mr F. P. T. Strube, one of the two brothers who had first discovered gold on the Rand, was trundled out at municipal functions. Sir Abe Bailey, one of Rhodes' brotherhood of tycoons, still boasted the telegraphic address CORINTHIAN, JO'BURG. Mrs Marie Decker, whose husband had founded the *Transvaal Mining Argus* in 1887, liked to remember being paid for birth and death announcements in turkeys and potatoes.

Strains of boom and shanty-town, too, ran through Jo'burg, and made the old-school Afrikaner, in for the day from the countryside, hasten home aghast to his farm in the evening. Crime was common, and increasing. Many a white man kept his black mistress. The South African Labour Party had its headquarters in the city, and there were frequent strikes. In 1922 one of them had assumed such proportions that the army was used to put it down and 153 people had been killed: it had been organized by a Marxist Council of Action, and even now, on Labour Day, when the Jo'burg workers marched in procession to the Town Hall, predicants and politicians of the Volk were horrified to observe that Communist banners proliferated.

For the Boers had not abandoned Johannesburg. They still thought in terms of Redneck and commando, they knew that Jo'burg was theirs by inheritance, and they planned for the day when Afrikanerdom would take over the city. The Broederbond, the semi-secret society of Afrikaner activists, was busy there already:

and though in this, the biggest city of South Africa, there was no daily paper in Afrikaans, still the Afrikaner population was increasing steadily, year by year—increasing in influence, too, as clerks rose to managerships, as yokels became technicians, and the first Afrikaner investors put their boots tentatively in the door of the Stock Exchange.

Besides, when they surveyed the humming, grasping city, they could reassure themselves with the truth that if its society was alien, many of its philosophies were home-grown. Jo'burg hardly subscribed to the trusteeship ideal of imperialism. Liberal progress was scarcely a preoccupation at the Chamber of Mines, or in the drawing-rooms of the Rand Club. In Johannesburg as in Pretoria or Bloemfontein, the Boer view of race was paramount. By the 1920s South Africa had virtually abandoned any pretence to racial synthesis, and the country was fast moving towards the ultimate negation of the imperial trust, apartheid—the absolute separation of the races, which was seen by its academic progenitors as a philosophical solution to a human predicament, but was interpreted by the populace as a licence for *baaskap*, boss-ship. Nowhere was the idea more welcome than in Johannesburg, for the whole economic and social structure of the city depended upon that vast helot community of blacks: blacks mined the gold, blacks cleaned the houses and watered the gardens, blacks without franchise, without unions, without even the right to move freely about the country, enabled the citizens of Johannesburg to live in the manner to which easy money had long accustomed them.

The Boers had founded Jo'burg, and the Afrikaners saw it still as theirs. The shape they foresaw for South Africa as a whole was already visible in this city of the uitlanders. The Colour Bar Bill, then going through Parliament, would finally separate the races in Johannesburg's public places; the new Immorality Bill would deprive its citizens of an old imperial privilege, the right to sleep with partners of any colour. Mr Justice Stratford passed four death sentences in one week of 1928, one on 'a native called Jim', two on women. The civic authorities of Jo'burg were preparing to celebrate, for the first time, Dingaan's Day, one of the great festivals of the old Afrikaner republics, which commemorated the Boer victory over the Zulus at Blood River in 1838, but had nothing whatever to do with the British Empire.

So we see a city not, like Toronto, somewhat neurotically displaying its Britishness, but steadily deserting it. Once more the concentrated folk-will of the Afrikaners was proving too formidable for the flabbier convictions of the British. The Boers were on their way to winning the third and last of the Boer Wars, and Jo'burg had come a long way indeed since Lionel Curtis took office as its first Town Clerk in the brave days of the Kindergarten.[1]

5

So the Dominions diverged, and there was no pretending either that the scattered peoples of the British stock were always in harmony. British Governments at home were at best bored, at worst infuriated by the average Dominion leaders, Haig's 'second-rate sort of people', and often treated them disgracefully, ignoring them in their economic and strategic calculations, imperiously demanding their compliance when needed, and sometimes being downright rude. The only colonial leaders they really welcomed to their homes and councils were the cultivated statesmen of Afrikanerdom, and as the legends of Anzac and Vimy faded the Dominions became ever mistier, ever less interesting in the public mind.

> *Kangaroo! Kangaroo!*
> *Thou spirit of Australia,*
> *That redeems from utter failure,*
> *From perfect desolation,*
> *And warrants the creation*
> *Of this fifth part of the earth!*[2]

[1] But not much further, it seems to me, since the 1930s, except that its whites are richer, its blacks are angrier and its Afrikaners are now much more powerful. Johannesburg had no television until 1975, and its older citizens still call a cinema a 'bio'—short for bioscope.

[2] By Barron Field (1786–1846), a quarrelsome London litterateur who briefly practised law in New South Wales: Disraeli called him 'a noisy, obtrusive, jargonic judge', and when he wanted to write a biography of his

As for the colonials themselves, they viewed the British still with an astonishing ambivalence. On the one hand there remained a profound sentimental loyalty to the *idea* of England, running much deeper than mere snobbery, and constituting even in the 1920s a powerful political energy. The colonials were stirred by England, responding almost despite themselves to its age, its grandeur, its continuity, even its damp and misty climate. Susceptible Newfoundlanders felt a sense of pride even in speaking its name. New Zealanders never out of Ruatoria thought of it still as 'Home'. 'If you were to ask any Canadian', wrote Stephen Leacock, ' "do you have to go to war if England does?" he'd answer at once, "Oh no." If you then said, "Would you go to war if England did?", he'd answer "Oh yes." And if you asked "Why?" he would say, reflectively, "Well, you see, we'd *have* to." ' Most of the Governor-Generals were still transplanted English grandees, and few of the colonials objected: they preferred it, feeling that Lord Y, the Duke of C or General Z elevated the tone of the place by his presence and impartiality, even if he *was* a bloody pom or limey.[1]

At the same time they were, by and large, more realistic about the state of the Empire than were the policy-makers of the Mother Country. Even the New Zealanders, the most conformist of the colonials, complied with British wishes so obligingly chiefly because they were, economically, hardly more than an agricultural annexe of the United Kingdom. The Australians were often at loggerheads with British Governments over one issue or another. In 1930 their Labour Government even clashed with George V himself, when they insisted upon their own Australian nominee as Governor-General: the Irish had done the same thing, they said, but the King was shocked by the analogy—'Does Australia, with her traditional loyalty to the Throne, wish to be compared with Ireland, where, alas! a considerable element of disloyalty exists?'

---

idol Wordsworth, 'the poet', says the *Dictionary of National Biography*, 'begged him to refrain'.

[1] 'Pom', Australian, allegedly but to my mind unconvincingly derived from the letters POHM, 'Prisoner of Her Majesty', said to have been stamped on the clothing of transported convicts: 'Limey', originally American, from the lime-juice drunk on British ships to prevent scurvy.

Certainly the politicians of the Irish Free State, who now entered the company of the Dominions premiers, had not abandoned their republican aspirations, while even the most Empire-minded of the South African leaders, even Smuts himself, never forgot their republican origins. And through it all the Canadians doggedly pursued, year after year, their object of absolute independence within the Empire, formal and actual: by 1923 they had signed their first independent treaty with the United States (though the Americans checked with London first, just in case) and by 1927 there was a Canadian Ambassador in Washington.

Their interests differed widely. On race, for instance, while the British fitfully honoured the criterion of equal rights for all civilized men, the Australians and New Zealanders were concerned to keep all Asiatics out of their territories, the Canadians had allowed no Asian immigrants since the turn of the century, and the South Africans denied their vast black majority any rights of citizenship whatever. On defence, while the British were now concentrating their strength in the Middle East, the Australians and New Zealanders saw Singapore as the most important imperial base, while the Canadians looked most anxiously to the Pacific. On economics the British were anxious to balance their own industries with the raw materials of the Empire, while the Dominions were only anxious to industrialize themselves.

The Imperial Conferences which met in London, and once in Ottawa, regularly during the post-war years, though they were commemorated always in dutiful group portraits, wing-collared and pin-striped in the garden of No 10, were in fact full of acrimony and exasperation, some of their participants very much disliking one another. Tact was seldom the strong point of colonial politicians. In 1926 the Prime Minister of Australia, Stanley Bruce, went to Canada: but though he was welcomed with the dignity peregrinating Prime Ministers expect, he did not hesitate to say just what he thought of that Dominion. No proper British subject, he said, could have much patience with the place! Canada called herself independent and self-governing, but when it came to trouble she relied on others to get her out of it! Why, Australia was spending 17/2d per head on naval defence—Canada was spending *8d!* 'How can Canada today', asked this pugnacious guest, bidding farewell to his hosts before boarding his ship at Vancouver, 'possibly maintain that she is the equal of Australia, *my* great country?'

6

No wonder King George V, still the Emperor of India and of the British Dominions beyond the Seas, was often perturbed about his Empire. He had visited all its major territories, the first British monarch to travel widely in his Dominions, and he doubtless looked back nostalgically to the heyday of the imperial meaning— to his Indian visit of 1906, perhaps, when 14,000 people with 600 elephants escorted him on shikari, when a road fifty miles long was built to connect his two hunting camps, and when in a single day he and his party shot 39 tigers, 18 lions and 4 Himalayan bears.

He was concerned to restore some of that old splendour, and to preserve the strength of the Crown Imperial. A proposal that his four princely sons should become Governors-General of Canada, Australia, New Zealand and South Africa came to nothing: but instead the heir to his throne, Edward, Prince of Wales, went on a series of imperial tours. One of these took him to India, where he was boycotted by Gandhi, frequently booed by disloyal demonstrators, and assured by old hands that it was no place for a white man any more, but the most successful visits were to the white Dominions: and the flavour of these tours, their period jollity, their mixture of the buoyant and the defensive, the picture they offered of the handsome but unhappy young prince, freed from the restraints of Windsor and Balmoral, almost becoming a colonial himself—the memory of those royal holidays, preserved in many books and thousands of photographs, piquantly reflects the spirit of the colonial Empire during the years of transience.

This is hardly a king-to-be visiting his future Dominions, this is a young man seeing a new world. Gone is the stately progress of the Viceroys, calm beneath their panoplies, or the grave composure of George V himself, when he sat with his wife as in marble on the Coronation dais at Delhi. The Prince of Wales, heir to all this, wore his shirt without a tie, his trousers short, his cap a'tilt, his heart on his sleeve. It is true that he sailed in the battlecruisers *Repulse* and *Renown,* at 32,000 tons among the great warships of the world. It is true of course that he was greeted everywhere with pomp and eulogy, from the Official Odes of the Torontonians and the Sydney-siders to the drum-poems of the Ashanti:

> *Thy fellows proclaim thee a man*
> *Triumphant from the struggles of war.*
> *To the ruler of kings who comes,*
> *Who inspires awe in the greatest,*
> *Hail! Hail! Hail!*

Edward was fawned on everywhere, naturally, flattered at garden parties, curtsied to by a thousand Lady Eatons, bowed to by innumerable Sir Henrys, blessed by manly bishops in surplices and war ribbons, saluted by old soldiers in crutches and eye-patches. When his battlecruiser passed through the Suez Canal biplanes of the Royal Air Force escorted her to the Red Sea, and Indian troops of the Canal Zone base cheered her on her way. When she put in at Aden a huge banner greeted her beneath massed Union Jacks: TELL DADDY WE ARE HAPPY UNDER BRITISH RULE. The Maharajah of Bharatpur came to meet him in an open landau drawn by eight elephants. The King of Nepal gave him a rhinoceros, a baby elephant, two bears, a leopard, a black panther, an iguana, a python, several partridges and two Tibetan mastiffs.

All visiting royalty met these people, unveiled these monuments, accepted these gifts. The Prince of Wales's tours were different, though, for he represented in his own person an almost reckless break with tradition. He was glamour, visiting an Empire that was fast losing allure. He was modernity, honouring an ideal that was growing fusty. He seemed to be visiting his future subjects not in a spirit of authority, but almost with fellow-feeling, and so took to life on the Canadian prairies, indeed, that he bought himself a ranch in Alberta. The white colonials loved him, and many of the coloured subjects too, for he seemed a foretaste of emancipation, a young, fresh embodiment of an ancient legend. 'Never since the heroic days of Greece has the world had such a sweet, just, boyish master. . . .'

Daddy was less impressed. Like the Wembley Exhibition, the royal tours seemed to please people for quite the wrong reasons, and the King was much disturbed by the newspaper cuttings which reached him from his far dominions. Riding bucking broncos indeed! Flirting with colonial girls! Foxtrotting in the small hours! This was not the spirit that made the Empire—or rather, his Majesty perhaps corrected himself, for in point of fact it was exactly that, it was not the spirit that would keep the Empire British!

7

New realities must be recognized. It had been clear since Chanak that the white Empire was not exactly an Empire any more, but a group of independent Powers of more or less common origin and generally compatible policies. Federal solutions had been abandoned —in the short run the Dominions would always be outvoted by Britain, in the long run Britain would always be outvoted by India. Instead, in 1930, the assembled Prime Ministers formally approved a new device of the pragmatic British political imagination, a Commonwealth of Nations. Though the name was old, the idea was said to have been perfected by Arthur Balfour, and it was full of sophistry. Britain and her white Dominions, it was decreed, were 'autonomous communities within the British Empire, equal in status, in no way subordinate to each other in any aspect of their domestic or foreign affairs, though united by a common allegiance to the Crown, and freely associated as members of the British Commonwealth of Nations'.

This proposition aroused long and intricate discussion. It was like a debate among the Schoolmen. Was the Crown indivisible, or could one man be King of England and King of Canada? Could the King give his signature to opposing Acts from different countries? Could he indeed be at war with himself? Since nobody was subordinate, could anybody be expelled? Mr Hughes of Australia called it 'an almost metaphysical document', and King George was much disturbed by it—'I cannot look into the future', he wrote, straining his grammar to the limit, 'without feeling no little anxiety about the continued unity of the Empire.' The New Zealanders felt the same, and did not subscribe to the new statute.[1] Even the Canadians waived the right to alter some aspects of their own constitution, investing it still in the British Parliament at Westminster.[2]

Though nobody had ever succeeded in binding together the disparate parts of the Empire, its self-governing Dominions, its Indian Empire, its dependent colonies, its mandated territories, its protectorates and Treaty States and condominiums—though nobody had managed to make a rational structure of it, still the overriding authority of the Crown, exerted through an imperial Parliament at Westminster, had in the past provided a recognizable unity. 'The British Empire', is how all the imperial delegates were jointly described, on the roster of the League of Nations. Now even that indeterminate formula was discarded. The King was still the King of all the Dominions, but separately: as the irrepressible Hughes remarked, the King of England was no longer the King of South Africa, but the King of South Africa was also King of England. In future the Governor-Generals, the King's representatives in each Dominion, would no longer be nominated by the British Government, to represent not only the Crown but the imperial authority itself: in future they would be chosen by the Dominions, and would have no contact with the British Government in London, only with the King.

In many other ways too the Statute disintegrated the imperial whole. Now Dominion Parliaments could pass legislation in direct contradiction to Westminster. Even the grand machinery of the imperial law, all its multifarious courts culminating in the Judicial Committee of the Privy Council in London, was now shorn of its certainty: the Dominions were free to withdraw from it when they wished, and so the Empire stepped down from the most truly splendid of its postures. A man could still say *Civis Britannicus Sum*, but he could no longer look with certainty to those remote and impartial justices in Westminster: *Civis Britannicus* my foot, a judge might soon say in Bloemfontein or Montreal, and there would be an end to it.

The British had tried hard, since the death of Queen Victoria, to give substance to a mystery. Now they gave mystery to a sub-

[1] Until 1947.   [2] Where it remains in 1977.

stance. The British Commonwealth of Nations was cloudy from the start. At the time many people claimed to see the Statute of Westminster as a final charter of imperial development, and Balfour himself described it as 'the most novel and greatest experiment in Empire-building the world has ever seen'. But it was really an admission of failure. The Empire would never be a super-Power now. Its Roman aspirations were abandoned: the only overseas delegates ever to sit in that Tribune of Empire, the House of Commons, were those from Ireland. As the Empire's diverse parts matured into independent nations, first the Dominions, then India, then inevitably the great African and Asian colonies, so the British Empire would cease to be among the prime movers of the world, and Britishness would hive off once and for all in separate and often conflicting patriotisms.

8

It soon began to happen. Within a couple of years the Irish Free State took advantage of the Statute of Westminster to abolish the oath of allegiance to the Crown. Within four years Catholic Ireland was proclaimed 'a sovereign, independent and democratic State', and before long it was a Republic and not a member of the Commonwealth at all. The British, though dismayed by so bold and contrary an interpretation of their vision, expected no better of the Irish, and had no choice but to acquiesce: but reluctant to admit that even an Irishman could opt out of being British altogether, they decided to classify citizens of the new Republic not as un-British exactly, but as 'non-foreign'.[1]

---

[1] George V died in 1936, his last words being variously reported as 'How is the Empire?' or 'What's on at the Empire?', and these arrangements surprisingly survived the abdication of his successor, Edward VIII, later in the same year: the Dominions passed their own acts of acceptance, one by one, so that the new monarch, George VI, became King of South Africa on December 10, of Great Britain, Canada, Australia and New Zealand on December 11, and of Ireland on December 12. The abdicated King, reconstituted as the Duke of Windsor, became Governor of the Bahamas during the Second World War, but otherwise had nothing more to do with the Empire which had welcomed him so rapturously a few years before.

CHAPTER SIXTEEN

On Technique

ON October 4, 1930, when the imperial Prime Ministers were about to assemble in London to conclude the Statute of Westminster, when Gandhi was about to celebrate his sixty-first birthday in prison at Poona, when the chiefs of the Al 'Alawi were still squabbling in the Hadramaut and Feisal I still ruled in Baghdad, there took off from its mooring mast at Cardington in Bedfordshire the airship R101, the largest and most expensive flying machine ever built in Britain. This was an imperial occasion too. Ever since the end of the Great War the imperialists had been planning to link the greater components of the Empire by the thrilling new medium of the air. It could revolutionize the very nature of the Empire, they thought, as had the advent of the steamships in the previous century. The idea that the King-Emperor might be in London one week, Canberra, Bombay or Vancouver the next, offered altogether new imperial prospects, and visionaries of the Rhodes kind already foresaw an All-Red Route of the Skies binding everything into a new cohesion.

Fortunately there was to hand an invention awesome enough to match the grandeur of the conception. Few people thought the aeroplane could ever master the prodigious distances of the Empire, but the airship, the rigid dirigible, certainly could. The Germans had done marvellous things with their Zeppelins during the war, sending one as far as Khartoum in an attempt to supply the German forces in East Africa, while the British R34, modelled on the German pattern, had become in 1919 the first aircraft of any kind to make a double crossing of the Atlantic. The airship was the Liner of the Future! It could be at once the means and the symbol of the imperial revival. In 1924 Ramsay MacDonald's Labour Government officially adopted an Imperial Airship Scheme, meant to provide a regular passenger and mail service along the principal routes of the Empire —to India, to Australia, to Canada, to South Africa. One day, the enthusiasts prophesied, flotillas of great dirigibles would be serenely sailing through all the imperial skies, saluting each other as they passed over Suez, or magnificently perceived in the dawn of mid-Atlantic.

The R101, which was built by the Government itself, was to inaugurate the service with a return flight to Karachi, in India, via Egypt. Its construction had been one of the great national efforts of the post-war years. The airship sheds at Cardington, the Royal Airship Works, were the biggest buildings in the British Empire, and the design team, it was said, had 'lived and worked like a religious community intent upon their single purpose'. So proud of the project was Brigadier-General C. B. Thomson, the Secretary of State for Air, that when he was elevated to the peerage in 1924 he took the name of the airship works for his title, and became Lord Thomson of Cardington. The airship took six years to build, and was full of new all-British ideas—diesel engines instead of the usual petrol, steel framework instead of aluminium, new kinds of valves, harnesses for the gasbags, fabric dope. She was propelled by five engines and was designed to carry 100 passengers at speeds up to 70 miles an hour.

'She', because everything about her was consciously shiplike, as though the Empire Airships were in direct line of succession to the British India liners, or P and O. Passengers bunked two to a cabin, with portholes to heighten the nautical effect, and the crew wore neo-naval uniforms (though most of them were in the RAF). There was a promenade deck with deck-chairs. The passengers' lounge, embellished with gilded pillars and potted palms in the Cunard style, was pictured in artists' impressions at the height of a tea dance, with young blood and flappers waltzing among the wicker chairs, and distinguished seniors watching benignly from the sofa. The wireless transmitters were as powerful as any ocean liner's, with a range of over 200 miles, and a radio-telephone that could carry conversations up to 100 miles. The R101 was, so the publicity men said, one of the supreme examples of British inventiveness, in the line of the great Victorian constructions.

On the ground the arrangements for the Airship Scheme were just as elaborate. Immensely expensive mooring masts were erected at Ismailia, half-way to India, and Karachi, where another gigantic hangar was also built, and across the Empire other sites were prepared. Work started on a base at Montreal, terminals were surveyed at Durban and Perth, in Ceylon and on the island of St Helena in the South Atlantic. Weather stations at Malta, Ismailia, Baghdad and Aden collaborated to begin a meteorological service more thorough than anything conceived before.

It was Lord Thomson who conceived a fitting moment for the launching of the project. The Imperial Conference of October 1930 was to be a particularly important one: it would be attended by all the Dominion Prime Ministers, plus representatives of India, and upon its talks, as we have seen, the future shape of the British Empire would depend. How grand, thought Lord Thomson (who had been tipped as a possible Viceroy of India), if the biggest airship on earth could make its maiden voyage along the most imperial of all the imperial routes, to bring the Secretary of State for Air himself direct from the Empire's frontiers to the conference chamber in Westminster! So it was arranged. Lord Thomson would fly with the R101 to Karachi and back, his return to coincide with the start of the conference. On the way the airship would land beside the Suez Canal, the lifeline of Empire, for a celebratory banquet at Ismailia—just as, over so many generations, British warships had flown the flag and offered hospitality in the imperial ports of call.

Everything was hastened to this end, and by October 4 the airship was ready for the flight. The publicity was terrific. The Prince of Wales himself drove up to Cardington to inspect progress, and the papers were full of the excitement of the project. 'As I set out on this journey,' said Lord Thomson at a farewell ceremony, 'I am reminded of the great hopes that have been pinned on this magnificent ship of the air as a link with the furthest corners of that everlasting entity, the British Empire. . . . This is the Empire link of the future, and I set out now to prove that the air and the far corners of the earth are ours to command.'

It was a grey cold night, but at Cardington the vast silver shape of the R101 was ablaze with light. Searchlights played upon it, and there were rows of bright lights from the passenger quarters and the control cabins, and green and red navigation lights on the fins. The headlights of hundreds of cars, too, illuminated the scene, so that from a distance the great hulks of the airship sheds, the resplendent silver of the airship at its tower, the crowds milling about the field, the stream of lights pouring out along the country roads from Bedford, gave it an other-wordly air, of fantasy or nightmare.[1]

Fifty-four men boarded the R101 that evening—12 passengers, all officials except Lord Thomson's valet, 42 crew, from Flight-Lieutenant Irwin the captain to J. Megginson the nineteen-year-old cabin boy. By 6.30 they were all aboard, and by 6.45 the airship's engines, spluttering one by one into life, were thundering at the mooring mast. Slowly the ship backed away, as the crowd sang 'Land of Hope and Glory' below, and the cars around the airfield flashed their lights in farewell. Rain began to fall, and a gusty wind blew across Cardington as the R101, making a last slow circuit of the Royal Airship Works, flew heavily away to the south.

By 9 o'clock the R101 was crossing the coast near Hastings, and Irwin was preparing his route down the line of the Rhone to the Mediterranean. The passengers, after a cold meal, soon went to bed, and the radio operator sent a reassuring message home. 'After an excellent supper', it said, 'our distinguished passengers smoked a final cigar, and having sighted the French coast, have gone to bed after the excitement of their leave-taking. The crew have settled down to a watch-keeping routine.'

The watch-keeping was not all routine, though. The weather worsened, strong winds blew up, the great airship rolled and pitched, and was sometimes blown sideways. From time to time the crew inspected the gasbags, clambering around the ropewalks that criss-crossed the interior of the airship's envelope: everything in there creaked, groaned and hissed as they worked, the gasbags themselves creepily squelched themselves into new shapes, as the pressure shifted inside them, and the metal chains that supported them clanked and chafed in the darkness. The night outside was very

[1] Still easily to be evoked, for the Cardington airship sheds survive, and seen especially from the low hills to the south, look like two gigantic barrow-graves.

black. Only occassionally did a dim light show from the ground below, through the drifting cloud and the rain.

Soon after 2 in the morning the radio operator exchanged messages with the airport at Le Bourget, and confirmed that the R101 was approaching Beauvais, a market city some eighty miles north of Paris. She was flying very slowly now, shaken all over by the wind. Soon afterwards an engineer in one of the nacelles, slung beneath the body of the airship, looked through his window in the darkness and saw an astonishing thing. Protruding grotesquely out of the rainy mist, only a few yards from the airship, was the humped roof of a building, a massive grey object stuck about with pinnacles and queer gargoyles, very old, very stark. For a few seconds he saw it there, as the airship laboured by, and then it was lost again in the rain and the dark. It was the roof of Beauvais Cathedral.

The engineer scarcely had time to tell his companion when the airship gave an abrupt lurch, dropped, recovered, dropped again, and with a colossal judder was suddenly still. There was a pause: then suddenly a tremendous breaking roar, like the lighting of a million bonfires, a frenzied ringing of bells, a clatter of feet along the companionways, and the shout of an officer somewhere—'We're down, lads! We're down!' In a moment the R101 was a mass of flames. She had covered some 300 miles of the 3,652-mile route to India.

2

The R101 had hit a low hill on the southern outskirts of Beauvais, and was destroyed in a matter of minutes. All but eight of her complement died at once: two more died of burns. Lord Thomson, Flight-Lieutenant Irwin, Sir Sefton Brancker the Director of Civil Aviation, the representatives of the Indian and Australian Governments, all died, and were soon forgotten. So absolute was the catastrophe that the Empire Airship Scheme was abandoned at once, and the Empire never did see the grand spectacle of the dirigibles dipping their ensigns in the empyrean. It was an immense blow to British pride. Worse still, it was a revelation of British failure in that most basic of imperial elements, technique.[1]

[1] Nothing was left of the R101 but a pile of steel, presently used to make

The story of the R101 was seen by the contemporary public as a heroic tragedy. 'Another band of pioneers have sacrificed their lives on the long track of the Empire's advance', wrote Sir Samuel Hoare in *The Times*, and the Imperial Conference, by now in session in London, passed in sad silence a resolution of condolence. Behind the tragedy, though, there lay a record of ineptitude. The building of the R101 had looked an imperial enterprise in the great tradition; in fact it possessed little of the daring confidence and common-sense that had characterized British technology in the nineteenth century. Compromise, makeshift, amateurism, plagued the work from the start, and the designers were hamstrung by political interference. The airships' engines were too heavy and too weak, the new valves were found to be more ingenious than practical, the new dope set up a chemical reaction, so that the whole envelope had to be scrapped. Even the galley with its Electric Stove proved inadequate —a party of MPs lunched on board the airship at Cardington once, but the meal they ate was surreptitiously prepared by RAF cooks on the ground. When it was found that the gasbags were chafing against the steel girders of the ship, wads of padding were simply tied around the metalwork. Far more drastically, when they discovered that the airship's payload would be no more than 35 tons, instead of the 77 tons predicted, they simply cut the entire airship in half, inserted an extra gasbag, and joined it together again.

It was only two days after the completion of this major surgery that the R101 set off on her maiden voyage. She had made only eight flights, all in good weather, and she had flown only once since the insertion of the new gasbag, which had increased her length by a

---

kitchenware in Sheffield, and the RAF ensign, which now hangs in the church at Cardington. The bodies of those killed were taken to Cardington too, and lie in a common grave within sight of the airship sheds. On the road from Beauvais to Paris an impressive monument records the sad roster, from Brigadier-General the Right Honourable Lord Thomson of Cardington, PC, CBE, DSO, to J. W. Megginson, Galley Boy, and James Buck, Valet. On the actual site of the crash, though, a mile or two away over fields and woods, there is a less grandiose memorial. It is a stumpy concrete pillar, half-hidden in the woods, and you can reach it only by scrambling through scratchy undergrowth and pushing aside the hazel branches. All it says is: LE DIRIGEABLE R 101, 5 OCTOBRE 1930.

quarter. Her engines had never been tested at high speeds—it was suggested that the test flight might be the maiden voyage itself. But all the time, as these slipshod preparations continued, as one fault after another was hastily mended, as the confidence of the crew gradually waned, and it became clear to the Cardington team that they were being rushed into a haphazard venture for political reasons—all the time the inexorable light of publicity shone upon the airship. There was not a child in the kingdom who did not know the R Hundred-and-One would soon be sailing, and scarcely an adult who did not think of the project as a great technical triumph. 'They're rushing us,' Irwin told visitors to Cardington in the week before the flight. 'We're not ready, we're just not ready. . . .'

This then was the truth behind the great enterprise. The Royal Commission of Inquiry did its best to minimize the meaning of the disaster, but the crash of the R101 was an indicator of the British condition. Technique had been the truest foundation of British power, and often the actual cause of Empire: Britain had truly been the workshop of the world, and the British Empire had eagerly seized upon each new product of technology—even the distant and inessential colony of Mauritius had got its first electric street lights in 1893, its first cinema in 1897, its first X-ray machine in 1898, its first motor car in 1903, its first telephone in 1912.

Since the Great War, though, the nation seemed to have lost its touch. The world was moving out of the age of steam, and the British, who had been masters of the greased piston and the mighty boiler, did not adapt so easily to the age of the automobile and the aircraft. They were not made for it. Unlike Americans or Arabs, Englishmen did not respond as it were by instinct to the new machineries. Their genius for improvisation did not run to the gob of chewing-gum, or the string of twisted fencing-wire, by which other peoples mended the broken carburettor or fixed the dead transmitter.

In the Crown Colony of Gambia schoolchildren used to sing a melancholy little ballad:

> *It was sad when the big ship went down,*
> *It was sad when the big ship went down,*

On Technique

Husbands and wives,
Little children lost their lives,
It was sad when the big ship went down.

They were remembering the loss of the *Titanic* on her maiden voyage in 1912, and there were some people who, in a superstitious way, traced the British technical decline to that older tragedy. Others remembered Beatty's *cri de coeur* at Jutland, as he watched his ill-designed battlecruisers inexplicably blowing up around him: 'There's something wrong with my bloody ships today.' Economists argued that the nation had over-extended itself once and for all in the war, by becoming at once a naval, an industrial and a continental military power, and it was true that since 1918 British industry had consistently lost ground against its foreign competitors.

The British were paying for their old success. The overwhelming superiority of their Victorian technique had made them complacent, even timid about new ideas. They were slow, or at least reluctant, to grasp the meaning of the aeroplane or the motor car—in August 1914 the British Expeditionary Force, the striking force of the British Army, possessed 80 cars and 15 motorcycles. In this as in so many ways, their social structure retarded them, too. Even now, Britain was still geared to a rural hierarchy, in which country landowners were pre-eminent. Industry was hardly for English gentlemen. Not many of the country's best brains graduated from Oxford and Cambridge into industry, and there were no equivalents in England of the great technical universities of Germany and Switzerland. British industry stuck, by and large, to what it knew best: and what it knew best was the technology of the preceding century, when Made in England had been the universal hallmark of quality, and hardly a French steam-cock or an American lathe had reached the limitless markets of the Empire.

It was not that invention had failed. British scientists remained remarkably resourceful, and there was no shortage of striking prototypes: in the 1930s Britain held the world speed records on land, on sea and in the air. But the nation seemed to lack the flair, the will or perhaps the incentive to translate ideas into solid achievement. Habit and long success made the British look at the twentieth century through nineteenth-century eyes, and the immense machine of Empire itself, with all its systems, all its schedules, even much of its equipment, remained essentially a Victorian device, coal-fuelled, steam-hauled.

3

By and large, the imperial specialities of Victoria's day were the specialities of George V's. The great age of railroad building was over, but still here and there in the Empire the railway engineers were at work, and the lines they laid, the trains they despatched, could still be spectacular. The most dashing of all trains were surely 'The Silks', the non-stop expresses by which the Canadian Pacific Railway hurtled the Chinese silk consignments from Hong Kong across Canada to the Atlantic ports and so to the London markets. The stateliest was unquestionably the white official train of the Viceroy of India, twelve carriages for one man, with its quarters for aides and secretaries and servants and bodyguards, its great crested locomotive, and the guard of honour which, during troublesome times in India, lined every mile of its way across the sub-continent, a soldier every few feet along the railway track, presenting arms one by one as the train went by. The most gentlemanly trains were certainly the splendidly serviced sleepers of the Sudan Railway, like travelling clubs of finesse, which ran down the tracks laid by Kitchener's engineers during his advance to Omdurman, and deposited their passengers at last in the cavernous bedrooms of the Grand Hotel, Khartoum (where the laundry lists quoted prices for Jodhpurs and Nannies' Uniforms). The most forbidding train was perhaps 'The Ghan', which ran across the inexpressible wastes of the Australian outback from Port Augusta to Alice Springs, held up for weeks at a time by frequent flooding, bumping and swaying horribly on its rocky roadbed, and named ironically in honour of the Afghan camel-men who first laboured along that uninviting route.

And the most romantic of railways, at least in conception, was certainly the Cape-to-Cairo line by which Cecil Rhodes had hoped to establish his British axis north and south through Africa. By the 1920s this had actually become possible. South Africa—Rhodesia—

Tanganyika—Kenya—Uganda—Sudan—Egypt—Rhodes might have done it had he lived, under the Union Jack all the way. It never happened in fact, but it had its own great memorial anyway, for the centre-piece of the enterprise, the work which meant most to Rhodes himself, was the bridge by which the railway crossed the Zambezi River, on the northern frontier of Rhodesia. This was nothing like half-way from Cape to Cairo, but it had a symbolic quality about it that everyone felt. The river was one of Africa's greatest: beyond it lay the legendary country of the Great Lakes and the Nile sources; and here, heralded across the bush by its perpetual rainbow, there fell into the great chasm the waterfall which Dr Livingstone himself had named for the Queen-Empress long before. At Victoria Falls many elements of Empire were magnificently expressed, and there it was that Rhodes had decreed his railway should cross the Zambezi. 'I should like', he observed, with his odd mixture of the banal and the poetic, 'to have the spray of the water over the carriages.'

The bridge was built after Rhodes's death, and it would have been much easier to cross the river a few miles upstream, where it was narrower: but his wishes were honoured, and the railway crossed the Zambezi within range of the spray. Nobody lived there then. The Africans would not go near the falls, and all around was dense bush, swamp and forest. In this place, where almost nobody could see it, one of the great imperial artifacts was erected. The gorge was 350 feet deep, and Ralph Freeman, the chief designer, bridged it in a single steel span, at once a *tour de force* of engineering and a gesture of grand romance. It was a combination of arch and girder, and it carried two lines of track across the ravine. By the 1920s there was a rest house beside the falls, and a little community was established on the north bank, to look after the tourists.

Not everyone paid much attention to the bridge, so dazzled were they by the falls; but to stand on the river-bank when the Bulawayo train steamed across the gulf was an unforgettable experience of Empire. Against the tremendous thundering background of the waterfall, its rolling spray, its reverberating cannon-cracks, the seething mass of its water, the bridge stood there defiantly, almost frail, with its slender arch curving gracefully across the void. The noise of the water was so deafening always, and the scale of the scene so immense, that at first one often failed to notice the train when, with black smoke billowing, it came puffing heavily out of the forest: but when it cautiously felt its way across the girders of the bridge, when the spray really did splash on its windows and sizzle on its locomotive boiler, when the passengers inside rushed as one man to the windows to see the falls through the streaming glass, then what allegories were invoked! Here was Empire still! Here was Rhodes! Here was Kipling's old power of Steam and Fire![1]

4

The greatest bridge of the 1920s was the Sydney Harbour Bridge, another of Ralph Freeman's designs. As an engineering work this was undeniably impressive. For more than a century people had talked about bridging the narrowest part of the great sea-inlet known as Sydney Harbour, to connect the original nucleus of the city with its growing suburbs on the northern shore, but the plan took time to crystallize. Not until 1924 did the well-known British firm of Dorman Long, after submitting seven alternative schemes, win a contract to build it, their tender quoting a craftsman's price of £4,217,221 11s 10d.[2] The bridge was a single-span steep arch structure. It was built to withstand gales of 250 knots, and carried six lanes of roadway, four railway lines and two footpaths across a main span of 1,650 feet. It was then the longest single-span bridge in the world, and the biggest arch ever made.

There has probably never been a more talked-about bridge, nor one that was to become more universally known. It was hailed, especially in Britain, as a triumph of the imperial technology. It

---

[1] The bridge has acquired different meanings now, for by the 1960s it had become the border between white-dominated Rhodesia and the black republic of Zambia, formerly Northern Rhodesia, and so one of the racial frontiers of the world. When the Prime Minister of South Africa met the President of Zambia in 1975, they held their talks in a railway carriage half-way across.

[2] The 11s 10d representing, it was coyly suggested at the time, the contractors' profit.

took seven years to build, and year by year the newsreels and magazines carried pictures of its progress, the vast concrete pylons going up, the steep arch hanging ever further over the water, the 'creeper' cranes high on the unfinished steelwork, against the blue glittering background of the harbour. A deep-water wharf was especially made to handle steel shipments from England. More than 150 men worked for five years quarrying the necessary granite. Sydney Harbour Bridge was 'one of the Great Engineering Wonders of the world', said Arthur Mee's *Book of the Flag*, and when it was finished in 1932 it really did present a striking image of Australian strength, its silhouette becoming a national symbol familiar everywhere on earth. Younger Australians viewed it with a mordant affection, traditionalists saw it as a proud figure of the imperial connection—

> *There is a land where floating free,*
> *From mountain top to girdling sea,*
> *A proud flag waves exultingly;*
> *And freedom's sons the banner bear,*
> *No shackled slave can breathe the air,*
> *Fairest of Britain's daughters fair—*
> *Australia!*

Yet in the somewhat lumpish structure of the bridge itself, with its great white towers monumental at each end, later generations might see signs of the imperial ageing. It was a very big bridge, but it lacked spirit. It did not soar. Even Arthur Mee's *Book of the Flag* did not call it beautiful. It had none of the colossal affront of the bridges, often hideous but never weak, by which the British had bridged the rivers of India in the Victorian century. Its function was disguised, as far as possible, in a tentative traditionalism. Its street lights were mock-lanterns. Its presiding pylons were vaguely Egyptian in style, like something from the Wembley Exhibition, or war memorials, but they served no technical purpose, being there simply for effect. The Sydney Harbour Bridge was a true product of the imperial thirties, when the British Empire was larger than it had ever been before, but rather less tremendous.

Sydney made a great occasion of its opening—it was easily the most prominent structure in the city, and for thirty years would remain the only Australian building that the average visitor would remember. In March 1932, the Governor of New South Wales, representing the King, unveiled a commemorative tablet, and the State Premier prepared to cut the tape. A royal salute was fired. The harbour ferries sounded their sirens. Most of the Royal Australian Air Force flew in formation overhead. But before the Premier had a chance to declare the bridge open, a young man in elaborate military uniform, wearing many decorations, rode on horseback past the Governor's stand, past the mounted guard of honour, past the astonished Premier himself, and drawing his sword from its scabbard, in one slash cut the ribbon and shouted in a loud voice: 'On behalf of decent and loyal citizens of New South Wales, I now declare this bridge open.' He was a member of a political protest group called the New Guard. They arrested him at once, replaced the ribbon and started all over again.[1]

5

Dams were another old British speciality, and the expertise of imperial hydrologists had been translated from British India, where 13 per cent of all the cultivated land had been irrigated by the British, to the imperial dependencies along the River Nile. The British relished the symbolism of these works. Records of the rise and flow of the river had been kept for 900 years: controlling the Nile was a charge handed across the centuries, from the river-lords of the Pharaohs to the Ministry of Public Works.

The Empire controlled the whole of the White Nile. Egypt's river service was run by Englishmen; Uganda was a Crown Colony; the Sudan, though ostensibly an Anglo-Egyptian Condominium, was in effect a British possession, headed by a British Governor-General and run by a corps of British administrators generally

[1] The harbour bridge was transformed in the 1960s by the construction beside it of the most beautiful building in the old overseas Empire, Sydney Opera House (by the Danish architect Joern Utzon), whose wing-like white sprawl at the water's edge provides a perfect foil for the now elderly formalism above.

thought to be, especially by themselves, the best in the Empire. This unprecedented unity meant that for the first time the flow of the Nile could be regulated logically in the interests of all the riverain people (and some further away too, like the Lancashire cotton magnates). Dams could be built in one country to benefit the people of another, and water could be fairly divided between them all.

The British had begun work on the Nile at the turn of the century, their purpose being to transform flood irrigation, which depended upon the amount of each year's river flood, into storage irrigation, which would provide insurance against a poor year's rainfall. They set about it spaciously. Enormous reports were printed. Splendid maps were made. The river was measured, recorded and analysed as never before, and a whole new vocabulary entered the imperial lexicon: Timely and Untimely Periods, Century Storage, Flush Irrigation. Just as Victorian Englishmen had devoted their whole lives to the study of the Ganges basin, or the irrigation of the Punjab flatlands, so now Britons in the Egyptian and Sudanese services spent all their days thinking about the Nile.

By the 1920s their works were inescapable. Hundreds of miles of canals had been cut or restored, many new barrages had been built, and vast new areas had been opened to irrigation. There were some unhappy consequences to these works—the spread of bilharzia, the exhaustion of the soil, the concentration on a single crop, cotton. Still they were among the chief achievements of the imperial decline, and the irrigation engineers became great men in Egypt, were greeted by crowds of grateful *fellahin* when they made their inspection tours, and blessed by imams in mosques.

The greatest of all their works was the Aswan Dam, which was to be after successive enlargements the bulkiest of all the British Empire's artifacts. Imperial publicists called it the Eighth Wonder of the World, and with its immense line of dressed masonry, its massive buttresses and frothing line of sluices, set stupendously in the dun desert around the first cataract of the Nile, it did have a truly classical grandeur.[1] The Aswan Dam stood at the head of the entire

[1] Though it was originally reduced in size, at a sacrifice of 1,500 million cubic feet of water storage, to avoid flooding the Ptolemaic temples at Philae. 'The State must struggle and the people starve', commented Winston

Egyptian irrigation system, for behind the dam a great reservoir extended upstream to Wadi Halfa, and out of this reserve the engineers could release water in dry periods to provide perennial irrigation all over Egypt. Downstream eight smaller barrages distributed the water among the irrigation canals, and the engineer in control at Aswan had it in his hands to destroy the life of Egypt (or alternatively, Egyptian nationalists suggested, to divert the entire flow of the Nile direct to England).[1]

Upstream the Sudan was less dependent upon the river, and it had been agreed that from the middle of January to the middle of July each year all the natural flow of the river was to be allowed to go to Egypt. The Anglo-Sudanese hydrologists, though, saw profitable possibilities of irrigation in a tract of land called the Gezirah Plain, which lay between the Blue and the White Niles above Khartoum. The country was unprepossessing, and was inhabited by people of awkward migratory habits—sometimes there, sometimes not, farmers in the rainy season, Omdurman drapers or railway track-layers when it was dry. The British resolved nevertheless that this should be the site of an experiment on the Anglo-Indian scale, the deliberate creation of a cotton-growing community dependent entirely upon artificial irrigation.

Churchill, aged twenty-four, 'in order that professors may exult and tourists find some place on which to scratch their names.'

[1] The designer of the Aswan Dam was Sir Benjamin Baker (1840–1907), consulting engineer for the Forth Bridge and the first London tubes, and inventor of the vessel upon which, in 1877, Cleopatra's Needle was floated from Egypt to the Thames Embankment. He is buried splendidly in the village churchyard at Idbury near Burford, in Oxfordshire, beneath a masonry structure, like an open pyramid, evidently intended to recall both his Egyptian associations and his cantilever genius. His dam, incidentally, contained $1\frac{1}{2}$ million cubic metres of masonry, compared with the Great Pyramid's $2\frac{1}{2}$ million, but its successor downstream, the High Dam built by the Russians half a century later, is claimed to be the biggest thing ever made. Long after the end of the Empire British hydrologists retained their links with Egypt, and Mr H. E. Hurst, author of the monumental *Nile Basin*, was still a consultant to the Egyptian Government in the 1960s (in *Who's Who* for 1977, when he was in his ninety-seventh year, his address was recorded as Sandford-on-Thames, Oxford, his club as the Gezira Sporting Club).

It took them six years to settle the land rights, renting the whole area from its owners, then re-allotting it in thirty-acre tenancies. It took them three years to build the Sennar Dam, by which 3 million acres of land was eventually to be made arable. But by the end of the 1920s the Gezira Scheme was one of the showpieces of the Empire, and the men of the Sudan Civil Service assiduously took visitors to see it. It was just as they wished Empire to look. It was a mixture of public and private enterprise: the Government Irrigation Department controlled the water, but two private British companies supervised the scheme itself, and the tenant farmers not only took 40 per cent of the profits of their cotton, but also grew food and fodder crops for themselves. It was replete with social organizations and improving systems—child welfare classes, domestic science schools, courses in civics and economics, village councils, educational films and a highly responsible newspaper. As one of its creators characteristically observed, it was 'a great romance of creative achievement': an object lesson too, for anybody, given rulers of equal merit, could do the same—'vision and foresight', as he generously added, 'are not qualities peculiar to the Government of the Sudan.'

So, less than a century since Speke had discovered its source, less than half a century since Sir Garnet Wolseley hoisted the Union Jack over Cairo, the British had mastered the White Nile. Even as the Empire began its retreat, they had schemes for much larger works—dams at the Great Lakes, a canal to bypass the Sudd, huge new developments at Aswan, a series of new barrages in the Sudan —but they already controlled the river more absolutely than any of their predecessors. From the remote control points, high in Uganda, by which their hydrologists estimated the flow of the water, to the last Delta barrages, almost within sight of the Mediterranean, they had laid their skills upon the whole length of the Nile. Never again, though they did not know it, would they have such matter for experiment, or such truly imperial tasks to perform.

6

These were traditional concerns of Empire. When it came to the internal combustion engine, and all that went with it, the approach was less sure. Everywhere people associated the British with trains, bridges and dams, but most of the first cars in the Empire were foreign, French, German and American manufacturers having gained an early lead over the British. American cars dominated the Australian and Canadian markets from the start, and even in India the first car of all was French—the Maharajah of Patiala's De Dion Bouton, whose licence plate number was O. The British built cars for the small easy roads of their own islands, and until the end of the Empire never did master the tougher imperial markets. Rover's Indian and Colonial Model of 1907 hardly swamped the far-flung highways, and when in the 1920s Morris Motors introduced a new model actually called the Empire, they sold *four*.

The motor car never blossomed in the imagery of Empire. One remembers Lawrence's Rolls-Royces hurrying north from Aqaba, or Dyer's sinister armoured cars outside the Jallianwallah Bagh; one glimpses a Governor emerging from his *porte cochère* in his dowagerly Humber, or a visiting Prince whisked from the quayside in his especially imported Daimler. Symbolically, though, aesthetically, the imperial modes of transport remained pre-automobile: the white shuttered train, the gubernatorial barouche, the elephant, the bullock-cart, the double-decker trams surging through Sydney, the street-cars steamed up from their coal-stoves in the snowy streets of Toronto. The one memorable motor-route of the British Empire was established by New Zealanders, the brothers Nairn, in the new territories of the Middle East: and even its buses were built in America.

The Nairns, who had gone to the Middle East with Allenby's Mounted Corps, astutely realized that the new British dependency of Iraq, so long immured in the obscurity of the Ottoman Empire, would now need swifter access to the west, and in 1922 they established a desert bus service linking Baghdad with Damascus, the capital of Syria, and so with the Mediterranean—a journey of some 500 miles along unsurfaced tracks. As Waghorn had been to the Overland Route to India, as Cunard was to the Atlantic steamship crossings, as P & O were to the Suez voyage, so the Nairn Brothers and their buses became to the new empire in the Middle East. They carried the mails, and until the advent of air services, all the most important travellers in and out of Baghdad. Everyone knew them. 'Taking the Nairn' became a familiar part of speech, and one of the sights of the day was the spectacle of a big Nairn six-wheeler sweeping across the gravel desert, marked by the great plumes of its dust, on its overnight journey between the capitals. The Nairns had several competitors over the years, but they out-drove and outlived them all, remaining familiar institutions of the Middle East when the Empire itself had gone.

The only stop on the journey was at Rutba Wells, almost halfway, in the very depths of the Iraq Desert. This was a fortified oasis and customs post, surrounded by barbed wire and a high wall built of sand-filled petrol cans. There were a few mud huts outside the gates, a police barracks and a rest house inside. It was an awful place. Not a trace of green could be seen, except for a few patches of grass about the wells; all around the desert extended drab and brown, and behind the fort the dried-up bed of a river, which came to life only after rains, meandered away into nowhere. The tracks of camels and goat herds came out of the desert to converge upon Rutba, and the wide beaten path of the Baghdad to Damascus road passed outside its gates, to disappear over the horizons east and west.

Nothing much happened there. A group of Bedouin might lollop in, to settle with their beasts and black tents in the shade of the petrol cans. Oil company lorries might stop for beer or petrol. Sometimes a convoy of RAF trucks arrived, on its way to Azraq, Amman, or Habbaniyah, and occasionally even private cars appeared, chauffeur-driven generally, and generally in convoy, to discharge their dehydrated diplomats or car-sick Levantines briefly and unhappily into the rest house. Most of the time, though, Rutba simply sweltered and dozed behind its barbed wire, buzzed about by flies, ranged about by pi-dogs, with only the blare of the police post radio to break the silence.

But once a day, regular as clockwork, far in the distance there appeared the cloud of dust that heralded the arrival of the Nairn bus, and the oasis shook itself into life. The café tables were flicked more or less clean, the cook was woken up, the petrol men loitered over to the pumps, the lemonade was put on ice, and even the police and customs men tipped their hats back from their eyes, stretched themselves and delicately picked their noses. Soon one could hear the agonized growl of the bus as it changed gear for the approach, and its dust came billowing ahead of it over the walls, and suddenly there it was rumbling and juddering at the gates—high, hot and thick with dust, with its two Arab drivers already clambering from their cab, and through the sealed windows of the passenger compartment tired lined faces looking out.

Every kind of person tumbled from the Nairn bus, when it arrived at Rutba Wells: Bedouin sheikhs in spacious hauteur, Baghdad Jews with thick beards and wide black hats, swaggering Kurds from the north, Persian wives in the prison-tent of the *burkah*, Iraqi Army officers stony in khaki drill, American oil technicians, Egyptian politicians with significant brief cases, Italian priests, Greek grocers—and aloof among them all, red from the sun, rather sweaty, in crumpled linen jackets or RAF serge, separated from all these companions by race, function, taste, history and prejudice, from each other by class, rank, preference and diffidence, there travelled the British, whose presence in that desert had sponsored the equipage in the first place, but who already seemed the least assertive of its passengers.

7

As to the air, the British never did establish the supremacy aloft which was so long theirs at sea. As one might expect, they were adventurous pioneers of flying. Englishmen were the first to fly an aircraft across the Atlantic, the first to fly to Australia, the first to fly over the Himalayas, the first to fly air mail, and they established speed records in all directions. After the war they realized the uses of air power in imperial government, too—the security of the Middle East was in the hands of the RAF, and as early as 1931 two battalions of troops were airlifted from Egypt to a trouble-spot in northern Iraq.

But there was something laborious, or even reluctant, in the application of all this initiative to the everyday business of air transport—a subconscious desire, perhaps, not to hasten the air age, in which all competitors were starting from scratch. Imperial Airways, the airline which did in the end launch an All-Red Empire Air Service, was born in 1924, and even then it seemed faintly anachronistic. If the splendour of steam could have been transferred to the air age, Imperial Airways would have done it: as it was, its plodding biplanes and flying-boats, nearly always slower, nearly always better upholstered than their rivals, sailed the skies with a distinctly maritime dignity, their captains talking of ports and moorings, coming aboard and going ashore, just as though they were in fact navigating the steamships of the imperial prime.

The airline was heavily subsidized, and as a carrier of the Royal Mails enjoyed semi-official privileges, punctiliously maintained by the management. If an Imperial Airways aircraft made a forced landing in imperial territory, its captain was authorized to stop any passing train and oblige it to take on the mail-bags. When the airline began scheduled services to Iraq, the Iraqis were induced to dig a furrow right the way across the Iraqi Desert, to guide its navigators more conveniently into Baghdad. Though it operated in Europe too, this was essentially an imperial airline, and by the end of the 1930s it flew to most parts of the Empire. There were routes to Egypt, India, Iraq, South Africa, Singapore and Australia—'buckling the Empire together', as Churchill put it, and greatly changing the life-styles of the imperial administrators.[1]

It was not a very efficient airline. Its pomposity was a joke among competitors, and even the British themselves often found it too formal and officious—so many of them preferred the KLM service to Singapore that the Dutch complained they couldn't get seats on their own aircraft. The imperial air routes, which looked so impressive on the maps and murals, turned out to be, if you actually tried to fly them, less than handy. Flights were unpunctual, staging-posts were uncomfortable. A traveller to South Africa in the early 1930s had to change six times, flying in five different types of aircraft: a traveller to India had to travel by train from Basle to Genoa, before

[1] 'Am I responsible or are you', a senior official asked his pilot, dubiously beginning a flight to Baghdad, 'for seeing that this machine is not overloaded?' 'That will have to be decided at the inquest.'

flying on to Rome, Naples, Corfu, Athens, Tobruk, Alexandria, Cairo, Gwadar and, after seven nights and four changes of aircraft, plus a Swiss wagon-lit, at last to Karachi.

All this was a far cry from the lost dreams of the R101, and Imperial Airways did not really get into its stride until the All-Up Empire Air Mail Scheme of 1934. Under this plan letters with an ordinary 1½d English stamp, the normal surface rate, were delivered by air mail to any Imperial Airways destination.[1] It was another attempt to unite the Empire by technology, and it gave the airline an inspiriting sense of purpose. To handle the new traffic Imperial Airways ordered, direct from the drawing-board, twenty-eight new flying-boats, called of course the Empire Class, and destined to set once and for all, in memory as in imagination, the tone of the imperial air services. They were not very fast aeroplanes, they were not altogether reliable—only one in three of the Karachi flights arrived more or less on time—and they had their share of trouble. One collided with an Italian submarine, one dived into Lake Habbaniyah, one sank in the Hooghly River, one was blown up by an exploding fuel barge at Southampton and one was permanently stuck in the mud in a lake at Tonk. Nevertheless they became a familiar and beloved part of life for thousands of Britons.

Like the R101, the Empire flying-boats tried hard to be ships. They had sleeping bunks, inhabited in the advertisements by elegant husbands in spotted silk dressing-gowns, bobbed wives in crêpe-de-chine. They had a smoking cabin, and a promenade deck, and the lavatories had tiled floors, and the seats had legs of chromium plate, and the chef in his galley wore a chef's hat, and the captain and his first officer, high in their cockpit above the Mooring Compartment, were splendid in blue, gold rings and medal ribbons. When the Empire flying-boats alighted with a lordly swish on the Nile at Cairo or the Hooghly at Calcutta, flags proudly burst from the cockpit roof—the Blue Ensign of the Royal Mail, the crested emblem

---

[1] Before the Great War the imperial postage rate had been 1d—

> *The stately homes of England*
> *Shake hands across the sea,*
> *And colonists, when writing home,*
> *Pay but a penny fee.*

of Imperial Airways. It was all very imperial, very self-conscious, and perhaps rather touching.

The home base of these imposing aircraft was Hythe, in Hampshire, and the British public were now indoctrinated with the idea of British supremacy in the air. Bradshaws, suppliers of railway timetables to His Majesty the King, brought out their first Air Guide, with useful tips for imperial air travellers (it was advisable to take a dinner suit, but otherwise no special clothing was required, the cabins of the airliners being closed and heated). The 'Speedbird' image, the blue-winged flash of Imperial Airways, became one of the best-known of advertising symbols, and the airline diligently fed its passenger lists to the newspaper columnists (though their selection was sometimes less than exciting—in September 1935, for instance, they could only suggest the owner of the *Karachi Daily Gazette* and the Rumanian Secretary of State for Air). In the brilliant posters of Imperial Airways the Empire boats flew gracefully over the Pyramids or settled on limpid blue seas beneath palm fronds, attended by lithe natives in canoes.

It all sounded very novel and up to date: in fact the air services were thoroughly traditional, not simply in their manners, but in their systems too. For Imperial Airways as for P and O, the hub of imperial communications was the eastern Mediterranean. In Egypt the air routes diverged, one going north-eastward into the new Anglo-Arab empire, one down the Cape-to-Cairo corridor to South Africa, one following the steamship routes across the Red Sea to India. Based upon Alexandria was the Royal Navy's duty destroyer, detailed from the Mediterranean Fleet to patrol the imperial air routes. Based upon Crete was the company's motor-yacht *Imperia*, with full servicing equipment for the Empire boats. Even Corfu, the Ionian island once thought indispensable to the security of Britain's eastern routes, came into its imperial own at last, so that the local cricketers, looking up from the game bequeathed to them by the Empire a century before, might often see the old flag breaking from *Canopus* or *Cambria*, *Caledonia* or *Capricornus*, as the boat for Alexandria settled on the emerald lagoon of Gouvia.

Flying the routes of Empire gave many Britons the same proprietorial sensations they had so long enjoyed from their liners, watching through sea-slapped portholes the passing of the imperial fortresses. 'I was an Imperial passenger, a proud title,' wrote Sir Edward Buck, CBE, of a flight through Africa, and indeed he felt easily at home. Among his fellow-passengers were Lord and Lady Chesham 'who have interests in Swaziland', and Mr F. Kanthack, CMG, formerly of the Indian Irrigation Department. At Entebbe he had time to make a brief visit to Government House, to congratulate the Governor on his new appointment to Nigeria, and to hear some of His Excellency's recollections of Sangor, in the United Province of India, 'where he served as Assistant Remount Officer in 1917'. They spent a night at Juba, where the paragons of the Sudan Service had created a model village in the bush, with its clean hostel, its well-planned native housing, its mission church and its spick-and-span District Commissioner rising courteously from his desk to greet his evening visitors. They had a couple of hours in Khartoum, where Major Barker showed them round the Zoological Gardens—Lady Chesham was much amused by the injunctions at the gate, which forbade her either to bring her donkey into the gardens, or to spit at the animals there.

Next day they spotted the Aswan Dam, the Empire's pride, as Sir Edward, Mr Kanthack and the Cheshams enjoyed a game of bridge, and when they landed for the night at Luxor Lady Chesham insisted upon a visit to Tutankhamun's tomb, whose replica she remembered so vividly from the Wembley Exhibition. So at last on the fourth day they descended thrillingly over the Pyramids to Cairo, where the Union Jack still billowed serenely over Kasr-el-Nil, and passengers on the port side of the aircraft caught a glimpse of the High Commission gardens beside the river. A line of uniformed Imperial Airways officials awaited them when they alighted upon the brown airfield at Heliopolis, snapping into a salute, as they stepped out into the warm sunshine, for all the world as though those passengers were being piped ashore from an imperial war-ship.

The 1935 African schedule allowed a week from Cape to Cairo, but never for a moment did the aircraft have to fly over foreign-controlled territory, or find itself out of sight (for it never rose higher than 10,000 feet) of more or less imperial soil. What was more, if passengers wanted it Imperial Airways would land at many another imperial outpost along the route: at Sheraik or Kosti, for instance, at Victoria West or Moshi or even at Mokia (though even in 1935, hardly anybody wanted to get off at Mokia).

8

It did not come naturally. Younger nations grew up with internal combustion: the British Empire was too old to find it easy, and for all the publicity of Imperial Airways, the British public were much more excited by the launching in 1934 of the greatest of all their Atlantic liners, the *Queen Mary*.[1]

One innovation which did fire the British imagination was something more mysterious, something almost spiritual it seemed: Wireless. It was, of course, a transforming agency for the Empire, and at one time the British hoped to make it as much their own as undersea cables had been in the previous century. Before the Great War Fisher had wanted to make it a world-wide Government monopoly, shared only with the Americans—'It's VITAL for war! The HOURLY developments of Wireless are *prodigious!* You *can't cut the air!* You *can* cut a telegraph wire!' He pressed the idea upon the Imperial Conference of 1911, and the assembled Prime Ministers did plan a series of wireless relay stations, throughout the Empire, which would effectively have dominated the world's communications systems.

The war prevented it, but wireless came to mean much to the British Empire, and its mystique powerfully attracted the British. Kipling even wrote a short story about it, more than he ever did for the Empire Flying Boats. When Freya Stark went to an English Mission service at Hamadan in Persia in 1930, the preacher likened the Lord himself to a Radio Receiving Station, tuned into the world's

---

[1] Now surviving all the flying-boats, alas, as a degraded tourist spectacle at Long Beach, California. Legend says she was to have been called the *Queen Victoria*, but when Cunard officials told King George V they wished to name their liner after the greatest of all English queens, 'Oh', said His Majesty, 'my wife *will* be pleased'—so *Queen Mary* it had to be.

prayers,[1] while at home the Archbishop of Canterbury, exalted by the possibilities, wanted to know if he had to leave his windows open to receive signals. The British Broadcasting Company, founded in 1922, began regular Empire broadcasts ten years later, intended to 'keep unshaken the faith the British nation has in its Empire', and the King himself said that wireless could 'work the miracle of communication between me and my people in far-off places'.

For most imperial citizens of the 1930s indeed, the wireless evoked above all the image of the King-Emperor himself, presenting his annual Christmas broadcast to his peoples across the world. This was the one occasion in the year when Empire and modernity truly coincided, but even so it was technique not in the cause of power or development, but for old time's sake. How Queen Victoria would have loved it! The chimes of Westminster relayed so magically around her Empire, the faint suggestive crackle on the loudspeakers, as though Buckingham Palace were even then being plugged into the system, the plummy accents of the 'announcer' (wearing, as everybody knew, evening dress for the occasion), the theatrical moment of absolute silence, and then, heard at that very instant in home and office, ship and barrack, kraal, hill-station, rubber estate and trading post across the British Empire, the thick bearded voice of His Majesty, speaking very carefully, as if to make allowances for the younger members of the family. 'Another year has passed. . . .'

[1] 'Terrible sermon,' Miss Stark commented.

CHAPTER SEVENTEEN

Art Forms

SINCE the Great War art had generally been hostile to grandeur, and so there was lost the last chance of a definitive imperial art form. Too late! Now there never would be such a thing as a British Imperial style, in art, in literature or even in architecture. The epic was past its climax, and nobody had commemorated it epically.

It was not that creative artists were necessarily hostile to the imperial idea—at least until the Great War few spoke out against it, and even the maverick Irishman Bernard Shaw believed that, in the absence of a world government, the British Empire was best qualified to rule the backward communities of the world. But they were seldom fired by it, either. No English Camöens arose, to celebrate the grand adventure—Tennyson went back to Arthurian legend, when he wanted a theme of chivalry and heroism, and Hardy preferred the wars against Napoleon. Even colonial artists failed to exploit the splendid story of their origins. Oliver Goldsmith II, born in Canada, hardly emulated his great-uncle's success with *The Rising Village*, his colonial successor to Auburn, while the muses did not immediately respond to the Australian William Charles Wentworth's attempt to win the Chancellor's Medal at Cambridge—

> . . . grant that yet an Austral Milton's song
> Pactolus-like flow deep and rich along,
> An Austral Shakespeare rise, whose living page,
> To Nature true, may charm in every age;
> And that an Austral Pindar daring soar
> Where not the Theban eagle reached before.

It showed how transient was the British taste for glory, how shallow perhaps the imperial instinct itself, that the most lasting artifacts of the British Empire were mostly green, gentle and quirky things, gardens and pleasant tropic cities, novels without heroes, limericks, wry ballads, echoes and suggestions.

2

Between the wars, almost for the first time, artistic intellectuals looked at the Empire speculatively in its decline, and dealt with it ironically. Aldous Huxley looked at India, and was reminded of the old man of Thermopylae, who never did anything properly. George Orwell looked at Burma, where he had been a policeman, and thought it all second-rate. 'I wouldn't care to have your job', an American missionary once remarked to him, observing the scarred buttocks of a Burmese suspect in the police station, and the remark cut Orwell not as insulting, but as contemptuous—'so *that* was the kind of job I had! Even an ass of an American missionary, a tee-total cock-virgin from the Middle West, had the right to look down on me and pity me.'

Out of these attitudes came one undoubted masterpiece, E. M. Forster's *A Passage to India*. Of all the novels written about the Empire, except possibly *Kim*, this was the most influential: two generations found their view of imperialism affected by it, if not actually formed, and anyone who read it found that the scenes of imperial life never seemed quite the same again. Forster wrote as a *half*-insider. He had been private secretary to the Maharajah of Dewas State Senior, in which capacity he was once preposterously photographed, wearing a long-skirted spotted gown and a sort of oriental tam-o'-shanter, in very English lace-up shoes against a painted background of flowers and mullioned windows. But he was anything but Anglo-Indian, only a life-long college man translated briefly into the Indian environment, and in a thoughtful, melancholy way, half-enchanted by it.

A Passage to India is not really about imperialism, but about human nature playing itself out against an imperial background, between people of different origins thrown together by the imperial chance. In particular it is about the specifically British imperial technique that was summed up by the image of The Club—the deliberate enclavity, the hanging-together. This was an ugly system, based as it was upon racial awareness and arrogance, but it was undeniably effective in sustaining the brazen bluff that lay at the heart of the Empire. Forster recognized this—'we're not pleasant in India,' as one of his characters says, 'we've something more important to do'—and in exposing the idea of the Club in its sadness and falseness, he did not frontally attack it.

Nor did he attack imperialism in the abstract. *A Passage to India* is not an ideological tract—nationalists often did not know what to make of it, and all too often its Indian characters really do seem incapable of running their own affairs. Forster was in India shortly after the Amritsar Massacre, and the book is full of allusions to that event, but still he was not repelled by the principle of Empire, the spectacle of one nation governing another, but by the personal implications of imperialism, the sham alienations it fostered, the hypocrisies, the apparently unbridgeable gulfs. Perhaps Forster did not worry himself much about the political meaning of it. He was looking deeper, and in his tentative, inconclusive way treated the imperial phenomenon as a Greek dramatist might handle a fourth Fate, as an ever-present imponderable decreeing and ensnaring the lives of human beings.

Yet he sees it too, paradoxically, as something transient and inessential. Render unto Caesar, he seems wryly to be saying as his book reaches its famous ending. 'Why can't we be friends now?' asks the Indian of the Englishman. 'It's what I want. It's what you want.' But no, India herself answers, not yet—'and the sky said, "No, not there." ' It would come in time, sooner than Forster could have dreamt as he finished his book in 1924, but better not to hurry it.

3

There was an imperial folklore of sorts, a corpus of popular art that had coalesced over the generations around the theme of Empire. Much of it was Kiplingesque, for the one period when imperialism impinged upon the popular consciousness was pre-eminently Kipling's period; his verses, tales and characters entered the public vocabulary, and were often transmuted into proverbs, like comedians' catch-phrases. 'You're a better man than I am, Gunga Din', 'But that's another story', 'East is East and West is West', 'You'll be a Man, my Son'—all these and many more, the very stuff of the imperialist philosophy, went into the language, and were bandied about in pubs and prize-giving speech-days as if they were immemorial saws. Every drawing-room baritone sang Oley Speaks' stirring setting of 'The Road to Mandalay', every church-goer knew the solemn verses of *Recessional*, which gave to the Empire the most hackneyed, and later the most mocked, of all its epithets— 'far-flung':

> *God of our fathers, known of old,*
> *Lord of our far-flung battle-line,*
> *Beneath whose awful Hand we hold*
> *Dominion over palm and pine. . . .*

Around the real thing, too, there assembled a mass of neo-Kipling ballads like 'The Green Eye of the Little Yellow God', songs like 'Pale Hands I Loved, Beside the Shalimar,' and a whole body of literature, given a home by magazines like *Blackwood's* or *The Strand*, which came to possess a stylized unity, as folk-tales do. *Blackwood's* published, in the 1930s, twelve volumes of such imperial mythology, under the generic title *Tales from the Outposts*, and they included *Speech Day in Crocodile Country, Ode to One of the Old Indian Troopships, My First Execution, The Left Hand of Abdullah the Beggar* and *A Solo Flight from England to the Gold Coast in Cirrus-Moth G. EBZZ*. Here are the opening lines of *Khyber Calling!* a novel by 'Rajput' (Lieutenant-Colonel A. J. E. Dawson):

I was donning my Blue Patrol jacket for dinner in mess when Firoze Din, my orderly, suddenly said: 'Huzoor, a Gurkha mule-driver was killed by Pathans this afternoon.' I made no answer, but grunted. . . .

Imperial humorists contributed prolifically to the form. There was never a shortage of them. Skilful or heavy-handed, subtle or naive, the satirist was the familiar of every imperial station, first to last, scribbling for the local magazines like the innumerable pseudonymous comic writers of Anglo-India, breaking into the bestseller lists like C. S. Jarvis of the Sinai or Arthur Grimble of the Gilbert and Ellice Islands. Of them all the funniest, and the most revealing, was

Lord Edward Cecil, whose book *The Leisure of an Egyptian Official* is the one classic of the kind. Cecil was no artist perhaps, no historian either, but his outrageous set-pieces of imperial farce, his gallery of characters dominant or subject, translated many of the imperial attitudes, prejudices too, into universal terms.

Unexpectedly, too, Hollywood powerfully propagated the imperial myths. The yarns of Empire were so tremendous, the settings so colourful, that inevitably the film industry, seeking an occasional alternative to cowboys and Indians, seized upon sahibs and savages. Many an old stalwart of Warner Brothers or MGM was to be seen in the 1930s leading his sepoys into the jaws of the Khyber, or limping blood-stained out of the African bush. Here Victor McLaglen guides The Lost Patrol through the burning sands of Mespot, here Gary Cooper and Franchot Tone, dashingly disobeying King's Regulations, rescue the colonel's son from his Pathan torturers and so save the honour of the Bengal Lancers. Walter Huston played a wistful Rhodes against Oscar Homolka's Kruger, Ronald Colman turned the rapacious Clive into a matinee idol, Spencer Tracy, lifting his hat politely, greeted Cedric Hardwicke at Ujiji with that incomparable one-liner, 'Dr Livingstone, I presume?'[1]

British film-makers treated the Empire more as a morality play. *Sanders of the River*, for example, made by the passionately Anglophile Hungarian Alexander Korda in 1935, was a lavish rationale of the White Man's Burden: Sanders, played by Leslie Banks, was beloved and respected by his simple African charges, and as his canoe was paddled along inky tangled creeks by faithful oarsmen, led by Paul Robeson, the cry of 'Sandy the Strong, Sandy the Wise' rang repeatedly through the imperial forests. *The Four Feathers*, taken from a novel by A. E. W. Mason, himself a wartime secret agent, was a thrilling justification of imperial attitudes, so thrilling that the official film censors actually thanked Korda for making it, and critics of the anti-imperialist Left felt obliged to point out that it had been

[1] The film critic Pauline Kael, writing in *The New Yorker* in 1976, observed cautiously of these stirring old movies that they induced in their audiences 'pride in the imperial British gallantry . . . despite our more knowledgeable, disgusted selves'.

financed by the Prudential Assurance Company, capitalist lackeys of racialism.

Away in the overseas Empire the folk-art was more spontaneous, and seemed to spring more directly from the collective imagination. Sometimes indeed its origins were forgotten. Nobody could remember the first appearance of the monocled English policeman portrayed by the wandering Indian players of the United Provinces, or the top-hatted character called The Lord who, in the folk-plays of Corfu, represented the last echo of British authority in the Ionians. The Br'er Rabbit stories were pure imperial folk-art, for they were a mingling of Ashanti lore with Algonquin Indian spirit-stories, the two being brought together by the transportation of slaves to Canada.[1] More often the lore was invented by professionals, sometimes to fulfil a need in a country without traditions, but had been blurred by time and usage until it came to seem organic.

Every white colony had its tall stories, its familiar rhymes, its ballads and its legendary characters. Sometimes real people had been metamorphosed into myth—Rhodes, for instance, became nearly more than human in the Rhodesian memory, and the Anzac soldiers of Gallipoli achieved a similar kind of apotheosis. The Canadian Mounties entered the folklore almost as a matter of principle, Ned Kelly the Australian bandit entered it *faute de mieux*, there being a paucity of alternative domestic heroes. In India there really were peasants who worshipped images of long-dead British administrators. In Sarawak the first White Rajah had become a dimly imagined embodiment of perfection.

And properly enough the most vividly remembered totem of any, till the very end of the Empire, remained Queen Victoria herself. Her grand image survived all her successors, and though as the years went by her statue was removed from all too many parks and plinths, her memory remained, if inexact, lastingly potent. Most folklores have their supreme being, god, hero or arbitrator, the hovering presence behind their stories and enactments—Ansansi the Spider, Heitse of the Hottentots, Zeus, Haroun al-Rashid.

---

[1] A conjunction listed as the fifty-third of *1,001 Reasons for Being Proud to be a Canadian* (Toronto, 1973). My favourite Reason is No 35: 'Canada has an almost square shape.'

Unmistakably the presiding genius of the imperial lore was the Great White Queen, whose presence was all-pervading, inherent to the lays of primitive bards, not altogether amused by Lord Edward's more provocative anecdotes, and present always, out of camera, in the Hollywood spectaculars of Empire.

4

Some of the imperial settlements, especially those that deliberately reflected functions of the imperial mission, had come in their maturity to possess the quality of art. Bombay in India was one, Penang in Malaya another.

Bombay had been acquired by the British in the days of Charles II, long before the birth of the Indian Empire, but it was the imperialism of the High Victorian age, inspired by a masterful and self-righteous set of convictions, which made it a great city. In its nineteenth-century plan, as in distant English prototypes, the British municipal virtues were exemplified in stone, and all its monuments stood testimony to their appointed values. Enterprise, for instance, the first and most fundamental of the imperial qualities, was magnificently embodied in Bombay's Victoria Terminal, the headquarters of the Great Indian Peninsular Railway, which provided a symbolic centre-piece for the entire city. It was one of the supreme memorials of the railway age anywhere, part Oriental, part Gothic, all unmistakably imperial, carved about with crests, emblems and gargoyles, with stained glass windows like a cathedral's and brass-bound teak doors like a palace's, and gigantic lions guarding its central staircase. The Punjab Mail and the Delhi Express snorted all the more purposefully for the splendour of its girdered roof, and it was only suitable that from its central tower there should gaze down upon the commuters the busts not merely of Queen Victoria and her Viceroy Lord Dufferin, but the Company Chairman and his Managing Director too.[1]

Order was embodied in the monumental Secretariat, overlooking the long Maidan beside the sea, from whose lordly galleries, shaded

[1] 'Proving the *bona fides*', a local guide-book says, 'of the well-known proverb "a thing of beauty is a joy for ever". '

by gigantic rattan screens, the imperial administrators could survey their passing subjects in the blazing sun below, or keep an eye on the cricket on the green. Down the road was Law, in the fabric of the hardly less overpowering High Court, and all around were the structures of Enlightenment: the University, supplied with a ceremonial tower by a public-spirited Parsee; the Prince of Wales Museum, with a dome copied in a scholarly way from the Bijaipur Mosque in Mysore, and a genial statue of the Prince himself, the future Edward VII, meditating in the gardens in front; the School of Arts and Crafts, an institution in the progressive spirit of William Morris and the pre-Raphaelites (and truly a memorial to its age, for here in the Director's House, still standing cool but dusty among its lawns of buffalo grass, trellis-shadows on its floors, hammocks beneath its banyans, was born the laureate of Empire, Kipling himself).

Healthy Pleasure was not forgotten in this architectural catalogue —*mens sana in corpore sano* was always an imperial motto. The promenade along Back Bay was developed Brighton-style, to give the citizenry of all races the benefit of fresh sea air out of the west, while parks and gardens proliferated. The Bombay Gymkhana Club was built not in the Gothic mode but in a rustic mock-Tudor, an affirmation perhaps of underlying pastoral values, and marvellously preposterous on the foreshore was the Taj Mahal Hotel, one of the great hostelries of Empire, so open-minded in its hedonism that even the great black jazz musicians of America were to be heard there.

And so to Profit, if not a virtue, at least an intent. The urge for profit was implicit everywhere in Bombay, and set the seal upon the city as a work of Victorian allegory: in the dockyards clanging and smoking on the eastern shore, in the grand Mint and the Cotton Exchange, in the portly old offices of Horniman Square around their gardens, in the queues of the taxis and the rickshaw men, in the ships that lay perpetually beyond the Gateway of India, and the dhows threading nimble passages among their anchorages. It was a shabby old city by the 1930s, but its patina of age and dereliction served only to bind its symbolisms together, and make it feel more than ever a category of art.[1]

[1] A category not everyone admired. G. W. Steevens, at the turn of the

Bombay was a picture in cracked and sombre oils. Penang, which reached its zenith in the 1930s, was a watercolour place, and stood above all for the serenity of Empire—for there were many parts of it that really did provide solace, for rulers and ruled alike. Penang had been British almost as long as Bombay, but since it suffered no social problems on the Indian scale, no terrible pressures of poverty or congestion, it had retained down the years a truly eighteenth-century elegance of manner, and looked like an imperial settlement in a diorama, or a quiet corner at Wembley. Penang Island was about fifty miles round, and lay three miles off the mainland coast in a blue and generally languid sea. A busy ferry connected its capital, Georgetown, with Port Butterworth on the mainland, but still it retained a pleasant island feeling, secluded, almost private. Nearly everybody liked Penang, and this affection was apparent in the look and feel of the island, which was green, and had agreeable beaches, and nice buildings, and often basked in a seductive sense of *dolce far niente*. Penang lived by the export of rubber and tin from the Malay forests, but it had a gently festive feel too, softened by the bland-ness, the sense of fading virility, that was so characteristic of the late Empire.

The great charm of the place, the circumstance which gave it its air of artistic composition, was its microcosmic completeness. With-in the perimeter of the island the whole pattern of imperial life in the tropics was conveniently displayed, as in an exhibition. On the southern shore, nearest the mainland, stood Georgetown. Here were the docks, and the Government offices, and the racecourses, and the

---

century, wrote that any Englishman would feel himself a greater man for his first sight of Bombay, but Aldous Huxley thought it 'one of the most appalling cities of either hemisphere'. Most of the Victorian and Edwardian monuments still stand, overshadowed now by huge new developments along Back Bay. They never built the great Processional Way which was intended to lead from the Gateway of India into the heart of the city, but the Government buildings are almost as grand as ever, the hammocks still swing in the Kiplings' garden, and the Taj Mahal Hotel has flowered in an enormous modern extension— when in 1956 King Ibn Saud of Saudi Arabia ordered a picnic lunch for 1,200 to take to the races, he got asparagus soup, paté de fois gras, smoked salmon, roast turkey, chicken, lamb, guinea-fowl, nine different salads and four desserts.

polo ground, and the Penang Club of course, and all the usual appurtenances of a colonial town in the tropics. A delightful esplanade sauntered by the sea, with a cenotaph commemorating the dead of the Great War, and grouped all around were the white City Hall, the Supreme Court, a nice Anglican church with a spire, and a marvellous old imperial cemetery littered with the broken columns, rotundas and veiled urns of the Empire's funerary tradition, its tombs often moulded together by age and dilapidation, and coloured by drooping frangipani.

In their downtown offices, built in the truest 1930s Georgian along King Street and Caernarvon Street, Pitt Street and King Edward Place, the imperial businessmen in neatly pressed shorts and open-necked shirts supervised the imperial accounts and took their fair share of the imperial spoils. The Eastern and Oriental Hotel— the 'E and O' to every imperial traveller—sprawled cheerfully beside the water beyond the Esplanade, old Fort Cornwallis crumbled away beside the sea, and all over the city, filling its streets in like a filigree, embroidering its edges, were the bazaars and booths and stilted houses, the fishing huts and tenements, the mosques, temples, monasteries and Chinese restaurants of the various indigenes.

And just along the road from Georgetown, thus concentrating the whole aesthetic of the imperial east into a few delightful square miles, there stood the island's hill-station. In India, where hill-stations were invented, hundreds of miles separated Calcutta from Darjeeling, Madras from Ootacamund. Here one did the journey in half an hour, by way of a bustling little funicular railway, completed in 1923—which, lurching and slithering upwards through several layers of tropical foliage, with a change of trains half-way, deposited you blithely on the sunny green upland of Penang Hill, 2,500 feet above the hot and bustling city.

The Governor had a bungalow up there, Bel Retiro, whose classic drawing-room views had been made familiar by many a beloved print, and there was a little village of attendant bungalows, for lesser officials, and a police station with a sergeant on duty to look after it all. Nearby was the Penang Hill Hotel, set among gardens: and on its verandah, looking across the lawn to the sweep of the distant sea, drinking a cool Malay stengah you could experience, as in some participatory theatre, the whole pattern of tropical empire, the leisure and the authority of it, the prospect of reward, the sweep of the eastern shore, the orchids, the remoteness, perhaps a little of the boredom, and possibly, if it were a Thursday afternoon, the tinkle of a piano slightly out of tune, the thin thread of Mr Ribiero's Goan violin, from the tea-dance in the lounge behind.[1]

5

The institutional art of the Empire was seldom a success, at least in the twentieth century, because its message was seldom very decisive. It was compromise art, appeasement art, and it lacked the punch of certainty. Kuala Lumpur, for instance, the capital of British Malaya, was scarcely an imperial capital at all, so assiduously were its offices supplied with domes and Moorish arcades, so closely did its railway station approximate to a mosque, so modestly was the cricket club tucked away amidst the surrounding arabesques. Canberra, the capital designed for the Australians by the American architect W. B. Griffin, was hardly more assertive, being a kind of half-cock Washington, all avenues and artificial lake, where it took an age to walk from one Ministry to the next, and nothing but the scale of the place seemed to have anything specifically Australian about it. In South Africa the most thoroughly imperial of all the imperial architects, Sir Herbert Baker, created a Government headquarters intended to seal once and for all, in stone, the reconciliation of Boer and Briton. High above the old Transvaal capital, Union Buildings looked out across the Apies valley with a noble air, but not much fire—absolutely symmetrical, absolutely balanced, its wings balanced, its columns and fountains balanced, its war memorials all aligned, tree facing tree, peony matching peony,

---

[1] All delightfully survives, rejuvenated by independence. Mr Ribiero, alas, was not playing when I was at the Penang Hill Hotel in 1975, but the funicular is still running rigidly to schedule, there are race meetings five times a year at the Turf Club, and the E and O has given birth to a lively disco called The Den.

cactus aligned with cactus, giving the whole ensemble a suggestion of grand but sterile ritual, like a permanent thanksgiving service.

But in the last generation of their power the British did build themselves one truly colossal self-memorial. The world might soon forget the messages of Canberra, K.L. or even Pretoria, but for better or worse, right or wrong, New Delhi, the capital of imperial India, would remain for ever an embodiment of the British presence. The imperialists intended it so, and they spent some thirty years putting the notion into practice. There were craftsmen in India who spent a working lifetime constructing this pyramid of the Raj: it was as though the British instinctively prepared for the end of their supremacy with an indestructible monument to themselves, like Ozymandias in the sand. The foundation stone of New Delhi was laid by King George V when he went to India for the Coronation Durbar of 1911; by the 1930s it was still only half-built.

Like those other capitals, New Delhi as a whole lacked conviction. Built on the hillock called Raisina, outside the walls of old Delhi, it was splendid, it was beautiful, but it was faintly at odds with itself, as though it was not altogether sure what it was supposed to commemorate. It was too late for arrogance, too soon for regrets, and one could not be sure whether these structures were intended to be the halls of an eternal dominion, or whether their architects foresaw, as the sublimation of their art, Indians themselves sitting in graceful succession upon these thrones of power. When New Delhi was started, the New Imperialism was hardly over, the Great War had not happened, and if the Empire had lost its aggression, it had certainly not lost its complacency. Twenty years later the fire had left the imperial idea, and Gandhi was able to squat on the viceregal floor, as one senior official remarked of his visit to Irwin, eating 'some filthy yellow stuff without so much as by your leave'. No wonder the great new capital lacked *insolence*.

The architects of New Delhi were the ubiquitous Baker, who had spent almost his whole working life in the Empire, and Edwin Lutyens, who had hitherto had little to do with it, and it is revealing that the weakness of New Delhi lies mainly in Baker's work, the true magnificence in Lutyens': as though the one architect were faltering in his imperial convictions, while the other had none.

Though they were old acquaintances, they were temperamental opposites. Baker was shy and un-social, and disliked the formalities and pretensions of Anglo-Indian life: 'Ned' Lutyens was witty and gregarious, was married to the daughter of a Viceroy, and relished every garden party and gymkhana. For twenty years they worked in an anomalous and later uncongenial partnership—his Bakerloo, Lutyens called it—each spending some six weeks of each year in India. English masons went out too, and the consulting engineers were British, but most of the work was done by an army of Indian labourers, men and women, working under able Sikh contractors to whom the building of New Delhi was a career in itself.[1]

The setting was solemn, and the structures of the new capital looked out across a dun plain littered with the abandoned fabrics of predecessor empires. This was deliberate, for New Delhi was intended to evoke the historical consequence of Delhi—the Rome of Asia, as Murray's Handbook used to call it, where a dozen dynasties had risen and fallen into decay. The main axis of New Delhi ran exactly east and west, but a secondary alignment, 60 degrees off, connected Raisina Hill directly with the great Jami Masjid mosque, the masterpiece of the Moghuls in the heart of old Delhi. Like a Chinese city governed by the principles of Feng Shui, it obeyed injunctions apparent only to its creators, and incorporated esoteric messages of its own. It was ornamented everywhere, for instance, with clusters of stone bells, meant to suggest the alleged Indian prophecy that when the bells of Delhi rang no more, the Raj would come to an end, and it was replete with improving texts and symbolisms, like the nauseating injunction to the indigenes that was affixed enormously to the Secretariat: *Liberty Does Not Descend To A People. A People Must Raise Themselves to Liberty. It Is A Blessing That Must Be Earned Before It Can Be Enjoyed.* There were Meanings all over New Delhi, and far away down the ceremonial approach, Kingsway, ornamental pools, tracks and shrubberies seemed to suggest astronomical implications, like the mystic lays and circles of the

[1] And to one of whom, Sir Sobah Singh, I am indebted for much in this chapter. Sir Sobah asked Lutyens to build a house for him, too, not far from the job. 'I don't build houses', replied the architect, 'I build palaces'—and in the consequent building Sir Sobah and his family live palatially to this day.

ancients. New Delhi was intended to be, so its architects said, neither British, nor Indian, nor Roman: simply Imperial.

The capital proper was conceived in three parts: the Secretariat, the circular Assembly Chamber, the Viceroy's palace. Around it was laid out a garden city for the chiefs, bureaucrats and feudatories of the Indian Empire. The richer princely States had their own palaces, surrounded by a Princes' Park closed to the public, and radiating from several hubs ran the domestic streets of the civil service and the military, each planted with a different kind of tree. The Commander-in-Chief, the second most powerful man in India, had his own palace, due south of the Viceroy's, and half-way along the axis to the Jami Masjid a big shopping plaza was built, providing a sort of junction between India of the Raj and India of the people. Kingsway ran for two miles straight as a die from the Viceroy's palace towards the great ruined citadel of the Purana Qila, and there were plans to complete the conception with a ceremonial boulevard along the river.

The Secretariat buildings, facing one another at the western end of Kingsway, were Baker's work. They were grandiose but weak, and managed in a curious way, rare in monumental architecture, to look rather *smaller* than they really were. Their immense colonnaded façades, eaved and turreted, and riddled with huge echoing courts and staircases, felt oddly insubstantial from close quarters, and Baker's characteristic progressions of vistas, arches seen through arches, columns ending arcades, belvederes and courts and galleries, worked less well in the changeable and dust-laden Indian light than they did in the clarity of Africa. The buildings looked what they were, institutional, and if their eaves and shady corners made them pleasantly cool in summer, generations of civil servants were to curse the name of Baker when the bitter Delhi winter reached through the unglazed staircase windows.

Beyond them on the hill, though—the Acropolis, Baker thought, to their Propylea—Lutyens built the greatest monument of the imperial architecture, the Viceroy's palace. Its presence was weakened by a squabble between the architects. It should have been approached as through a ravine between the Secretariat buildings, but Baker declined to reduce the gradient of the roadway, so that an unworthy bump in Kingsway obstructed the central view, and weakened the impact of the palace. Even so, it formed a tremendous climax to the capital. Lutyens conceived it as the destination of a colossal ceremonial way: away from the river boulevard to the east, up the uncompromising unwavering Kingsway, past the tall white figure of King George V in his cupola, under the Arch of India inscribed with the names of the imperial war dead, between the lofty slabs of the Secretariat, through the tall wrought-iron gates of the palace entrance, across the gravelled courtyard with its grave viceregal statues, up the immense ceremonial steps, through the arched loggia, and so, hardly deviating an inch from the start of the main east-west axis two miles away, directly into the throne room of the Viceroy of India, where the Crown's Anointed, seated with his lady in the sumptuous regalia of his rank, would greet you with a fitting condescension, and accept your humble courtesies.

In designing this unexampled palace Lutyens made no concession to the imperial falter. Perhaps he had not noticed it. He built the house as for an absolute monarch, a later Moghul—India, he said, had made him 'very Tory—pre-Tory, feudal'. Certainly he seemed to accept implicitly the old Anglo-Indian maxim that to rule India successfully a ruler must live magnificently. Grander and grander had risen the structures of Indian sovereignty, ever gaudier trundled the elephants of Empire, until in New Delhi 'Ned' Lutyens, the urbane and puckish creator of English country houses, built upon Raisina Hill a palace fit for a Sun-King.

It was bigger than Versailles in fact, and perhaps more powerful, for it was more compactly structured. It was 600 feet square, more or less, and including its twelve enclosed courtyards, covered four and a half acres. Lutyens designed it all, down to the 130 chairs of the State dining-room—it was said to be the largest project ever undertaken by a single architect. It was not at all like his neo-Georgian fancies of Kent or Sussex, with their gentle brickwork and unassuming entrances, its style being set by a rather bizarre central feature, a vast shallow dome of faintly Byzantine bearings. In fact this was curiously derived from the celebrated Buddhist stupa at Sanchi in Bhopal. In adapting it Lutyens cut off its base, lifted the whole of it bodily, so to speak, as he might remove the dome of

St Paul's, and deposited it, complete with its surrounding balustrade, on top of a squat square tower.

The rest of the house was just as boldly hybrid. If it was like a gentleman's country house in some ways, it was like a despot's castle in others. Its comfortable private quarters were balanced by immense offices of State, and its wings and courtyards were linked by interminable bazaar-like corridors, marble-arched and marble-floored, whose traffic of liveried servants, aides and hurrying bodyguards gave the building the feel of an arcaded oriental city of its own. The garden was part rose-bed English, part water and clipped trees in the Moghul style, and all through the courtyards and passages cobra-fountains spouted their water into the sunshine, and lines of elaborate lamps swung heavily in the breeze.

The Viceroy's private quarters were like rooms in the stateliest but most comfortable of English country houses—the Lutyens vernacular enormously magnified. The baths were Romanly lavish, the lavatories had the very latest kind of flushing system, there was a handsome loggia suitable for conversation pictures and a classic country gentleman's library of fiction, travel, history and biography. The Durbar Room was circular, lofty and awesome, with a floor of porphyry and columns of yellow jasper. The grand central staircase was open to the sky. The Council Chamber was decorated with a fresco, covering one entire wall, showing the imperial air route from London to New Delhi, complete with hurrying flying-boats and biplanes, and a puffing foreign locomotive, somewhere in the Alps, admitting that in the 1920s even Viceroys had to come part of the way by train.

Deep below stairs laboured the servants. Nearly 6,000 of them manned the great house, and they lived in a township of their own, just out of sight in the viceregal estate beyond the gardens. More than 400 of them were gardeners, and fifty of *them* were boys whose only job was to scare the birds away from the vegetables. In the basement of the palace, running the entire length of the building, a great household community was always at work. The pot-cleaners perpetually scoured their great copper pans, the chicken-pluckers squatted among their feathers, the linen-men stood guard over their cavernous linen-cupboards, the dish-wallahs laboured up to their ankles in washing-up water. The kitchens, hung picturesquely all about with skillets, skewers and choppers, were equipped with the latest English electric ovens, and generally presided over by a French chef. The stables housed a regiment of cavalry. English mechanics tended the three Rolls-Royces in the garage, sewing-men sat at their machines in the Tailors' Shop, numberless pheasant hung in the Game Room. There was a Tinman's Room, and a tent store, and away at the north end between the carpet godown and the tiffin rooms, the ever-busy Viceroy's Press clanked away at its menus, court circulars, seating plans and confidential agendas—His Excellency's visit to the dentist that morning, or the arrival of Sir Hector and Lady Edgington-Shore (from Calcutta, to the Minto Suite).

From the start opinions varied about New Delhi. Visitors who saw it first at moments of great ceremony or delight, when the crimson-jacketed lancers were parading in the Great Square, when the howdah'd elephants majestically advanced along Kingsway, when the handsome young ADCs, elegant in navy drills and sky blue lapels, welcomed one so charmingly to the ballroom, and the Viceroy's house was a pageantry of uniforms, saris, ball gowns and decorations, scented with orange blossom from the garden and roofed by the stars above the great open stairs—people who saw it first in its lordly moments were generally overwhelmed, and thought the capital a worthy crown to a tremendous enterprise. On a stifling dusty day when nothing much was happening it seemed less impressive, and then most observers thought it too big, or too pompous, or not Indian enough, or not British enough. One Secretary of State for India, Edwin Montagu, even wondered if it was sufficiently *large*. Perhaps it was at its best, though, seen through eyes of gentle irony, as Forster might have seen it—early on a spring morning, say, when the dewy haze was beginning to clear, when the heat was starting to shimmer in the air and the old walled city was astir with the sun—when the early-morning riders were trotting along Kingsway, and the first buses rumbling shakily towards Connaught Place. Then if you stood far back from New Delhi, down by the river, say, with the cracked monuments of other empires lifeless all round, and looked back over the dun and green expanse of the Kingsway gardens, glistening here and there with dew and lily-pond, then its proud distant bulk up there, towered, domed and flagged in the morning, looked a thousand years old itself, and perhaps, in that fallacious morning light, a little dishevelled already, like the dead tombs and fortresses all round.[1]

6

But epic, never quite. Suppose the Victorians had built this, the one supreme temple to the British idea of empire! Then what marvels of assurance we might have seen on Raisina Hill, what glinting flamboyances of skyline, what a tight-packed elaboration of pride, stern and imperturbable above the Jumna! Too late! The flare of the imperial confidence had been too brief, too illusory perhaps, and the only epic of Empire lay in the memory of the thing itself, in the surprise and the effrontery of it, in its own images of labour, service, swank and avarice, and in its effect, for good and for ill, upon the lives and manners of mankind.

[1] The ensemble of New Delhi has adapted easily enough to the republican style, though the names of its imperial creators are still commemorated beneath the cupolas of the Secretariat approach, where the martins like to nest: Architects, Sir Edwin Lutyens, Kt, KCIE, Sir Herbert Baker, Kt, KCIE; Chief Engineers, Sir Hugh T. Keeling, Kt, CSI, Sir Alexander Rouse, Kt, CIE. For years the old viceregal portraits remained in the palace, now the home of the President of India, and though they have now gone a bust of Lutyens still stands on a landing of his great staircase. When I was kindly shown around the kitchens in 1975, I was impressed to observe that though the cooks downstairs were busy smoking yellow pomfrets over charcoal braziers, scented with wheat grain, in the housekeeper's room upstairs there was a well-thumbed copy of Escoffier's *Modern Cookery*, 1928.

CHAPTER EIGHTEEN

Stylists

THE British Empire lacked charisma now. By the 1930s it was more benevolent than it had ever been, more idealist in an unassertive way, more sympathetic to its subjects, less arrogant, more humane: but it was becoming a somewhat dowdy presence in the world. Even for its practitioners it had often lost its tang—'in comfortless camps,' as Orwell wrote of them in 1935, 'in sweltering offices, in gloomy dark bungalows smelling of dust and earthspoil, they earn, perhaps, the right to be a little disagreeable.' There was a reason for this. Nobody had properly analysed the root causes of imperialism, but everybody recognized the prime characteristic of the phenomenon: aggression was necessary to its spirit, whether it be aggression for bad or good causes, and when an empire lost its aggressiveness then the excitement of imperialism itself, for better or for worse, was lost.

In this chapter we meet some men who, in this prosaic matinée of the imperial performance, acted still with a sense of style.

2

In Kenya was Lord Delamere. Kenya *was* Lord Delamere. The British had acquired that delightful country, with its high pastoral downlands, its flamingo lakes and its tropical seashore, in the 1880s, and by building a railway through it from the Indian Ocean to Uganda, had opened it up to European settlement. They had always loved the place. Its rolling downs were like a freer, grander Wiltshire. Its fauna was noble, its tribespeople were handsome, its soil was fertile, much of its farmland was more or less free for the taking. It was Crown land. Lord Kitchener, looking around for an agreeable investment in the days of his stardom, acquired a few thousand Kenyan acres, and by the 1920s many other Europeans had done the same. Most of them were British, including a sizeable minority of aristocrats, but many more were Afrikaners, who trekked up with their families from the south.[1] Of all the white settler colonies, Kenya was the most stylish, and the highland country north of Nairobi, much of it fenced off for European ownership, became the most desirable ranching and farm land of the overseas Empire. Kenyan life demanded hard work, rough living and real risks: but for a fit and adventurous man of the right disposition, and preferably the right connections, it was the imperial dream fulfilled.

Officially it was governed by the Colonial Office. A Governor sat in his handsome palace in its garden outside Nairobi, well-paid, well-servanted, with an elaborate official administration to interpret his wishes—more than a thousand men of the Colonial Service, in the most coveted posting of all. There was an elected legislative council, in the usual Crown Colony pattern, but all executive power was in the Governor's hands. To him, and to his masters in Whitehall, Kenya was an African country. One day, one distant day, it would be returned to the black people, when they had been educated to run it themselves: in the meantime the Crown, represented in the field by His Excellency and his thousand diligent officials, would look after it for them.

This was not the view of the white settlers, and the leader of the settlers was the most vigorous, impatient and imaginative man in Kenya, Lord Delamere. He had first come to the country in 1898, leading an exploratory expedition out of Somaliland. Seeing for the first time the glorious landscapes that lay beneath Mount Kenya, rippling with grasses in the sunshine, and inhabited only by genially co-operative blacks, he had fallen in love with the place, and after an unsuccessful attempt to settle down on his family estates in Cheshire,

[1] Some hundreds of them to the Uasin Gishu Plateau, in the west, which in 1903 had been offered by the British to the Zionist Movement as a National Home. What the Jews spurned, the Boers cherished, and the area became virtually an Afrikaner province. For years its principal town, now Eldoret, was known simply by a leasehold number: Sixty-Four.

came back to Africa, aged thirty-three, and obtained a 99-year lease on 100,000 acres of Kenyan highland.

Delamere was as tough as nails. He arrived on his new lands on a stretcher, having been injured in a fall, and all his life his fortunes alarmingly fluctuated. He was always in money trouble. With his young wife he lived for many years in a mud hut, curtained with sacking, and he tried almost everything on his property. He tried sheep, cattle, ostriches, wattles, wheat, pigs, oranges, tobacco, coffee. He built a flour mill, and bought a chain of butchers' shops. He never did get out of debt, but his ranch Soysambu became a famous showplace, his merino sheep won prizes everywhere, he evolved the first successful East African wheat, and he walked through his wide acres with all the pride of a man whose roots are in the soil, and whose land is his by right.

For he believed firmly in that right. He thought this was country ordained for the white man, and he wanted to see a large European farming community controlling Kenya for ever. In 1921 the settlers of southern Rhodesia, resisting Colonial Office attempts to liberalize the colony, had persuaded the British Government to give them a limited form of self-government, by which they confidently hoped to sustain white supremacy indefinitely. Delamere wanted to achieve the same in Kenya, and he dreamt of a chain of British white-dominated settlements running down Africa from Nairobi to Salisbury, linked in some federal arrangement as a huge East African dominion. He arranged conferences of delegates from Kenya, Tanganyika, Northern Rhodesia and Nyasaland, which met three years running in different places, and discussed magnificent dreams of sovereignty. The first was held in a deserted mission house in southern Tanganyika, 450 miles from the nearest railway station, 800 miles from the nearest shopping centre. It was not for nothing that Delamere was called the Rhodes of Kenya.

He was a man of powerful convictions, or prejudices—'part politician,' a Labour MP thought him, 'part poseur, part Puck'—and so resolutely did he practise what he preached that by the 1930s he was indisputably the great man of Kenya. He had been there longer than almost any other European. He had suffered and succeeded there. He knew exactly what he wanted for the country, and he cut through the liberalism of colonial officialdom as a knife through margarine. He believed at once in the natural superiority of the white man, and the qualities of the noble savage, and he thought the idea of eventual black sovereignty no more than a mischievous or sentimental day-dream.

Down the years he had watched the Governors come and go, like a constitutional monarch surviving all his Prime Ministers—'the uncrowned King of Kenya', thought Sir Edward Grigg, one of them, 'to whom all the settlers looked up for leadership'. Officials prepared themselves nervously for his visits, hostesses vied for his company, and the Legislative Council, in which he represented the Rift Valley, the heart of the settler country, was dominated by his arguments. He not only influenced the policies of Kenya Colony, he also powerfully affected its manners. He was not a handsome man, with his small figure, his rather hawk-like face and his slightly cauliflower ears, but in his youth he had been distinctly racy. He dressed eccentrically then, in peculiar hats and disreputable clothes, and he grew his hair outrageously long, down to his shoulders. He had a taste for violence, his temper was foul, he indulged himself in bad language, strong drink and practical jokes, and he loved every sort of high jink and irreverence. His had been a subaltern's, hunt ball kind of fun—shooting out street lamps, midnight rickshaw races, throwing oranges at the windows of the Nakuru Inn or shutting the manager of the Norfolk Hotel inside his own refrigerator, when he declined to serve more liquor.

In his maturity he was more sober, devoting himself to his farms and his politics, but by then he had become a legend, and the reputation of his youth attached itself to the colony, many of whose settlers liked to think of themselves as lesser Delameres, hard-riding, high-spirited, fast-living, well-bred individualist adventurers, there to stay.

3

They were a small community—in 1934 there were some 9,000 white people in Kenya, 1,500 of them farmers—and though some of them would hate to hear it said, essentially provincial. Nairobi, their capital, had started life as a shunting-station on the Uganda railway, and was now a curious mixture of the drab, the boisterous and the snobby. Its architecture was hardly distinguished, its roads were unpaved, its Africans lived in petrol-can shanties and its many Indians had created an untidy bazaar quarter of their own. But its rollicking New Stanley Hotel was always full of boisterous young men in Stetsons and plus-fours, and on the edge of town was the Muthaiga Club, encouched in lawns and creepers, which was the cosy bastion of the settlers, refusing membership to blacks and browns, and begrudging it to officials. There were three banks in town, and a synagogue, and a masonic temple, and a cantonment of the King's African Rifles. The air was marvellous, for Nairobi stood 5,500 feet up, and the brilliance of the light and atmosphere gave life in the town a hectic and often bawdy feeling. Scandals of one sort and another were common, and from time to time the London popular papers printed exposés of Nairobi life, full of titles and wife-swappings.

It was generally agreed that this was destined to be a great capital. Government House itself had been consciously designed by Herbert Baker, Rhodes's favourite architect, to become one day a viceregal residence, where the King's surrogate could hold sway over a great new African dominion. Where the parties differed was over the character of Greater Kenya. The Colonial Office foresaw an African India, the indigenous blacks gradually acceding to power under Whitehall's kindly tutelage, the European settlers relegated to the role of a planter class like any other. Lord Delamere and his friends held quite different views. They stuck to convictions long discredited at home, concerning the nature of civilization and the hierarchy of race, and they envisaged the East African Dominion as a species of Virginia, gracious and spirited, where a gentlemanly white ruling class would hold power in perpetuity. In this Lord Delamere was no more reactionary than most of his contemporaries. He had come to Africa at a time when black civilization seemed hardly more than a joke—'blank, brutal, uninteresting, amorphous barbarism', is how the first Commissioner for the East Africa protectorate saw the native cultures in 1900. Delamere had virtually created Kenya; he had written the style of the country upon what appeared to be an empty page, and it was beyond his imaginative powers to revise the conceptions of a lifetime.

He was bound to lose. In 1922, when the settlers were particularly incensed about Indians in Kenya being put on a common electoral roll with Europeans, there were plans for a rebellion on the Ulster pattern—communications to be seized, the Governor to be gently kidnapped, the imperial Government obliged by force to accept the Delamerian view. But it never happened, and by the end of the decade the settler community was already an anachronism. Most of Delamere's cherished criteria were out of date even then. His opinions on race, on responsibility, on laissez-faire, were all robust survivals from the previous century, and more and more people were wondering if the white man had any right to settle in Africa at all.

Delamere's last political act was a mission to London, in 1930, to protest against the Labour Government's declaration that Kenya was primarily an African territory, where 'the interests of the African native must be paramount.' He must have known he would fail, fighting as he was so clearly against the tide, and he returned to Kenya depressed, sick and disillusioned, dying in the following year. They erected a statue of him in the middle of Nairobi, at the junction of Delamere Avenue and Government Road, and they buried him on his own land at Soysambu in the highlands, where the pink mass of flamingoes murmured and meditated at the lake's edge, and across the open range the old adventurer's merinos peacefully and profitably grazed.[1]

4

In London was Frederick John Dealtry Lugard, first Baron Lugard of Abinger in the county of Surrey, who had been a mercenary of

[1] The statue has gone, and Delamere Avenue is renamed for Jomo Kenyatta, but the New Stanley and the Muthaiga Club flourish still, and I think Lord Delamere might rather like the racy and somewhat garish flavour of modern Nairobi, black though its rulers be. His dreams of a great new Dominion never, of course, came about, Baker's Government House now providing a lordly headquarters for the President of the Republic, but in the last decades of the Empire the separate territories did co-operate in many services, from pest control to East African Airways.

Empire, and a pro-consul, and was now one of its few theorists. He had spent a lifetime in the imperial cause, starting in India, ending as a member of the Mandates Commission of the League of Nations, but he made his name, and evolved his philosophies of Empire, in a colony which he more or less created, Nigeria.

This was a very different kind of Africa. Nigeria was unarguably Black Man's Country, sweltering tropical country, where few white men much wanted to work, still less to settle. With its 350,000 square miles it was much the largest African possession of the Crown, but hardly the most serene. There were at least 100 different Nigerian tribes, each with its own language, and they were by no means all fond of each other. Some, in the north, were Muslims, ruled by the elegant Fulani Emirs, whose bodyguards wore chain mail and blew fanfares on horn trumpets. Some were boisterous pagans of the seashore, who ate slugs and practised necromancy. There were great tribal federations like the Ibos and the Yorubas: there were infinitesimal clans, lost in the rain-forests, who lived by the hunt and the barter, and knew nobody else.

Long after the British had established themselves on the West African coast, they understandably stayed clear of the unnerving Niger hinterland, contenting themselves with a trading establishment at Lagos, the chief coastal town. The stimulus of trade took them tentatively into the interior, often against their better judgement—of the forty-eight Europeans of the first Niger trading expedition, thirty-nine never came home. By the end of the nineteenth century flag had followed trade, and partly for profit's sake, partly to keep the French and Germans out, the British Government presently established protectorates over the whole country, and in 1914 converted it into the biggest of all their Crown Colonies. Its first Governor-General was Lugard.

Lugard was one of the few professional imperialists to evolve an ideology. In 1922 he published a widely read and almost fulsomely admired rationale of imperialism, *The Dual Mandate in British Tropical Africa,* for which Nigeria had provided the laboratory. The views it expressed came just at the right time, for they offered at once a sop to the liberal conscience and a prod to the wavering imperialists. Lugard offered sensible, generous reasons why the

British Empire should continue to exist in Africa at a time when self-determination was all the rage. His 'dual mandate' implied that Empire was good for everyone: it was a mandate to swop skills for resources, to the equal benefit of the subject peoples, the ruling Power and the rest of the world. 'The tropics are the heritage of mankind and neither on the one hand has a suzerain power or right to their exclusive exploitation, nor on the other have the races which inhabit them a right to deny their bounties to those that need them. . . . The merchant, the miner and the manufacturer do not enter the tropics on sufferance or employ their technical skill, their energy and their capital as "interlopers" or as "greedy capitalists", but in the fulfilment of the mandate of civilization.'

If Delamere was behind his time, Lugard was ahead of his. He was the archetype of the paternal imperialist, and paternalism was to prove the ante-penultimate phase of the British imperial process—the fifth age of Empire, perhaps.[1] Lugard had spent most of his life

---

[1] *First go the Spies,*
Bowing and presenting plausible credentials,
And then the honest Traders with their ribbons,
Striking an excellent bargain, one for two,
Tin for gold. Hard on their heels, the Conquerors,
Full of just cause and grievance, flying fierce flags,
Speaking of hinterlands. The Consuls next,
Elephant-borne or vastly palanquined,
Grandly distributing the Queen's command,
Hanging disloyal miscreants now and then
After due process of law. And then a shift,
Into a worthier Schoolmaster's pride,
Of benefits bestowed and gratitudes,
Like pedagogues regretfully reprimanding
Pupils of wasted promise. The sixth age
Adopts the shuffled posture of Apologist,
All ready now to see the other side,
Opening the door for last month's criminals,
Bending the knee to vassals once in awe,
Giving the titles and the booties back
With murmurs of goodwill. Last scene of all,
Sail from the scented shore the Abdicators,
Back to their distant island, small and damp,
Sans guns, sans gold, sans flags, sans everything.

in Africa: he had fought, explored, adventured and administered there, and he loved the continent and its peoples. Never for a moment, though, did he regard the African as his equal, any more than Delamere did. The typical African, he once said, was a 'happy, thriftless, excitable person, naturally courteous and polite, full of personal vanity, with little sense of veracity . . . his thoughts are concentrated on the events and feelings of the moment, and he suffers little from apprehension of the future or grief for the past'.

It was as a kind of prefect that the Briton should live among these people, teaching them the rules (even some of the dodges), encouraging them to honour the school code, and preparing them for the day when they might win First XV colours themselves, and walk across the quad with their hands in their pockets. Furse's young men, Lugard thought, were perfect for the job—English gentlemen whose only apparent passion was for fair play. There was nothing possessive to these attitudes, he always emphasized. The Briton in tropical Africa was not there to stay, only to guide and cherish: not to settle, only to spend a few terms there, or perhaps a long vacation.

Lugard's influence was immense, for like Delamere he spoke with utter assurance. The system of indirect rule he evolved for Nigeria was in fact nothing new—it had been the British system among the Indian States for generations—but he made it *sound* new, he intensified it, and it was associated always with his name. Under it the Nigerian chiefs ruled their own people as they always had, but subject to the supervision of the Sixth Form, in the person of the Empire, Frederick Lugard and his District Commissioners. By these means they would acquire the best of western methods and values, while preserving their own cultures and prestige. It was a humane and practical system, but it was hardly less anomalous than Delamere's opposing philosophy. As the enlightened Lugard sought to conserve African civilization, all the more vigorous of the Africans did their best to escape it. If Delamere was a reactionary in one kind, Lugard was seen as archaic in another. The Kenya Africans would be serfs: the Nigerians, exhibits in a folk museum.

Even Lugard did not recognize that whatever the white man did in Africa would be wrong. As in India, so in Africa, sooner or later his very presence there would be seen as an intrusion. No flicker of doubt, though, seems to have crossed the pro-consul's mind as he dispensed his wisdom to the world. Lugard was another small man, and a nice one. Everyone liked him, even his critics. There is a photograph of him escorting a party of African chiefs on a visit to the London Zoo, and while he himself looks properly headmasterly, with his umbrella and his watch-chain, his voluminously robed companions, gleaming black and very serious, look as though they trust him absolutely, and would not in the least resent it if, muttering something about it hurting him more than them, he gave them six of the best with his rolled umbrella.

In the field he looked rather more peppery, and during his pro-consulship visitors sometimes found him daunting in his fame, especially when there stood at his side, slightly taller and no less formidable, the celebrated Lady Lugard, née Shaw, formerly Colonial Correspondent of *The Times* and an exceedingly clever woman. Lugard was one of the half-dozen men of the twentieth-century Empire who could claim, if not to have created a country, at least to have transformed it, but he was dominant chiefly in a philosophical sense. He had enjoyed a warlike youth, but in his maturity his presence was not pugnacious, only persuasive, almost reproachful. He seemed to be expecting the best of the British Empire, and to be slightly hurt when he got less.

Nor did he die, like so many of his peers, disillusioned. On the contrary, he remained convinced of his theories to the last, universally honoured, sustained in his old age by book royalties and lecture tours, loaded with honorary degrees and commemorative medals, chairman of the International Institute of African Languages and Cultures, he a peer, his wife a Dame, successful in all he did—and like so many Empire-builders, whether *post* or *propter hoc,* childless.[1]

5

In Cyprus there was Ronald Storrs, who as Oriental Secretary in

---

[1] As soldier, explorer or administrator he had served in India, Afghanistan, the Sudan, Burma, Nyasaland, Kenya, Uganda, Bechuanaland, Hong Kong and Nigeria, and he died at Abinger in Surrey in 1945.

Cairo during the Great War, had been one of the progenitors of the Anglo-Arab empire. So acute was he then in his use of Arab nuance that old King Hussein thought he must be a Muslim himself, and certainly of all the imperial activists of his time Storrs was the most polished, some would say the most exquisite. There he sits on his tasselled sofa, smoking a Turkish cigarette and reading a poetic broadsheet, and no one would guess that he was a favourite of Lord Kitchener, a protégé of Lord Cromer. Long-limbed, rather effete in appearance, with a delicately clipped moustache and a volatile, mannered conversation, he looked less like a pillar of Empire than a more than usually worldly dramatic critic, perhaps, or a fashionable museum curator. He was a classicist, an aesthete, something of a snob, with a preference, if not for the great, at least for the gifted. 'I see I shall have to struggle', he wrote in his diary after a journey up the Tigris with some more ordinary imperialists, 'against developing into a Prig: but the whole paraphernalia of whiskies and sodas, the plugging and lighting up of great briar pipes, the bubbling and sucking, the pointless gusty sigh of relief (from what?), the halting oracular uttering of common-places!'

In no other period of imperial history would Storrs have become an activist of Empire, but it happened that when, in 1903, he came down from Cambridge the Egyptian Civil Service was recruiting British volunteers: and this small corps of urbane enthusiasts offered him his life-long ambiance. Physically he was an Englishman absolute, without a drop, he liked to say, of any other blood: but he was a Levantine by nature and by taste, and it was in the eastern Mediterranean that he found himself.

He had no sense of Empire. He did not want to rule people, and was never subject to the *frissons* of the imperial mission. This is how he recorded his first visit to Bombay, the Gateway of India: 'Walked up and down the city with Said, who assumes that all big shops must be branches of similar establishments in Cairo. Alas, bought some books. Went into Cathedral and attempted to test organ, apparently good, but found bellows switch under lock and key. Very ugly church with interesting *basse epoque* monuments of Colonels who braved "inexorable Sultans" and died young for their pains. Lunched with Jukes, and his pretty Jane Austen wife in the splendour and opulence of the Yacht Club, to which we have nothing in Egypt *simile aut secundum.* . . .'

He was an instinctive cosmopolitan, and a sybarite. He wallowed in life, in the true Levantine style, delighting in the company of artists and musicians, writers and actresses, eating well, keeping self-conscious and over-literary diaries full of foreign phrases, collecting *objets d'arts*, moving charmingly through a succession of poses and enthusiasms, the ideal dinner-guest, the perfect flat-companion. He was an aesthete, but a shrewd and capable aesthete, and he reached the climax of his career as Governor of Jerusalem, the first Christian satrap of the Holy City since the Crusades.

His first administration, as Military Governor, was mixed and colourful. His assistants included, so he claimed, a bank cashier from Rangoon, an actor-manager, a Glasgow distiller, an Alexandria cotton-broker, a taxi-driver from Egypt, a Niger boatswain, two schoolmasters and a clown. With this eclectic staff he ushered Christian rule into the Holy City, later spending six years as Civil Governor too, and with it he rode the tricky whirlwinds of Empire in Palestine—the perpetual squabbles of sect and faith, the duties of Mandate, above all the implicit threat of Zionism.

Storrs' approach to these problems was essentially a-political. He viewed Jerusalem less as an imperial responsibility, less even as a Holy Place, than as a work of art, as his comrade-in-spirit Ian Hamilton viewed a battle, and he governed it with his adept and fastidious imagination. He cultivated all the communities. He made friends with the religious leaders, the archaeologists, the scholars. He founded the Pro-Jerusalem Society, an early exercise in civic conservation, and himself designed the lovely tiled plates, made by Armenians, which designated the streets of the Old City in English, Arabic and Hebrew. Storrs married while he was Governor of Jerusalem, and he loved the place always, feeling like Allenby that it represented the apex of his life. 'There are many positions of authority and renown within and without the British Empire,' he wrote in his memoirs, 'but in a sense that I cannot explain there is no promotion after Jerusalem.'

Alas for Ronald Storrs, promotion came his way, and never again did he find a post so suited to his temperament. In 1926 he was appointed to govern Cyprus, and there he was faced with more brutal realities. In Jerusalem he had succeeded by art and subtlety, by his own flair for the oblique and the elliptical. In Cyprus he must deal with Greeks and Turks whose methods of debate were murderous or arsonical. Cyprus had been effectively British since 1878, when Disraeli acquired it from the Turks as an outwork for Egypt, but it had never been a success. Its first British ruler had been Sir Garnet Wolseley, its first surveyor had been the young Herbert Kitchener, but by the 1920s it had decayed into an improvident and often incorrigible backwater. It was torn by a perpetual antipathy, between the Greeks who were the noisy majority of the islanders, and the Turks who were its stoic minority. In the days of the Victorian Empire, when an Englishman's mere presence was often enough to restore order, these fractious neighbours had been generally subdued: by Storrs' day, when the mystique of Empire had been sadly weakened, the island was habitually on the edge of chaos.

It was not that the British were unpopular. In the fifty years of their rule they had restored the island from desolation to a degree of fruitfulness, with irrigation works, new forests, roads, and all the usual material benefits of imperialism. They had even revived the wine industry, one of the oldest in the world but almost extinguished under the Muslim regime of the Turks. The trouble was that the Greek majority did not wish to be British citizens at all, however beneficial the Empire, but wished to be united with the Greek mainland, while the Turks viewed such an event with violent misgivings.

Poor Storrs, this was not his metier. With a somewhat desperate charm he did his best, presiding sympathetically over the Legislative Council, encouraging tourism and archaeology, patron to the Survey of Rural Life in Cyprus, joint editor of *The Handbook of Cyprus* (Jubilee issue, to mark the fiftieth anniversary of British rule). He persuaded Queen Mary to give fifty books to the Nicosia Public Library. He attended the Chess Club once a week, and even went to the Boxing Tournament. He nagged the Colonial Office to give more money for agricultural research. He tried genuinely but ineffectively to reconcile Greeks and Turks through the medium of scholarship, music and the arts, as he had momentarily brought together Jews and Arabs, Muslims and Christians, Orthodox and Catholic in Jerusalem.

Always, though, simmering beneath the surface of Cyprus, he felt the pressure of its old resentments, and heard the Greek call, more insistent every year, for *enosis*, union with the motherland. He was not happy in the island. He despised many of his British subordinates, he detested the provincial narrowness of the imperial way, and he came to loathe the treacherous practitioners of Enosis. In 1928 he had a breakdown, and it took him eight months to recover: when he returned to Cyprus, it was to discover that charm was not enough, in geo-politics, and that even the best-intentioned imperial dilettante must have his enemies.

One night in 1931 he was dressing for dinner after spending a very characteristic evening, preparing Christmas presents for his staff—'I remember writing Gunnis's name in Enlart's *Art gothique en Chypre*, which Enlart himself had given me'—when a crowd of several thousand people stormed the gates of Government House, shouting *'Enosis! Enosis!'* They broke all the front windows, and threw burning sticks and material into the house. The Riot Act was read, in English and in Greek, and a volley was fired by the police, but by the time the crowd had dispersed into the night, the house was in flames. It was made of wood, and in ten minutes Storrs had lost everything: all his books and all his letters, his Steinway and his mother's violin, his Homer won as a prize at Charterhouse, his Byzantine ikons and his Sienese primitive—his wooden hawk from ancient Egypt, his Greek torsos and his Armenian trays of beaten copper. He was not a strong man, physically or even perhaps morally, and though he did his best to draw happy conclusions from these dismal events, still he left the island saddened and disillusioned—'as for the things that went in Cyprus, perhaps they were taken because I cared for them too much. . . .'

He achieved no more. He was a man of Empire only by force of circumstance, and he was not really sympathetic to the imperial ethos, or born to rule in distant places. His last appointment was the most truly imperial of all, Governor of Northern Rhodesia, but he was out of his element. Now in his fifties, he hated every moment of it—'almost overwhelmingly disagreeable', he found life in the village-capital of Lusaka, *sans* libraries, *sans* antiquities, with a heritage altogether alien to him and an imperial protocol so rigid that in a population of 1½ million the only black man whose hand he was permitted to shake was the Paramount Chief of Barotseland. Furse's empire was not for Storrs. He hated the philistinism of it, he hated the arid white society, perhaps he even hated Africa. After only two years he retired to England, and there he lived happily enough, if disappointedly perhaps, ever after, lecturing, being charming to people, gracing innumerable societies, and writing his ornamental autobiography *Orientations*, the least imperial, most enjoyable and best selling of all imperial memoirs.[1]

6

In South Africa was Jan Christian Smuts, the only colonial statesman to become a world figure, or for that matter to be accepted on his own merits in the highest councils of the Empire itself. There was no questioning *his* style. His presence was elegant, his manners were delightful and he was generally assumed to have the qualities of a sage or prophet. With his neat grey beard and slim figure, the straightness of his posture, his calm grey-blue eyes and gentle high-pitched voice, he did not so much dominate company as attract it magnetically to his person. Yet there was a hint of the equivocal to him, and though to the end of his life the British honoured him, and accepted him as one of the great men of the age, his own people viewed him with distrust. For he was a Boer always and unmistakably, and for all the breadth of his vision and experience, all his life he was trapped within his origins.

As it happened he grew up within the British Empire, in Cape Province, and went on a law scholarship to Cambridge, where he got a brilliant first and declined a fellowship. He renounced his British nationality after the Jameson Raid, but when the Boer War

[1] He died in 1955, but to this day there seems hardly a second-hand bookshop in England without a copy of *Orientations* on its shelves. Storrs wrote it half in Wiltshire and half at Axel Munthe's famous villa of San Michele on Capri.

ended worked wholeheartedly for reconciliation with the British Empire. The British greatly admired him, and during the Great War he was co-opted to the British War Cabinet in London. This was an unprecedented honour. Smuts stayed in England for more than two years, and became a great public figure—a Privy Councillor, a Companion of Honour, an oracle everyone consulted. He advised generals on their strategy, politicians on their arguments. He was offered a safe seat in the House of Commons. He was invited to command the Palestine campaign. He reorganized the air force —the true beginning of the RAF. He proposed himself as field commander of the American divisions then arriving in France. He devised the structure of the League of Nations, he largely invented the Statute of Westminster, and he did much to convince the British Government that independence was the only solution to the problem of Catholic Ireland.

All this time he remained a member of the South African Government too, and when he went home in 1919 he became Prime Minister upon the death of Botha. In that troubled backwater of the world his stature was cruelly diminished. He found that to counter the more extreme of the Boer Nationalists, who wanted only to revive their own Republics, his moderate South African Party was obliged to coalesce with the Unionists, the party of Rhodes, Jameson and the jingos. He put down the 1922 Rand riots with unyielding ferocity, antagonizing the labour unions, and next year he was defeated by an alliance between the Labour Party and the Afrikaner Nationalists. In the world outside he was an elder statesman, very nearly a prophet: in his own country he was, throughout the 1920s and 1930s, a political failure, at worst out of office altogether, at best a deputy to his own principal opponent in the coalition Governments of the slump years.

Smuts had a philosophy of his own, which he called 'holism', and which postulated, insofar as anyone could understand it, the essential unity of everything within an amicable universe. He sprang from a people, though, who believed in the single punch, the undiluted draught, black-and-white. His private life was complex beyond the imagination of most Afrikaners. His home near Pretoria, Doornkloof, was no more than a large tin-roofed bungalow, built by the British as an officers' leave centre during the Boer War, not at all pretentious and not very comfortable: but he lived there with his devoted family like a scholar and a prince, developing his theories of history, holism and diplomacy, surrounded by a wild clutter of books and papers, clambered over by many grandchildren, visited by every distinguished visitor to South African shores and exchanging letters with correspondents all over the world.

Yet he remained, like it or not, a Boer himself—Cambridge-educated, British-fostered, world-renowned, liberally minded, estranged from the Dutch Reformed Church, detested by the Afrikaner Republicans, but still a Boer of the Volk. And perhaps it was his Boerness which flawed this truly great man, and kept him from the very highest ranks of human achievement. Fundamental to Boerness was the question of race, and behind every episode of Boer history, behind the Great Trek, behind the wars against the British, behind the re-emergence now of Afrikaner nationalism, lay the inescapable truth that the white man was outnumbered in South Africa overwhelmingly by the black. The profoundest Boer intention was to maintain the supremacy of the white race, and it was this unchanging resolve, touching upon the rawest nerve of Empire, which made South Africa so awkward a limb, of the imperial body. The British Empire might not always live up to its own best principles, but still it was *au fond* a liberal organism. To deny a man advancement because of his colour denied its own highest convictions: trusteeship implied, if not immediate equality of status, at least the recognition that, with luck and good behaviour, all men might be equal one day.

This was the root difference between the British and the South African view of Empire, and Smuts was torn by it. He believed genuinely in the community of the world, but he could not quite bring himself, so deep were his folk-instincts perhaps, to believe in the community of black and white. He was 'Slim Jannie' to his fellow-Boers because he seemed too flexible: he was faintly suspect to his fellow-cosmopolitans because on this particular issue he seemed too rigid. It was Smuts who, by persuading Milner in 1902 to postpone the problem of the native franchise, ensured that it would be handled in the Boer, not the British, way. '*Allas sal regkom*', he used to say—'everything will sort itself out'—and in the racial context this meant that nothing would change. 'I sympathize profoundly with the native races whose land it was long before we came here,' Smuts once said, 'but I don't believe in politics for them. . . . When I consider their political future I must say that I look into shadows and darkness; and then I feel inclined to shift the intolerable burden of solving the sphinx problem to the ampler shoulders . . . of the future.'

This was foresight. He could not have solved the problem even if he tried; sphinx-like it remained; shadows were truly to cloud its resolution. But Smuts' evasion of it tempered his greatness nonetheless, and perhaps affected his style too. Even at his most majestic, there was something withdrawn about him. He saw the world in the grandest terms; he believed evolution itself to be only a symptom of cosmic brotherhood; his influence upon international events was never less than dignified, and sometimes noble; yet all around this famous figure, as he sat with his beloved and comfortable wife on his stoep at Doornkloof, there developed year by year the pseudo-philosophy of apartheid, which was one day to shatter his own dream of Commonwealth, and even threaten the unity of mankind. He was a man of his times, after all: a man of his race, too.

7

There were others, of course. There were wandering eccentrics like Bill Bailey of the Coconut Grove, whose refusal to leave his celebrated bar in Singapore immortalized him in a popular song. There were formidable individualists like Thomas Russell Pasha, the police chief of Egypt, who used to jump his favourite camel over the steeplechase course at Gezira, or his colleague Harry Boyle the Oriental Secretary, who was asked once by a total stranger on the terrace of Shepheard's if he was the hotel pimp—'I am, Sir,' he replied at once, 'but since I am enjoying my tea interval perhaps you would direct your inquiry to my deputy over there'—and off the stranger went to Sir Thomas Lipton the tea magnate, who was taking tea nearby. There were great merchants like Antonin Besse of Aden, an *entrepreneur* out of the Arabian Nights, who lived in delicate splendour in his palace in the Arab quarter, refusing to join the British Club but sending his hides, spices and coffees in his own ships across the world.

There was Charles Vyner Brooke, last of the great imperial freelances, who had succeeded his father and his great-uncle as hereditary Rajah of Sarawak, and against all the trends of Empire, remained the despotic ruler of the land, appointing his fifty British officials personally, recruiting his own army, and accepting White-hall's authority only in matter of foreign relations. There was the aberrant soldier Orde Wingate, Royal Artillery, the furious son of Plymouth Brethren parents, who had adopted the Zionist cause while serving in Palestine, had organized his own night squads of young Jewish guerillas to combat Arab sabotage, and was to be encountered loping around Palestine with his devoted desperadoes, a rifle in his hand, grenades tied to his belt and dirty gym shoes on his feet.[1] There was Captain A. T. A. Ritchie, the Old Harrovian Game Warden of Kenya, who fought in the Great War with the French Foreign Legion, drove about in a Rolls-Royce with rhino-ceros horns affixed to its radiator, and so adored all living things that his house was full of stray wild animals, and he even made friends with individual fish. There were saint-like missionaries, unconven-tional engineers, flotsam and jetsam characters everywhere, keeping hotels in Suez, sailing steamboats up the Gambia, fossicking still on the deserted Klondyke, or pretending to imaginary pedigrees among the Melbourne matrons.

But they were the exceptions, the rebels even. 'We spend our time', wrote Freya Stark, contemplating the imperial British in 1933, 'creating a magnificent *average* type of Englishman, the finest instrument in the world: none of our education sets out to produce great men.' So it was. The Empire was withdrawing into the first postures of apology, and it needed the estimable average, Santayana's just and boyish masters, not the great men or the heretics.

[1] The Special Night Squads, officially disbanded by the British in 1939, became units of Haganah, the Zionists' own defence force, and since this in turn developed into the Israeli Army Wingate is still honoured by the Israelis as one of the founders of their military strength.

CHAPTER NINETEEN

Memsahibs and Others

AN allegorical factor in that withdrawal was the advance of women. Empire had generally been unkind to women (to British women, that is, for the women of the subject races had often benefited greatly, being relieved from the necessity of self-immolation or barbaric surgery, spared the hazards of slavery or tribal war, and occasionally persuaded to regard their husbands as less than god-like after all). Empire marched uneasily with the feminine principle. 'We're not pleasant in India,' says Heaslop the City Magistrate in *A Passage to India*, 'and we don't intend to be pleasant', but his mother disapproves. 'God has put us on the earth to be pleasant to each other . . . and He is omnipresent, even in India, to see how we are succeeding.' Now, in the 1930s, the principle of being pleasant to people was gaining ground on the principle of being strong, and Mrs Heaslop's view of Empire was imperceptibly leading to its dissolution.

In theory the British Empire was staunchly feminist—the first women in the world to get the vote were those of the Isle of Man, the next were those of New Zealand.[1] In practice it was a purveyor of female heartbreak, relative or complete, temporary or life-long. The imperial memorials were full of women's sadness: dead babies, fatal pregnancies, early widowhoods, disease, loneliness, terrible boredom, perpetual separations, the bicker and envy of small isolated communities, the gnaw of homesickness and regret. The Empire's wars had deprived millions of mothers of their young sons. The adventure of Empire had destroyed countless families, and

[1] Unless you count *Y Wladfa*, the Welsh settlement founded in Patagonia in 1865, where everyone over eighteen had a vote from the start. But that, as Kipling would say, is another story. . . .

condemned thousands more to unhealthy alien climes, where for a couple of centuries the memsahibs and others tried to recreate, with their flowered curtains and their drooping cottage gardens, some sad verisimilitude of home.

To add insult to melancholy, it had become popular to blame women for the Empire's decline. The imperialist attitude to the sex had detectably changed over the years. In mid-Victorian times the memsahib and her kind were sacrosanct, figures of porcelain perfection, above and apart from ordinary human functions, to be cherished tenderly in life and Mourned Perpetually in death. By Kipling's day the memsahib had become a more fallible figure, and his Mrs Hawksbee, bitchy, snobby and conniving, became the fictional archetype of the imperial female. The ladies' drawing room at the United Service Club at Simla was nicknamed 'the snake-pit', and the doughty novelist Maud Diver, a bold supporter of General Dyer and an imperialist to the core, felt obliged to write a book specifically to restore the damaged reputation of her sisters.

With the 1920s an element of bantering contempt entered the Empire's attitudes to its women. 'The hen-house' was now the epithet for ladies' annexes in the imperial clubs, and women's efforts to fulfil their own potential, physically and intellectually, were generally greeted with affectionate and patronizing amusement. Later still their effect on Empire was represented as baleful. Until the women came, it was suggested in clubs and bar-rooms around the world, the Empire had been invincible. The British male, as everyone knew, was frank, fearless and no respecter of persons—the natives knew where they stood with him, an Englishman's word was his bond and all that. Besides, there was nothing like a bit of black to cement race relations, was there? Living dictionaries, as they used to say! It was the damned memsahibs who wrecked it all. 'Mark my words, old boy, if the women had never come we'd still be on top of the world. What will it be now? No no, old boy, it's my round . . . *Bearer!*'

There was something to this beery myth. Many women, like many men, were devoted to the country of their adoption—often their countries of birth indeed, especially in Anglo-Indian families. Few women, though, were imperialists. The Empire offered them so little. They were denied most of its chances and half its stimulations, and if it is true that they were often more race-conscious than their menfolk, equally they were much less ambitious to rule other peoples. The notorious absurdities of the memsahibs, satirized for so long by male writers, were generally only the frustrated expressions of unhappiness, fear, homesickness and waste—for surrounded as she generally was by servants, denied any real responsibility, the woman of Empire often felt herself to be no more than an ornamental, and progressively more *un*-ornamental, supernumerary. Nobody wanted her to be too clever, still less politically concerned. Any attempt to break from the herd would damage her husband's career. All too often the climate, the society and the way of life sapped her desire to be beautiful. Male values were supreme almost everywhere in the British Empire, and even the family role, even motherhood itself, was forlornly disrupted by constant moves and partings.

How sad it was! The girls who flocked bright-eyed to India, Malta or Egypt, 'the Fishing-Fleet' as they were cruelly called, all too often returned sad-eyed and sallow, when thirty years later their husbands reluctantly retired at last, and they settled alien and out of touch at Guildford or Yelverton, all their imperial pretensions crumbled, all their memories rather a bore. No wonder, in their exiled prime, they were often supercilious and overbearing: there was purpose to their husbands' imperialism, fun too very often, but there was not much satisfaction to their own.

2

Yet out of the generalizations, a thousand brilliant exceptions spring, for if the Empire did not offer much to the average woman, it was fertile in opportunity for the maverick, the solitary, the rebel and the visionary. The imperial adventure sharpened the outlines of exceptional men, and even more did it bring into focus the gifts of remarkable women, for the very fact of their independent presence on the frontiers marked them out as special.

Some were women who, for all the restrictions of their status, actively responded to the imperial idea: some were women who passionately opposed it. It was after all a matriarch's Empire, and Victoria herself had always been able to see in it womanly terms, a gigantic family strewn around the globe for whom she was a universal Earth-Mother. It was true that she never ventured deeper into her domains than Ireland, but she was with all her dear peoples night and day in spirit, which was in many ways just as demanding as going there in person. The evangelical aspect of Empire held a true appeal for many women not only because, like Victoria, they wished to be good, but also because the mission stations of India and Africa offered them rare chances of active service in the field. Many an adventurous Englishwoman found her fulfilment looking after lepers in Bengal, teaching Dinka children in the southern Sudan, nursing sick Eskimos, vainly trying to persuade the Ashanti towards a Truer Light, or even, like the indefatigable Mrs Dorothy Brooke of Cairo, rescuing from their miseries the derelict quadrupeds of the fellahin.

Often in their pursuit of the good, they came up against the mighty. Smuts was not being altogether flattering when he called Emily Hobhouse 'the eternal woman': having been Kitchener's scourge in the Boer War—'that bloody woman'—she went on to nag Smuts to distraction towards racial enlightenment, and was one of the few people who presumed to offer moral guidance to Gandhi. Lady Anne Blunt, who raised her magnificent Arab horses almost in the shadow of the Pyramids, and was often to be seen careering across the desert in full Arab costume, was a staunch ally of her husband Wilfrid, the gadfly of Empire, and disconcerted all the Cairo hostesses by her wilful combination of radicalism and aristocracy. Who could be more aggravating to Authority than stumpy Annie Besant, henchwoman of the Mahatma, founder of the Theosophist cult, socialist, atheist, strike-leader, Indian nationalist, imprisoned for subversion but becoming in the end, in a triumph of will over circumstance, President of the Indian National Congress? Or who more infuriating than the handsome Countess Markievicz, née Gore-Booth, who was born a favoured child of the Anglo-Irish Ascendancy, admired by Edward VII, but who went on to command the Irish rebels in St Stephen's Green during the six days of the Easter Rising?

These were rebels all, but there were many more who influenced events obliquely, within the system. There were women who flatly refused to go to the Hills, and by staying with their husbands in the sweltering plains of the Indian summer, helped to shatter the myth of female uselessness and fragility. There were women like Helen Younghusband, who furiously advanced her husband's interests whatever the opposition—she was, it was said, 'inordinately proud of him, and despises the whole race of officials'. While Lutyens was building the Viceroy's house his wife Emily was organizing crêches for the women labourers on the site—the first in India. While Frederick Lugard was presiding over the destinies of Nigeria his formidable wife Flora, confidante if not fellow-conspirator of Rhodes and Jameson, was powerfully propagating his ideas to captive visitors in the Gubernatorial drawing-room.

There were women who flew aeroplanes around the Empire, like Amy Johnson. There were women who treated Empire as a holiday, like merry Sarah Wilson in her Mafeking dug-out, or as a protracted nature ramble, like the painter Marianne North, who spent fifteen years depicting the imperial flora, including five species she discovered for herself. The Irishwoman Daisy Bates deserted her husband but became the most influential friend of the Australian aborigines, while Lady Florence Dixie, daughter of the Marquess of Queensberry, was one of the earliest champions of the Zulus. Mary Slessor, raised in a Scottish slum, spent nearly forty years as a missionary on the Niger coast, and ended up as British Vice-Consul in Okoyong. Clara Butt, a 6-foot 2-inch contralto, sang 'Rule Britannia' and 'Land of Hope and Glory' with such incomparable diapason that she became, in the first quarter of the twentieth century, the very voice of Empire.[1]

---

[1] She had actually suggested 'Land of Hope and Glory' to Elgar, just as another of the great imperial anthems, Samuel Liddle's setting of 'Abide With Me', was written for her. She was made, of course, a Dame of the British Empire before she died in 1936, but her singing was remarkable, says the *Dictionary of National Biography* carefully, 'for its broad effect rather than for its artistic finesse'.

3

Several great women travellers were sponsored by the fact of Empire—great not merely because of their journeys, but because of the use they made of them. They were seldom adventurers of the absolute kind, seeking excitement for its own sake. They went to achieve something, and the danger was incidental.

The brief travels of Mary Kingsley, for instance, influenced the course of history in West Africa. She was a marvellous traveller, but an even more enterprising political theorist. A doctor's daughter, a niece of Charles Kingsley the novelist, she had no formal education at all, and until she was thirty stayed at home and helped to look after the family: but when her parents died in 1892, and she was free to pursue her own fulfilment, she sailed away alone to Africa, a late Victorian spinster, to collect zoological specimens and study ancient cultures. Her fame was established by two tremendous journeys, both completed within four years, across great slabs of forbidding country almost unknown to Europeans before—to Kabinda and Matadi on the Congo, up the Ogowe river through the perilous rapids of N'Ojele, across the cannibal Fan country, around the obscure Lake Ncovi, to the dim island of Corisco, and finally to the summit of Mungo Mah Lobeh, the Great Cameroon, 14,000 feet above the sweltering desert of the Wouri!

Even the names of her itineraries sounded scarcely lady-like, and indeed of all the imperial explorers of Africa Mary Kingsley was the most astonishing. She was the born free-lance. She had no rich institutional supporters, but paid her way by trading in palm oil and rubber as she wandered. She was not a beautiful woman, and preferred to wear, so she said herself, 'elderly housekeeper's attire'— Africans, puzzled by the effect, frequently called her 'Sir'. Many people she instantly antagonized—'I do not worship at the shrine myself', said Lugard, and the mandarins of the Colonial Office came to detest her. Many more, and especially Africans, she instantly charmed, with her zest, her courage and her long, funny face.

She was an imperialist of a particular and then unfashionable kind. At a time when Empire was presenting itself to the world as a privilege and a mission, Mary Kingsley saw it as Trade. Especially in West Africa, she thought, trade should be its *sine qua non*, and she struck up an unexpected alliance with the great Liverpool merchant houses which, having been deprived of their West African slave trade half a century before, now dealt in the cocoa and copra of Nigeria and the Gold Coast. Mary Kingsley used to boast that some of her own ancestors had been slave traders, and she had nothing but contempt for the windbag pretensions of the New Imperialism. She despised missionaries, too, loathed Little Englanders, and called herself 'a hardened, unreformed imperial expansionist'.

Hers was Rhodes' view of Empire, 'philanthropy plus five per cent', pursued with a sense of practical style, and a dignified respect, without hypocrisy or condescension, for the native civilizations. She had strong views about race. The African, she allowed, was probably inferior to the white man, but what was much more important, he was *different*. His virtues and vices were arranged in a different way, and he was no more an undeveloped white man than a rabbit was an incipient hare. The African approached life more spiritually, and if he was to be educated, he must not be forced into European modes of thought. He was not a mere cipher, to be manipulated as Empire willed. He was, Miss Kingsley provocatively thought, very likely the archetype of World Man.

Commerce was to be the bridge, by which the Empire reached the truth about these peoples. She viewed the project romantically. 'You great merchant adventurers of England,' she once reproached the sober and bourgeois business community of Liverpool, 'you great adventurers must pull yourselves together, and become a fighting force, and a governing force!' Traders were the true experts of Empire, knowing more, and having a more genuine stake in their territories than any colonial officials. Traders had no wish to change the natives, who were perfectly profitable as they were. Traders had no fancy notions, as Mary Kingsley put it, 'about the native being a man and a brother'. The trading instinct was the key to progress—to dominion, stability and power.

All this made many enemies. 'Miss Kingsley', wrote *Concord*, a pacifist magazine, 'is a very unwomanly woman. . . . Her language is tainted with the demoralization of frontier life. . . . Her politics are bad, her economics worse, and her morals, in regard to these public concerns, worst of all.' At the same time her robust sympathy with West African traditions, even down to slavery, horrified the imperial evangelists, and her championing of indirect rule, long before Lugard's conversion, did not please the Whitehall supremacists. Purist eyebrows were raised when, invited to lecture to the London School of Medicine for Women, she chose as her topic 'African Therapeutics from a Witch Doctor's Point of View'.

But she was only ahead of her time. Indirect Rule was to become an imperial orthodoxy, economic influence would one day be the respectable euphemism for imperialism, the world was presently to share her view of *négritude*, and her vastly entertaining travel books were to amuse and instruct West African administrators for the rest of the Empire. She did not live to see these consequences, for all her adventuring lasted less than a decade. When the Boer War broke out she went to South Africa to nurse sick Boer prisoners of war at Simonstown. From them she picked up enteric fever, and she was buried, as she had characteristically asked, at sea.

4

And most remarkable of all was Gertrude Bell. We have already seen her from a distance—in a fox fur and a flowered hat at the Cairo Conference of 1922—for out of her journeys she had become, not a critic of Empire, but one of its pillars. She was a more obviously brilliant woman than Mary Kingsley, but less surprising. The daughter of a celebrated Durham ironmaster, she was the first woman to win a first class degree in modern history at Oxford, and her earliest travels were proper peregrinations in the tradition of the Grand Tour, round the world once, climbing the Alps, staying with diplomatic representatives in embassies here and there. But she learnt Persian, and so acquiring a taste for the Islamic east, went to Jerusalem when she was thirty-one to learn Arabic. Soon she was deep within the Arab subject, linguistically, archaeologically, and presently politically. She travelled all over Syria, down the Euphrates to Karbala, northward through Mosul into Asia Minor, and finally, in the most ambitious of her projects, deep into the Arabian interior to the desert city of Hail.

The Great War gave purpose to these extraordinary experiences, for on the strength of them Gertrude Bell joined the Arab Bureau, then busily mobilizing British expertise about the Arabs. So she became one of the original British administrators of occupied Iraq, where as Oriental Secretary she played a key role in the enthronement of the Hashemites. For the last seven years of her life, in her fifties, she was the most powerful woman in the British Empire, and politically one of the most powerful in the world. She never married, remaining all her life faithful to the memory of her only love, a British officer killed in the war, but she never suffered the taint of the blue-stocking, for she was delightful in her emancipation. At Government meetings her opinion was generally decisive; at Arab gatherings sheikhs and even princes, anxiously peering across the assembly in the direction of Miss Bell, would await her nod or flickering eye before committing themselves to a course of conduct—

> *From Trebizon to Tripolis*
> *She rolls the Pashas flat,*
> *And tells them what to think of this,*
> *And what to think of that.*

She was full of paradox. She loved clothes and hats and textiles, but she was tough as nails, riding every morning, swimming the Tigris frequently. She was a prime agent of the policy which made Iraq a puppet-State of Great Britain, but she was genuinely concerned for the independence of the Arabs. A tall, sharp-nosed, angular woman, red-haired and green-eyed, it was the fact that she *was* a woman that made a legend of her, for temperamentally she was not of the heroic mould, and the proudest of her achievements was the creation of the Baghdad Museum—'like the British Museum, only a little smaller'. Yet briefly, perhaps luckily, a legend she became—'The Arabs', it said hopefully on the plaque they erected to her in the Museum, 'will always hold her memory in reverence'— and every traveller to Baghdad wished to set eyes on her, meet her at a party, or best of all take tea with her in her house beside the river, cluttered with books, shards, maps, flowers and new evening dresses just arrived from England.

There is an old photograph of a garden party in Baghdad which illuminates this reputation. Its central figure is King Feisal, dressed in military uniform with medal ribbons, and there are a few sheikhs in the garden, and many British officers in topees, and two or three Englishwomen in long dresses, wearing elaborate hats and carrying parasols rather like Lady Pellatt in Toronto. A faded string of flags hangs above their heads, and the party appears to be happening in a decaying botanical garden, with banked lines of flower-pots, and a tangle of palms behind.

A row of kitchen chairs and an elderly sofa have been arranged for the principal guests, and on one of them, half-turned from the camera, sits Miss Gertrude Bell the Oriental Secretary, next to the king. She is having her cigarette lighted by an attentive Englishman in a trilby hat, and she is wearing a dark shady hat with flowers on it (by Anne Marie of Sloane Street), a long pale dress with some sort of ornamental belt, and white high-heeled shoes. What is it about her that compels the eye? Why do we look at her, rather than at comfortable Mrs G, on the other side of the king, or Mrs H who stands so daintily with her husband the general on the right flank of the picture? It is not that Miss Bell is about to smoke a cigarette, not that she carries no parasol, nor even that she seems to sit in that rather bony, defiant way characteristic of progressive women of the day. It is the fact that, as everyone in that picture knows, as she undoubtedly knows better than anyone, she represents Power. In her slim cottony figure imperial authority is improbably concentrated. The plump sheikh behind looks at her rather apprehensively; the king waits politely for a light himself; and nice Mrs G, one cannot help surmising, is thinking to herself really, that woman does give herself airs.[1]

[1] But the strain of it all defeated her; she died in Baghdad in 1926, aged fifty-eight, and was buried in the British cemetery near General Maude.

CHAPTER TWENTY

Adventurers

HARD times had come for the British people, in the years of slump and unemployment: terrible times were approaching, as the world walked like a blind man once again towards the precipice of war. Fifty years before the possession of Empire had offered an element of circus, to make up for the shortage of bread, and to a minority of Englishmen still it provided the stimulus of challenge and response, keeping their minds perhaps, like healthy team games at school, off more debilitating topics.

In the past the adventure of it had been something more, and had been a principal urge towards sovereignty—

He was the first to venture, he was the first man to find!
Trusting his life to his rifle, groping ahead in the blind!
Seeking new lands for his people!—This is the end of the day,
A little mound on the mountain, a little cross in the clay. . . .[1]

Every adventurous taste was provided for then, in the boom days of Empire: the frissons of exploration, speculation, prospecting or war; the call of the pioneer's trek and the settlers' bivouac; the grand excitement of emigration itself, to a new life on the frontiers. Even the imperial proletarians, the private soldiers, often prided themselves on the challenging hardship of it all. They called the South African sun 'McCormick' but loved to boast about the heat of it;

[1] By Kingsley Fairbridge (1885–1924), a Rhodesian Rhodes Scholar whose Rhodes Emigration Society, founded when he was an undergraduate, aimed at the mass emigration of English slum children to the overseas Dominions. Its only success was a Farm School in Western Australia, where Fairbridge died, but there is now a Fairbridge Memorial College in Bulawayo, and he is counted among the heroes of early Rhodesia.

they called the North-West Frontier 'The Grim', but shared their officers' wry regard for its murderous Pathans; they absorbed affectionately into their folklore the horrific hazards of life on the Indian plains, where strong men were said to lose their potency through eating mango skins, and the noxious air of the Terai was composed of the breath of serpents.

The Empire had been a gigantic employment exchange for the adventurous. We see its solitary clients, for instance, penetrating disguised into Tibet or Turkestan, to be thrown into verminous dungeons by Khans of Bokhara, or strangled in Afghan market-places. We see them struggling over the Chinook Pass into the goldfields of the Yukon, or hastening on dromedaries towards the ambush of the Khalifa. We see Professor E. H. Palmer of Oxford murdered by Bedouin during an Arabian intelligence survey in 1882, or Bishop French of Lahore dying in Muscat after his forlorn attempt to proselytize the fanatic Omanis—'I cannot say that I have met with many thoughtful and encouraging hearers or people who want Bibles and Testaments.'

For many people the instinct of adventure was killed in the Great War. It died with Rupert Brooke and his poetry, was buried with the bitter epitaphs of the war poets, and was one with the million ghosts who haunted the schools and universities of England.

> *Who are the ones that we cannot see,*
> *Though we feel them as near as near?*
> *In chapel we felt them bend the knee,*
> *At the match one felt them cheer,*
> *In the deep still shade of the Colonnade,*
> *In the ringing quad's full light,*
> *They are laughing here, they are chaffing there,*
> *Yet never in sound or sight.*[1]

Many a Briton swore, when he returned from the trenches, that he would never go abroad again, let alone risk his life in any cause.

---

[1] By E. W. Hornung (1866–1921), the creator of that archetypal gentleman-cracksman, Raffles. I was sent the poem by Mr M. A. Nicholson of Eton, who wonders, from the evidence of such writing, if the stiff upper lip of the public schools was ever so stiff after all.

Security was the national aim, preferably without exertion, and thought it proved an illusory hope, still the average Englishman disregarded the imperial challenge now, and preferred to make the best of things at home.

Yet adventure was in some sense the deepest truth of Empire. The craving for excitement, the yearning to break out, was perhaps the profoundest of all the imperialist motives. It was not dead yet, and throughout these decades one may observe it still, held in trust so to speak, or latent, until history once again required it.

2

Some Englishmen still went to be pioneers. They were seldom pioneers in the rawest kind, for most of the Empire's arable land had been exploited now, and everywhere the reach of Government had brought order and restraint to the frontiers. There were plenty of communities, though, which retained some of the old fizz, in the second or third generation of their citizens, where newcomers still felt themselves to be frontiersmen, and looked back with relief, sometimes condescension, to more ordinary places left behind.

In the spring of 1927, for example, a group of 300 British families, escaping the hunger-strikes and dole queues of England, emigrated with Government help to Alberta, and so found themselves in an authentic frontier-town of the British Empire, Calgary. It was a cow-town, as in the movies, and by the standards of the emigrants, inconceivably remote, being three days by train from Toronto. Just getting there was an adventure. In summer the journey could be dramatic enough, when the sun probed relentlessly through the chinks of the window-blinds, and made the waiters sweat in their high-buttoned jackets, but in the winter it was terrific. When the station-masters at the prairie halts put on their astrakhan hats and fur coats, when the snow lay feet deep through the forests, and the conifers creaked and drooped with the weight of it, when the fish lay embalmed in their frozen lakes, and a man could get frost-bitten crossing the village street—when the ice-grey skies of winter, like gun-metal, lay glowering and magnificent over the prairies, then the newcomer from Salford, Hull or Manchester felt he was venturing indeed.

'Calgary coming up! Calgary ten minutes!'—and there it was now, the first dirt roads running into the outskirts of town, the first scattered houses clamped against the cold, a clutter of sheds and marshalling yards, a hiss of the brakes, a shudder, and there already at the door was the conductor with his little wooden steps, accepting gratuities. At once you were in the middle of town, for like most imperial frontier settlements Calgary was built around the railway station. It was very thin on the ground—just a huddle of dowdy buildings around the tracks, with the limitless prairie everywhere beyond—and even in 1927 it was still in Indian country. There were people alive who had signed Treaty No 7 with the local tribes, and the Blackfoot, Blood, Piegan and Sarcees, in their reservation down the Sarcee Trail, were still paid five dollars a head annually in Government stipend, 'for as long as the sun shines, the grass grows and the rivers run'. There was no suburbia to Calgary. There it stood, take it or leave it, with the prairie all around, the Indians down the road, and the railroad running on to the Rockies and the Pacific.

It was still a very British frontier town, scarcely alarming to our apprehensive newcomers. Life was brash by Canadian standards, but only in a boyish and endearing way. There was a red light district beyond Centre Street, one heard, where the cowhands found their comforts, and there were a few discreetly unprosecuted gambling joints, one was led to believe, in the little Chinatown towards the river: but for the most part, like many twentieth-century imperial adventures, the adventure of Calgary was fundamentally *wholesome*. The crooked policeman was unknown there, the roads were safe for all, and there had not been an illegal still in those parts since the Mounties cleared up the border bad-men half a century before.

Why, even the Prince of Wales had bought a ranch nearby, and visitors with posher credentials than ours could be sure of a genteel welcome. The grand families of Calgary, the Hulls, the Lougheeds, the Burns lived in enfilade well south of the railway tracks, training their guns upon each other and upon visiting celebrities. They had built themselves elaborate villas on the prairie edge, with verandahs and dormer windows and scalloped eaves, and sometimes they flew Union Jacks largely on their lawns. Old Country cricket was regularly reported in the *Calgary Herald*, the Palliser Hotel had often been host to Dukes and Senior Officers, the Ranchmen's Club was all cigars and mahogany, the schools faithfully honoured Empire Day.

Yet even Lady Lougheed was proud to boast herself a frontiers-woman—did not the family firm of Lougheed and Taylor announce themselves on their office door as 'Western Pioneers'? The great event of the Calgary year was the annual Stampede, when ranchers and stockmen, breeders and cowboys came to town from all over, but some of the swagger of that event lasted all year through. Calgary was full of characters, remarkable to immigrant eyes. Stylish rich ranchers, many of them American-born, strolled about in wide Stetsons and elegant boots, and gave a patrician glamour to the Ranchmen's Club. Wild itinerant gypsies drifted in, to camp with their horses and trailers beside the railway tracks. Blanketed Indians, still be-feathered sometimes, hung about town at every intersection. Gloomy Hutterite Anabaptists, in cotton blankets and black Ukrainian hats, came shopping from their prairie communes.[1] Often cowhands clattered ostentatiously through town, and there were Russian horse-buyers sometimes, and prospectors of varying ambition, and the usual imperial assortment of remittance-men, drop-outs and loiterers.

It *was* an adventure. Calgary was not a substitute for older societies, but an alternative: it was loyal to the Empire, none loyaller, but it offered a true rebirth for people of the British stock. Our 300 families did not find it easy living. The great depression hit Alberta too, and between 1920 and 1930 not a single new office building was erected in Calgary. Life was demanding still in the Canadian west, and now and then, one may imagine, the new settlers pined for the hugger-mugger of a back-to-back terrace, or wished they could chalk a bitter up again on the slate of the Red Lion. They mostly prospered in the end, though. The Calgary instinct for success, if it

---

[1] Where they were governed by an all-male council of functionaries, including the Hog-Boss, the Geese-Boss and the Bee-Keeper.

was dashed one day, was likely to be boosted the next. Every night, when our homesick Britons went to bed, they could see a flickering blaze in the northern sky. It lit up the whole prairie, like a violent aurora, and was to remain for many of them the most pervading memory of their whole adventure, so different was it from anything at home, so strange, so theatrical. It was the glow of the burning gases from the Turner Valley oilfields, the first of the Alberta oil strikes, and it hung there like a banner over Calgary, an earnest of gambler's luck.[1]

3

This was adventure of an organic kind, a last excitement of the migratory urge. Another old thrill of Empire, still fitfully available in the twentieth century, was the thrill of exploration. The Empire had been the greatest of explorers' patrons. Under its auspices half the world had been penetrated, surveyed, mapped and tabulated, sometimes for strategic or economic reasons, sometimes in the interests of pure science. The islands of the south seas, the sources of the Nile, the Himalayan massif, the Canadian north, the Australian outback—all these wild or inaccessible places had been brought within the cognizance of western man by the agency of the British Empire.

There were still a few regions to be explored in the imperial interest, and a handful of men seized their chances still. Their motives, like their temperaments, greatly varied, for the imperial purpose was itself diffuse now, and men could read into it what they wished. F. M. Bailey of Tibet, for instance, bespoke all that was most fun about the great adventure. He was a soldier, the son of a soldier, and he first went to Tibet with Younghusband in 1904— seven years later, when the Dalai Lama escaped from the Chinese revolution into India, it was Bailey who hit upon the plan of disguising him as a *dak wallah*, a postal runner, actually giving him

[1] And paid off in the end, Calgary now being, with a population approaching half a million, one of the most prosperous of the Canadian cities. Its shape is still recognizably the same, the Indians still live on their reservation, the Hutterites wander through, the Palliser and the Ranchmen's Club prosper.

the mail-bags, and so smuggling him through the Chinese frontier posts.[1] Thereafter Bailey lived a life of magnificent hazard. He learnt Tibetan, his examiner being Sarat Chandra Das, the Bengali agent who was the original of Hurry Chunder Mokerjee in Kipling's *Kim*, and after several more journeys within the forbidden country, feeding intelligence to the Indian Government, he evolved a grand ambition: to solve the mystery of the Tsangpo gorges.

Nobody then knew how the Tsangpo, the greatest river of Tibet, flowed through the north-eastern part of the country to become the Brahmaputra. Bailey prepared for years to find out. He perfected his colloquial Tibetan, he read every available book, and in 1913 he and an equally imperturbable colleague, Captain H. T. Morshead, secretly set off up the Dihang river, over the Tibetan frontier towards the ravines. Ostensibly they were defying their superiors' prohibitions, actually they were spies.

They succeeded triumphantly. Not only did they solve the problems of the river itself, but they mapped large parts of the Tibetan frontier for the first time, they produced a detailed report of the border peoples, they cracked the esoteric riddle of the Eared Pheasant, and they discovered the blue Himalayan poppy, *Meconopsis betonicifolia baileyi*, which was to augment the incomes of seedsmen all over the world. Bailey went on to still more exciting adventures, as a secret agent in Central Asia during the Great War, and most melodramatically of all as a spy against the Bolsheviks in Russian Turkestan after the revolution. But it was the great Tibetan journey which was his triumph, and would be remembered as one of the last great strokes of imperial skulduggery.

Everything Bailey did was vivacious, daring, and in the profoundest way innocent. Sometimes he disguised himself as a Buddhist monk, with a compass hidden in his prayer wheel—'the usual secret

[1] The late Lord Cawdor, the 5th Earl, sent me this story—'the only time on record when His Majesty's mails were carried by an Incarnate God'. Cawdor was an Indian adventurer himself, an old friend of Bailey's, and vividly recalled for me the reception such young bravos got from the Burra Memsahib of long ago, when they returned to the cantonment from jungle, frontier or indigo plantation. Where had they come from, she would ask, surveying them through, 'or more probably over', her lorgnettes, and then, speculatively, '*You were staying at Government House?*'

service agent's equipment', as he casually observed. Sometimes, menaced by murderous tribespeople, threatened by avalanches, pursued by furious messengers from Calcutta, he would pause to observe the habits of a hitherto unknown shrew (*Soriculus baileyi*, no doubt), or take a cutting of a new rhododendron (*Rhododendron baileyi*, for instance). Once, swept away by a sudden snowslide, he saved himself only by the skilful handling of his butterfly net. Nothing daunted him, or suppressed his dry humour.[1] When he got home from Central Asia he married the daughter of a peer and retired to Norfolk, writing accounts of his great adventures, and slowly growing, as he aged, more and more like a Tibetan sage himself. (As for his friend Morshead, the happy companion of his great Tibetan journey, he was, so Bailey blandly recorded in his memoirs, 'taking a peaceful ride one morning in Burma when he was murdered'.)[2]

At another extreme St John Philby of Arabia expressed a darker side of the imperial spirit, its complexes and neuroses. He was the son of a colonial philanderer (and became the father of a Communist spy), and he spent his life restlessly betwixt and between: between patriotism and rejection, between anarchy and authority, between duty and impulse, between pathetic uncertainty and overwhelming pride, between Islam and Christianity, between the spell of Arabia and the loyalties of Westminster and Cambridge, between one wife and several mistresses. In many ways Empire was his whipping-boy, for upon it he unloaded his burden of bitterness, and in reaction to it he found his fulfilment.

He began his life, a salt fish in fresh water, as a member of the ICS, but the Great War took him to Iraq as an official of the occupying Government there. He quarrelled at once with nearly everybody

[1] Which ran in the family. WARN BAILEY MASSACRE SADIYA, said a laconic cable from his father, delivered to him one day by runner from Chengtu —an unsatisfactory message, as Bailey commented, 'because it failed to say who had massacred whom, and why'.

[2] By the time I met Bailey in 1958, nine years before his death, it seemed to me that he had been physically Tibetanized by his experiences, for his cheek-bones were high, his eyes were slightly slanted, his skin was like brown parchment, and he even moved, it seems to me in retrospect, in an indefinably remote or monkish way.

because he disapproved of the imperial policy, believing that an independent Iraqi Republic should be established; and though he later went on to serve the Emir Abdullah in Transjordan, he was never reconciled to the British presence among the Arabs, or the alliance with the Hashemites. Instead he presently devoted himself to the cause of the Saudis, the Hashemites' most potent rivals, and became the greatest of all foreign travellers in their kingdom.

Philby had first visited the Arabian peninsula as a British civil servant, on a mission to Ibn Saud, but so enthralled was he by the personality of the king, so bewitched by the empty grandeur of the country, that he eventually left the imperial service and went to live there as a private citizen. He earned his living as the Ford agent in Saudi Arabia, but devoted his genius to the exploration of the desert. He was the first man to cross the peninsula west to east, the first thoroughly to explore its inner fastnesses. He mapped it all meticulously, kept exhaustive notes, and sent home to London box upon box of specimens, botanical, zoological, archaeological, geological—a colossal memorial to his own professionalism.

Yet he was never a happy man. He was never fulfilled by adventure itself, as men like Bailey were, and adventure offered him no simple purpose, as it gave the Calgary pioneers. He was a very clever man, but querulous: it was his misfortune to have started life as the servant of an Empire no longer sure of itself, for his career was one long unresolved squabble, not least with himself. He worshipped King Saud, an absolute monarch, with an adolescent fervour, yet came to loathe the British Empire for its despotism. He adopted Islam out of crass opportunism, yet constantly accused the Empire of hypocrisy. He was a revolutionary at the foot of the throne, a palace radical.

The older he got, the more Philby disliked the British, and vice versa, and the more alien he became to everything their Empire represented. All too often he was proved right, and he looks back at us from his many photographs, dressed generally in the full Arab mode, with a look of told-you-so: a foxy, conceited look, with a touch of sneer to it. Philby disassociated himself from the Empire, preferring private adventures to public: but no less eagerly the Empire disassociated itself from him, and this was perhaps because its officials recognized in him some awkward reflections of their own self-doubts, and some disagreeable truths about dominion.[1]

For self-doubt, self-recognition had now become part of the imperial condition, and not least among the adventurers. The cause was hazier now, they had no sense of civilizing mission, and their explorations were often, in the age of Freud and Jung, inner journeyings. Among the most introspective of the late imperial explorers was Henry George Watkins, 'Gino', who was born in 1907 and died in his twenty-fifth year. Watkins' most celebrated exploit, his journey to Greenland in 1930, did indeed have a public purpose, but he really did it for his own reasons, to test himself, perhaps to compensate himself for missing the Great War. He seemed to his contemporaries the beau-ideal of the professional explorer. He was handsome not in an English kind, but Nordically, fair, sharp-featured, symmetrical, and he prepared himself with an un-English thoroughness for a life of physical endurance. As an undergraduate he climbed in the Alps, learned to fly, and led an expedition to Spitzbergen. Then he spent a year in Canada, exploring and surveying among the fur-trappers of the north. He was very hard and lean, very demanding of his comrades, and we are told he used to *run* home through the streets of London after all-night parties.

He was a man made for war, condemned to live in a brief period of peace. In 1930, though, a perfect opportunity arose for him. The Empire Airship Scheme was about to be launched by the R101, and one of its destinations was to be Canada. It was suggested that the best route for the airships might be not across the 2,000-mile expanse of the North Atlantic, but rather by the Great Circle route to the north, by way of Iceland, Greenland and Baffin Island: it was shorter, and it was said that the weather, to which the great

---

[1] Philby was imprisoned in England during the Second World War for his anti-British attitudes, and in the end he was banished from Saudi Arabia too. He died in Beirut in 1960, and is buried in the Muslim cemetery in the Basta quarter of the city. His son Kim, before defecting to Russia three years later, wrote a just epitaph for his tombstone—*Greatest of Arabian explorers*—but the fifteenth edition of the Encyclopædia Britannica gives three times as much space to the son as it does to the father.

dirigibles were dangerously susceptible, was more stable. The least-known part of the route was the interior of Greenland, and an expedition was mounted to survey it and to record its weather conditions. The Royal Geographical Society and the British Government gave the expedition its support: 'Gino' Watkins, twenty-three years old, flyer, explorer, surveyor, leader-figure, was its ideal commander.

It was his one great achievement in life. He never married, and he was to die on a second visit to Greenland, two years later, when his kayak overturned in an icy fjord. The Arctic Air Route Expedition made a hero of him, and perhaps in a tragic way gave his life a richer meaning than most. Within its limits it was a perfect adventure, for it had a practical purpose but was attended by a certain symbolic brilliance. It was a young adventure, an adventure of young friends, concerned with modern matters, and led by this dazzling young Cambridge prodigy.

Watkins believed in living off the country. Though motor-boats and aircraft were used too, he proposed to travel across the Greenland ice-cap eskimo-style. He learnt all about dogs and sledges, and became a skilful handler of kayaks. He became a skilful handler of Eskimos, too, enjoying a succession of mistresses in the expedition's base camp on the coast, and indeed his whole approach to adventure was functional, realist and amoral. He regarded himself as a machine to be tested, and he treated his expedition as a series of technical experiments, dispassionately conducted, with danger itself accepted as a scientific quantity.

They established a meteorological station in the mountainous interior of Greenland, planning to relieve it every month from their base camp on the coast, with the object of recording a full year's weather. The weather itself was so terrible, though, with hurricane winds and visibility down to a few yards, that one member of the expedition, Augustine Courtauld, volunteered to stay at the station all alone from December to May. Long before he was relieved his hut was buried deep beneath the snow, and he had run out of fuel for heating or cooking: when the relief team arrived in the summer they could at first find no sign of him at all, only a slight mound in the snow where the station ought to be.

He survived in perfect health, but Watkins welcomed him back to base without surprise or emotion. His ordeal alone on the ice-cap were no more than an interesting trial of stamina and psychology: it was also very economical in manpower. Besides, was not the very discomfort of it a sort of satisfaction? On a plaque, painted in gold letters above the door of his hut, Watkins' philosophy of adventure was declared in a verse by John Masefield, as true a Poet Laureate for the declining Empire as Tennyson had been for its materialist prime:

> *The Power of Man is in his hopes;*
> *In darkest night the cocks are crowing;*
> *In the sea roaring and the wind blowing*
> *Adventure—man the Ropes!*

4

The most emblematic adventure, in these years of contrived excitement, was the protracted effort to reach the summit of Mount Everest, the highest mountain in the world. The Himalayan frontier was the most evocative of them all, and generations of Britons knew its seduction: the long heavenly treks through the rhododendrons, the myriad cuckoos of the foothills, the amiable slant-eyed indigenes, the dizzy rope bridges over unfathomable gulfs, the wood-fire smells and the roast potatoes, the heady local liquors, the obliging local monks, the flap of the prayer-flag and the creak of the holy wheel—and always beyond the trees and foothills, the celestial rampart of the snow-peaks.

Tremendously at the centre of these imperial delights stood Mount Everest, recognizably the greatest of mountains, from whose summit there habitually flew, high above the Tibetan frontier, a plume of white driven snow, like a defiance. For thirty years the British were the only climbers to attempt this ultimate peak, and their monopoly was purely a gift of Empire. Everest was a British preserve because it stood on the limits of the Indian Empire, the border between Nepal and Tibet passing across its summit, and in later years it became almost an Empire-substitute in itself, providing successions of sahibs, with their attendant bearers, a late adventure in the old style.

Everything about the Everest saga was imperial. The mountain was named, in 1849, for Sir George Everest, former Surveyor-General of India.[1] The first man to think seriously about climbing it was an Indian Army officer named Charles Bruce, who had travelled all over the Himalaya on military business and viewed the mountain with the speculative eye of the Great Game.[2] The first man to sponsor an expedition was Curzon—the plan was squashed by Whitehall, on the grounds that it might upset the Russians. The first man to get near the mountain was another Army officer, John Noel, who entered Tibet in disguise in 1913 and penetrated within forty miles of it.[3] The first President of the Everest Committee was Younghusband of Tibet. No wonder it was generally assumed that there were strategic motives behind the British affair with Mount Everest: in those days there certainly were, for either route to the mountain, from the Tibetan or the Nepalese flank, would take an expedition into sensitive frontier country never yet seen by Europeans.

The first full-scale reconniassance, in 1921, was indeed accompanied by a military survey team, led by Morshead, but the climbing parties that followed really were pure sport. They were the imperial impulse atrophied, in fact, or sublimated, for they had no purpose but themselves. Even the scientific usefulness of the Everest adventure, though it was repeatedly stressed for propriety's

[1] Before that it was called, more romantically I think, Peak XV. Modernists nowadays, including the Russians, like to call it by its Tibetan name, Chomolungma, which means either Goddess Mother of the World, or Lady Cow.

[2] The son of the great South Wales coalowner Lord Aberdare, Bruce went on to lead, in his late fifties, the first two climbing expeditions to Everest. He died in 1939.

[3] Noel went back to Everest too—his cinema film of the 1922 expedition was the first high-altitude film ever made—and fifty years later was still lecturing about it from his home in Romney Marsh, Kent. In 1927 he advanced a plan for lowering a man on a rope from an aeroplane to the summit of the mountain, whence he would find his own way to the bottom: he would need, as Noel convincingly observed, 'the stoutest pluck'.

sake, was really vestigial. 'I think the immense act', wrote H. W. Tilman more frankly of the 1938 expedition, quoting Chesterton, 'has something to it human and excusable: and when I endeavour to analyse the reason of this feeling I find it to lie, not in the fact that the thing was big or bold or successful, but in the fact that the thing was perfectly useless to everybody, including the person who did it.' The great adventurers of the Victorian Empire, who had performed *their* immense acts in the causes of progress and power, would have squirmed in their graves to read it.

On the face of things the Everesters (as they called themselves) seemed the very type of the Empire-builder: Englishmen of the upper middle classes, Furse men, who went to Everest in a spirit of uncomplicated bravado. All the Everest expeditions between 1921 and 1938 reached the mountain through Tibet, by permission of the now restored and compliant Dalai Lama, and so it happened that six times in the post-war years jolly groups of Englishmen crossed the Tibetan frontier by Younghusband's route, up the Chumbi valley along his old telegraph wire, and made their way across the wide gravelly steppe that lay beyond the Himalaya. The Everesters became well known along this esoteric route. A whole generation of Tibetan villagers grew up knowing them, and the great monastery of Rongbuk, almost in the northern shadow of Everest, became in the climbing season almost a British club, where the sahibs unbent themselves in the congenial company of the monks, and wrote up their journals beneath the flapping prayer-flags.

There they sprawl now in the statutory group photograph, 1921, 1922, 1924, 1933, 1936, 1938, among the tumbled scree of Base Camp or the shambled shrines of Rongbuk, dressed in a cheerful variety of climbing clothes, some casual, some exhibitionist, with school scarves, and old Army jerkins, and Tibetan boots, some in flapped astrakhan hats, some in caps bought long ago in Tyrol or Chamonix—bearded and dirty from the long march, but essentially clean-limbed, clean-living young Englishmen of the post-war generation, speaking a common language, sharing accepted values.

They were not of course really as straightforward as that. That was the projection. Some in fact were unhappy men, working out their disappointments, some were running away from more complex challenges, some were very conceited, some fiercely ambitious to be famous. Some were neurotics, some poseurs. One disappointed his comrades by packing his bags and disappearing the moment their expedition seemed to have failed—not, they thought, the proper team spirit. One presently took to running nude through the lounges of respectable Swiss hotels. This is how George Leigh Mallory, who died on Everest in 1924, wrote home about the adventure: 'I sometimes think of this expedition as a fraud from beginning to end invented by the wild enthusiasm of one man—Younghusband, puffed up by the would-be wisdom of certain pundits in the Alpine Club, and imposed upon the youthful ardour of your humble servant.'

Mallory, the author of that sublime bathos 'Because it's there', was variously described by his contemporaries and climbing colleagues as a 'stout-hearted baby', an 'unimaginable English boy', and 'the magical spirit of youth personified'—'George Mallory!' quivered Lytton Strachey. 'My hand trembles, my heart palpitates, my whole being swoons away at the words—oh heavens! heavens!' Howard Somervell, who reached 28,000 feet on Everest in 1924, stayed in India for twenty-two years as a mission doctor in Travancore, abandoning his career as a London consultant, becoming an admirer of Gandhi and an advocate of Love as the ultimate political force. Hugh Boustead, who went in 1933, was the first man to be pardoned for deserting the Royal Navy, having jumped ship to join the army in search of action in the Great War. Three Everesters became university professors, three became generals, one became a Conservative MP and one Governor of the Gambia.

So they were very varied men really, individualists, often of striking and unexpected talent. To the monks of Rongbuk, nevertheless, they doubtless looked all of a piece, and to the public at home too, which watched the successive Everest attempts with a less than hysterical interest, the Everesters seemed a homogeneous company. They were gentlemen-adventurers of the imperial frontiers. Their doings were reported in despatches exclusive to *The Times*. Their leaders were generals, or landed gentry. They talked Oxford English, and they liked to come down from the mountain, if possible, to champagne and chicken in aspic—or alternatively, to porridge and a cold bath. They were sahibs—a Party of Sahibs, as the Tibetans rightly said—and as such they were figures of Empire, a type becoming more remote, more misty, a little more risible as the decades drew on, but still recognizable as a category of Briton, lounging gregariously beneath the flag on meadow, scree or ice-flow.[1]

5

So long as the Raj lasted in India, they never succeeded in climbing Everest, but that was not the point.

> *For when the One Great Scorer comes*
> *To write against your name,*
> *He marks—not that you won or lost—*
> *But that you played the game.*

More than ever now, perhaps, when success no longer seemed a national privilege, the British cherished their heroic failures. Of all the Great War battles, the all-but-victory of Gallipoli most bewitched the British imagination, and of all the imperial adventurers, none stirred the British memory like Captain Robert Falcon Scott, courageous always, gentlemanly to the last, devoted, diligent, second to the South Pole, and dead.[2]

Perhaps it was the elegiac instinct: as though the British sensed that ahead of them lay one last great adventure of another kind,

[1] Some sixty Britons went to Everest with these expeditions. Four more flew over Everest in 1933, and one went to the mountain alone—Maurice Wilson, who believed that moral and physical asceticism could conquer everything, and perished all alone on Everest, thirty-seven years old and undeterred to the last, in 1934. Thirteen men died in all, four of them British.

[2] Since 1912, when the Norwegian Roald Amundsen beat him to the Pole by a month, and Scott with five companions failed to return to base. The most moving memorial to this epic, I think, is that erected by brother-officers to Lieutenant Henry Bowers, Royal Indian Marine, who died with Scott. It stands in St Thomas's Cathedral in Bombay, and shows in white marble bas-relief his cairn and cross in Antarctica—while all around the Bombay traffic rumbles through the jalousies, birds shriek and chatter under the eaves, and the dust of India hangs heavy on the sun-shafts across the nave.

unwanted, unprofitable but magnificent, in which they would sacrifice not simply their lives, but their greatness too. In 1935 T. E. Lawrence, having withdrawn into pseudo-anonymity as an aircraftsman in the Royal Air Force, riding his motorcycle fast through Dorset, crashed and killed himself. A rumour ran through the country then, an Arthurian whisper, that he was not killed at all, but that, spirited away by Authority to secret duties, he would reappear when the moment came, when Drake's Drum beat perhaps, to lead the British once more triumphantly against their enemies.[1]

For if the spirit of Empire flickered dimly among the British now, the sense of destiny lay there still. This was a people bred to great things, after all, feeling itself a great nation still, equipped socially and historically for high enterprise. The Britain of the 1930s, timid and insular as it seems in retrospect, did not feel itself to be provincial. On the contrary, most Englishmen, of all classes, believed Piccadilly Circus to be the centre of the world, and assumed as a matter of course that in any new historical movement, any change in the world's direction, the British would play a leading part. There was a part of the British instinct which hungered still for challenge: the collective unconscious, to use Carl Jung's recently fashionable phrase, seemed to demand an element of trial or sacrifice, so that the people were at their best in times of hardship. The imperial adventurers of the post-war decades may have seemed anachronistic, compensating for the tremendous past in often unnecessary derring-do, but they were perhaps more representative than they seemed, and there were ordinary working men in England then who looked back to the miseries of the Somme itself as the richest period of their lives.

The British were no longer an aggressive or even an expansionist people. The imperial urge was over, and they no longer felt it to be their duty, still less their privilege, to be the arbiters of the world. But they were still intensely patriotic, convinced that on the whole

---

[1] It was not so, as far as I know. Lawrence certainly appears to lie buried in the village graveyard at Moreton, near the scene of the accident, where his gravestone describes him simply as 'Fellow of All Souls College, Oxford'. I happened to go there on the anniversary of his death, in 1976, and the grave was covered with flowers.

their ways were superior to foreign ways, and still, as a rule of thumb, dedicated to Fair Play as the British approximation of the Sermon on the Mount.

If it was their weakness diplomatically, it was their truest moral strength: and so it led them in the end into the noblest of all their adventures, the most Pyrrhic of their victories, the last war of the British Empire.

PART THREE
Farewell the Trumpets
1939–1965

CHAPTER TWENTY-ONE

The Last War

HITLER went to war, wrote King George VI of England in his diary in September 1939, 'with the knowledge that the whole might of the British Empire would be against him'. Fair Play was at stake, besides much else. As the good and simple monarch added, the British people were 'resolved to fight until Liberty and Justice are once again safe in the World'. Political theorists would argue that the Second World War had economic causes, or sprang from the inequities of the Treaty of Versailles. To most Britons, though, as to their King, it was started by an evil man in pursuance of wicked ends, and the rightness of the cause was never in doubt. Hitler and his allies were Bad: in this context at least Britain and her Empire were Good.

So the Last War assumed, for the British, a heroic quality. They alone fought the three great enemies—Germany, Japan and Italy—first to last. They alone held the breach in the dangerous months after the fall of France in 1940. Once again, and for the last time, their fleets and armies fought across the world, and the imperial strongholds from Bermuda to Hong Kong stood to their arms. The Empire did not always fight well. There were sorry defeats, timid failures, constant muddles and recriminations. Generals were all too often in the Great War mould.[1] Still on the whole it was a grand performance in a noble purpose, a swan song of some splendour, and a worthy last display of the imperial scale and brotherhood. 'Alone at last!' ran the caption in a 1940 cartoon by David Low, and went

[1] At least in the early years: by the end Generals Alexander, Slim and Montgomery had all succeeded to a remarkable degree in enlisting the confidence of their troops, and Montgomery's name even went into the Australian language—'it's a monty' meant it was a certainty.

431

on to answer itself, for there behind the resolute Tommy, steel-helmeted on the Dover cliffs, extended in endless line of march the soldiers of the overseas Empire, massed once more. A cherished cable from the Caribbean was received in Whitehall that same summer. 'Carry on Britain!' it said. 'Barbados is behind you!'

2

This time it was unmistakably an imperial war. Hitler himself professed to have no quarrel with the British Empire, and would certainly have preferred the British as allies rather than as enemies. Like so many foreigners, he was seduced by history. He thought of the Empire as it had been, as history and legend made it still. He conceived of it as aristocratic, ubiquitous, imperturbable, immensely experienced. He admired the supposed consistency of British policy —'foreign policy become a tradition'—and the latent brutality and toughness that was, he said, common both to the national leaders of England and to the mass of its peoples. 'I do not wish the crown of the British Empire to lose any of its pearls, for that would be a catastrophe to mankind.' Common-sense, he said, suggested a free hand for Germany in Europe, for Britain elsewhere. Otherwise a great Empire would be utterly destroyed, 'an Empire which it was never my intention to harm'.

But the British Empire, when once the die was cast, would have none of him. Mackenzie King, the Prime Minister of Canada, warned him that if he ever turned against England 'there would be great numbers of Canadians anxious to swim the Atlantic', and so it proved. 'In that dark, terrific, and also glorious hour', wrote Winston Churchill in retrospect of 1940, 'we received from all parts of His Majesty's Dominions, from the greatest to the smallest. . . the assurance that we would all go down or come through together.' This was, besides being not absolutely true, by no means a foregone conclusion, for by 1939 the Empire was far more diffuse than it had been in 1914. Then the King had declared war, in a single proclamation, on behalf of all his peoples. Now he was several kings in one, and there was no formal alliance among his separate Dominions, no constitutional obligation to go to war at all.

Ireland indeed never did, and was represented in Germany throughout the war by an Ambassador accredited in the name of King George. The Australians and New Zealanders declared war at once, soon sending almost all their trained soldiers to the other side of the world, but the South Africans decided only after bitter Parliamentary debates which side they wanted to be on, while the Canadians declared war on Germany after Great Britain, on Japan before Great Britain, and never declared war on Bulgaria at all. As for the 400 million people of India, they were committed to war by the sole word of their Viceroy, a Scottish nobleman, who consulted no Indians on the matter.

Still, as the King said, 'the whole might of the Empire' was presently mustered—even the Catholic Irish volunteered in their thousands for the British Army. The British had no expansionist war aims this time. They hoped the Empire and Commonwealth would survive the struggle intact, but no more. They coveted no new provinces. Their purposes were as altruistic as political purposes ever are, sharpened perhaps by a sense of shame about the indecisive years between the wars. If they did not go into battle in the sacramental spirit of 1914, they had apparently not inherited the disillusionment of 1918. Few of the old imperialist issues were at stake now—the British were no longer Thinking Imperially, as Joe Chamberlain had urged them to—but many of the imperial impulses found a fresh fulfilment. That old yearning for excitement came into its own once again, and helped the people through the ordeals of blitz, battle and separation. The aggressive instinct resurfaced, legitimized by war. Racy characters of the imperial legend reappeared from clubs or offices, to rediscover themselves in commando raids, parachute drops or weird prodigies of intelligence. The nation came to life again, as it had in the ebullient years of the New Imperialism, in a last flare of the commanding spirit.

This time the very existence of the Empire was at stake. Germany might have no designs upon it, but Japan and Italy certainly did. The Japanese envisaged a Greater East Asia Co-Prosperity Sphere embracing all the British eastern territories, India to Hong Kong, and perhaps Australia and New Zealand too. The Italians let it be known that they wanted Malta and Cyprus as Italian possessions,

Egypt, Iraq and the Sudan as Italian protectorates, and Gibraltar as an international port.

At the same time Britain's chief foreign allies in the struggle were hardly less dedicated to the redistribution of the Empire. Nations seldom go to war for old times' sake, and in the Second World War the British Empire was fighting alongside allies whose ideologies were specifically hostile to the imperial idea. The United States had actually come into being in reaction to Empire, and still earnestly supported the principle of self-determination for all peoples, while the Soviet Union equated Imperialism with Capitalism as surely as the muscular Christians of the previous century had associated cleanliness with God. It was plain that when the war against Hitler's Axis was over, these ambiguous associates would see to it that the British imperial era was ended. Britain would obviously be bled white by the war, and had no chance of remaining one of the great Powers, when the world sorted itself out again after the battle. Even the staunch Dominions, loyally though they sprang yet again to arms, could hardly suppose that the British Empire would survive another such reshuffling of interests, and subtly shifted their attitudes and allegiances as the war progressed.

But the British put it all out of their minds, and fought on, once more, as though they meant it. Theatrical arts had always been essential to their imperial methods, and in this their last exhibition they excelled themselves. Their initial defeat, their expulsion from the European mainland in 1940, they presented as the triumph of Dunkirk. Their initial success, the Battle of Britain later that year, they shamelessly exploited and exaggerated. Hitler thought their war propaganda 'wonderful', and by their skill in it, by the magic of their historical reputation, they convinced half the world that they were uniquely brave in adversity, subtlest in perception, most unanimous in patriotism. The BBC was much the most trusted of the wartime broadcasting services, and the legend of the British patrician style, eccentric and assured, was assiduously cherished— handsome Lord Lovat at the head of his Lovat Scouts, General Carton de Wiart with his arm-sling and his black patch, or the astonishing Admiral Sir Walter Cowan, who had won the DSO with Kitchener in the Sudan in 1898, and got a bar to it when he was seventy-three for his services with the Royal Marine Commandos.

And persistently it was as an Empire that the British presented themselves. In fact the possession of the Empire vastly increased their burdens and anxieties, but they interpreted it as a mighty asset, so that it never occurred to their own people, whatever their enemies sometimes thought, that on balance they might be safer without it. On the face of it, as David Low had suggested, it was an enormous reservoir of manpower and materials. More than 5 million fighting troops were raised by the British Empire, and there was hardly a campaign in which imperial troops did not play a part, sometimes a predominant part. Australians and New Zealanders fought in North Africa, Italy, the Far East, the Pacific. South Africans fought in North and East Africa and Italy. Canadians provided half the front-line defence of England in 1940, a quarter of the pilots of the RAF, and a sizeable proportion of the invasion force that went back to the European continent in 1943. Indians, forming the largest volunteer army in history, fought almost everywhere, and volunteers from the remotest and most insignificant of the imperial possessions, from Ascension Island and the Seychelles, from the Falklands and British Guiana, somehow found their way across the oceans to the British forces, the flashes sewn to their uniform tunics bringing an uplifting suggestion of far-flung brotherhood to the wartime London scene. Thousands of Englishmen were trained to fly in Canada. Jan Smuts, twenty years on and Prime Minister of South Africa again, resumed his place in the imperial councils, and this time became a Field Marshal too.

At their head, from the summer of 1940, stood that incorrigible imperialist romantic, Winston Churchill. His view of Empire was flexible, but he played the old theme unblushingly. If Britain fell, he said, he would continue the war from Canada, and thither the fleets and armies of the Crown would sail with him—'I have not become the First Minister of the Crown in order to preside over the liquidation of the British Empire!'[1] When the apes of Gibraltar, traditionally the custodians of British rule, seemed likely to die out, Churchill personally saw to it that new stock was brought in from

[1] Though King George himself had decided to stay in England whatever happened.

Morocco, and his speeches and messages reverberated with the imperial idiom—'From all over the Empire and from the bottom of our hearts we send to the Armies of the Nile every good wish for the New Year'—'I am proud to be a member of that vast commonwealth and society of nations and communities gathered in and around the ancient British monarchy'—'If the British Empire and its Commonwealth last for a thousand years, men will still say, "this was their finest hour!"'

Churchill lived the imperial theatre. He knew that Powers, like people, are taken at their own valuation, and he behaved always as though Britain were a super-Power, the leader of a great disciplined Empire destined to last a millennium. He approached President Roosevelt the American, Marshal Stalin the Russian, not simply as personal equals, but as political equals too. He spoke as he might have spoken if the Victorian Empire were still at his back, in its unchallengeable zenith. He spoke in imperial hyperboles, saw with an imperial vision, and gave to the British themselves, for the last time, the feeling that they were a special people, with honourable duties all their own.

3

Without her equivocal allies Britain could never have won the war, and she could no longer defend her Empire single-handed. In the first months of the war the Channel Islands, the oldest overseas possessions of the Crown, were surrendered to the Germans without resistance. Thereafter many an imperial property was attacked, and many lost. British Somaliland, Hong Kong, Burma, Malaya, Singapore, many islands of the Pacific, were taken by the King's enemies. Egypt was invaded, Australia was bombed, Malta was almost obliterated, enemy submarines penetrated the harbours of Sydney, Muscat and St Lucia. Even the frontiers of India were crossed by enemy armies at last, fulfilling one of the Empire's oldest nightmares.

Just as, in 1915, the British sought to make a decisive assault by way of the Ottoman Empire, so a generation later they still saw the Middle East as the epicentre of their struggle. The Suez Canal had become, if not in fact at least in theory, the fulcrum of their power in the world: their oil supplies, their links with India, their dominance of the Arab land mass—all seemed to depend upon their power in Egypt. Before the war their only standing expeditionary force was designed to go, not to France, but to Egypt: at the most crucial moment of the Battle of Britain, when an assault on England seemed imminent, they sent 100 aircraft and an armoured brigade to the eastern Mediterranean. Whatever else might happen Egypt, threatened first by the Italians, then by the Germans, must not fall.

If we think of the imperial armies in the First World War, we think of France and Flanders and Gallipoli. If we think of them in the Second we think of them first, perhaps, in Cairo. Cairo in the 1940s was the last great assembly-point of the imperial power, the last place where, in a setting properly exotic, the imperial legions mingled in their staggering variety. Every kind of imperial uniform was to be spotted in Cairo, in the first years of the war. There were kilts and turbans and tarbooshes, slouch hats and jodhpurs. There were Kenyan pioneers, and Indian muleteers, and Australian tank crews, and English gunners, and New Zealand fighter pilots, and South African engineers. There were scholarly staff officers straight from their Oxford colleges, and swaggering extroverts back from secret missions in the Balkans, and rumbustious troopers of the sheep-station and the surfing beach ('I look to you to show the Egyptians,' General Sir Archibald Wavell told the Australians, when they first arrived in Cairo, 'that their notions of Australians as rough, wild, undisciplined people given to strong drink are incorrect'—but alas, their notions were all too often confirmed).

Cairo had been a pseudo-imperial city for sixty years, and although since 1936 Egypt had been nominally independent, and was officially neutral in the war,[1] the whole capital was now in effect a British military base. The British Army still occupied the dismal old Kasr el Nil barracks, in the centre of town beside the Nile, and the British Middle East Command had its headquarters in Garden City, conveniently close to the British Embassy. From Cairo, during the first years of the war, operations were directed all over the nearer

[1] Until February 1945. Two months later Egypt was able, as a belligerent Power on the right side, to become a founder member of the United Nations.

East. The North African campaigns, the battles for Greece and Crete, the reconquest of British Somaliland, the capture of Ethiopia, the invasion of Syria, the reoccupation of Iraq, the occupation of southern Persia, guerilla war in Jugoslavia—all were organized from Garden City. Aircraft passed this way to India and East Africa. The Royal Navy's Mediterranean headquarters was up the road at Alexandria.

For some years of the war Cairo was the military capital of the British Empire. A British Minister of State was in residence there. Dignitaries and celebrities were constantly passing through, Noel Coward to the Duke of Northumberland, 'Chips' Channon the diarist to Cecil Beaton the photographer or J. J. Astor the owner of *The Times*, and in 1943 Churchill, Roosevelt and the Chinese leader, Chiang Kai-shek, met in conference at Mena House beneath the Pyramids. In Cairo life was still richly lived. They served French wines and grouse in season at the worst moments of the war, they played polo most days over the river at Gezira, the brothels of Clot Bey flourished exceedingly and military offices, like all others, closed from one till half past five each afternoon.

> *We never went West of Gezira,*
> *We never went North of the Nile,*
> *We never went past the Pyramids*
> *Out of sight of the Sphinx's smile.*
> *We fought the war in Shepheard's and the Continental Bar,*
> *We reserved our punch for the Turf Club lunch*
> *And they gave us the Africa Star.*

4

It was in this crowded and frenetic city, full of power, indulgence and intrigue, that the most traditionally imperialist gesture of the war was made. Britain's paramountcy in Egypt had always been based on force—acquired by invasion in 1882, consolidated by the perpetual presence in the country of strong British forces. By 1942 the omnipotent British Agent and Consul-General of Cromer's day, the persuasive High Commissioner of the 1920s, though still inhabiting the same Residency beside the river, had matured into a British Ambassador in the person of the bluff and overwhelming Sir Miles Lampson. Lampson was a gigantic man, 6 foot 5 inches tall and 18 stone in weight, and since his arrival in Egypt in 1934 he had become an inescapable figure of Cairo life, closeted with Egyptian Ministers, calling on the King, dancing, shooting, riding, gambling at the Mohammed Ali Club, learning to fly at Heliopolis airport, driving from appointment to appointment in the enormous Embassy Rolls, the best-known car in Egypt.

He was an old-fashioned, straightforward, robustly patriotic imperialist, and when war broke out he saw his duties as essentially imperial. Though Egypt was officially neutral, it was absolutely subjugated to the British Empire's war effort: and though Lampson was ostensibly an Ambassador like many others, he was effectively the Satrap of Cairo. He believed that any means were justified to keep Egypt within the traces. Many Egyptians held different views, and prominent among them was the young King, Farouk, who was twenty-two years old and understandably impatient of British authority. He had been educated, like other kings of Anglo-Araby, to be a puppet: schooled in England and tutored by an Eton master, kept always beneath the eye of the British Embassy, he was only just an independent monarch at all, but held his throne on sufferance. He was, however, far from subservient. His entourage was chiefly Italian, at a time when the British were at war with Italy, he lived a life of louche dissipation in his Abdin Palace in the heart of Cairo, or at Monza beside the sea at Alexandria, and his devotion to the imperial cause was distinctly fragile.

Lampson, who habitually called him 'the boy', treated him first with condescension, later with contempt, and the King responded predictably, by snubbing him, deliberately misunderstanding him, or keeping him waiting for appointments. Lampson was enormous, and very strong: the King was smallish, and fat. Lampson was a man of the past, a survivor of an all-powerful Britain; Farouk was a young man of modernist instincts, amoral, materialistic, who had perhaps sensed more quickly than the Ambassador himself the decline of British power. The two men naturally detested each other, with equal reason on either side, and when in 1942 King Farouk determined to appoint a new Egyptian Government more amenable to his own views, but less acceptable to the British, Lampson decided it could not be tolerated. Rommel seemed about to break through the British positions in the western desert and advance upon Cairo: at such a moment Lampson felt it within his authority to depose the King of Egypt himself, if British interests demanded it.

It so happened that among the British officials in Cairo, working in the propaganda department, was Walter Monckton, the lawyer who had, six years before, drafted the instrument of abdication for King Edward VIII. He was now instructed to draw up a similar document for Farouk, the king of an ostensibly independent Egypt. Farouk was to be sent an ultimatum demanding the appointment of an administration more favourable to the British. He was given until 6 p.m. on the evening of February 2, 1942, to name Mustapha el Nahas Pasha Prime Minister, or else face the consequences—by which Lampson meant that His Majesty would be obliged to abdicate, and would be whisked away by the British Army to be held prisoner on a cruiser in the Red Sea.

Here was a return to form! The British had hardly acted with such peremptory decision since Victoria's day, and Furse's men would probably have been appalled. Farouk replied only with a letter, received at the British Embassy a quarter of an hour after Lampson's deadline, protesting against such a blatant infringement of Egyptian sovereignty, and so there was set in motion one of the last acts of imperial swashbuckle. Shortly before 9 that night, when Cairo was in full spate, the cinemas bright with neon lights, the night clubs opening their velvet-curtained doors, the red-capped military policemen warily patrolling the stews of Boulac, the pavement cafés thronged, the open-air cinemas booming above the traffic—through the bright and noisy city, smelling as always of dirt, jasmine, food and inadequately refined petrol, a convoy of British tanks, armoured cars and military trucks rumbled across town to take up positions around the royal palace.

At 9 o'clock precisely Lampson's Rolls arrived at the gates, and the Ambassador, accompanied by the commander of the British troops in Egypt and followed by a specially picked posse of armed officers, entered the palace. True to his code even then, Farouk kept him waiting for five minutes in an ante-room, but once inside the royal chamber, with the general glowering at his back, Lampson allowed no nonsense. He was the last in the domineering line of Clive or Cromer, with an oriental monarch in his power before him ('it doesn't often come one's way,' as he wrote in his diary next day, 'to be pushing a Monarch off a Throne.')

First he read the unhappy king a statement, very loudly, 'with full emphasis', as he later reported to the Foreign Office, 'and increasing anger'. It accused the King of assisting the enemy and thereby violating his commitments to Great Britain, described his attitudes as unfaithful, wanton, reckless and irresponsible, and said he was 'no longer fit to occupy the Throne'. Lampson then handed Farouk Monckton's abdication instrument, and told him to sign it at once, 'or I shall have something else and more unpleasant to confront you with'. Even as he spoke, they could hear the growl and clatter of the tanks outside the palace gates.

The abdication instrument was simple enough: 'We, King Farouk of Egypt, mindful as ever of the interests of our country, hereby renounce and abandon for ourselves and the heirs of our body the Throne of the Kingdom of Egypt and all Sovereign rights, privileges and powers in and over the said Kingdom and the subjects thereof, and we release our said subjects from their allegiance to our person.' It was typed on old British Residency paper from which the letter-head had been removed, and Farouk's first response was to complain about its sloppiness. He was answered only by an overbearing silence, and after a moment or two of indecision, cowed and frightened he picked up a pen to sign the abdication. But then he hesitated. Would Lampson give him one more chance? The Ambassador agreed, provided that the King immediately complied with the ultimatum. Farouk thereupon submitted, 'for his own honour and his country's good', and the Ambassador and his cohorts stamped away down the corridors of the palace, between the steel-helmeted British soldiers with their tommy-guns at the gates, past the tanks squatting watchful in the street outside, into their cars and back to their great Embassy beside the Nile.

'So much', reported Lampson, 'for the events of the evening, which I confess I could not have more enjoyed.' Certainly they seemed to have been successful—'I congratulate you warmly,' cabled Anthony Eden, the Foreign Secretary, 'result justifies your firmness and our confidence.' Nahas Pasha, summoned to the palace next day, formed a Government that remained loyal to the British connection for the rest of the war, even when the Germans advanced so threateningly towards Cairo that the British Embassy burnt all its secret papers and officers queued around an entire city block to withdraw their money from Barclay's Bank. What the longer effects of the affair would be, when the world returned to normal, nobody knew or even cared: the imperial instinct now was for survival, and the days of apology were suspended, as the saying was, 'for the duration'.[1]

5

The Middle East never did fall—when the war ended the British military position there was stronger than it had ever been. Nor was the Mediterranean, Mussolini's Mare Nostrum, ever denied to those British fleets which had frequented it since the end of the Napoleonic wars: and this achievement was largely due to another peculiarly imperial episode of the war, the defence of Malta.

The peoples of the overseas Empire were not, *pace* Churchill, so unanimous in their fidelity as they had been in the Great War. They were rather better informed now, they had other standards to judge by, and they had learnt to question the meaning of the war and the purpose of the imperial connection. Treason to the Crown, so rare in the First War, was not uncommon in the second. The Mufti of Jerusalem, for instance, the leader of the Palestine Arabs, went to Germany and formed a Muslim Army of Liberation. Aung San, one of the most prominent Burmese nationalists, went to Japan and formed a Burma Defence Army. Subhas Chandra Bose, *quondam* president of the Indian National Congress, went first to Germany,

[1] In fact the incident rankled ever after, and Anglo-Egyptian relations remained deplorable until the last claims to imperial privilege were abandoned. In the end it was the Egyptians themselves who forced the abdication of Farouk—in 1952, when they shipped him off to Italy in his own yacht (to be returned later, like an empty bottle). He died in 1965, a year after Lampson.

where he formed an Indian Legion, then to Japan, where he formed an Indian National Army. One in six of Indian prisoners of war joined his forces (which included a Gandhi Brigade and were known to the British as 'Jiffs'—Japanese Indian forces): Bose declared himself head of an Indian Government-in-exile, and became in the end a national hero.[1]

On the other hand the loyalty was often touching. Most of the hundreds of thousands of colonial volunteers joined the colours simply out of loyalty, to an idea, a heritage or perhaps a set of values: and this truly Victorian sense of duty was encapsulated best of all in Malta, a small group of islands in the middle of the Mediterranean which assumed not only a high strategic importance in the war, but a symbolic meaning too. As the main base of the Mediterranean Fleet Malta was made untenable when Italy entered the war, but it remained invaluable as a submarine base, from which to harass the enemy supply routes to North Africa, and it became almost indispensable as a propaganda station. Through the long years of setback, when islands fell and British armies retreated everywhere, Malta remained defiant in defence, furious in attack. If it had not been for Malta's submarines and aircraft, Egypt might well have fallen, and like the defence of Mafeking in the Boer War, the defence of Malta was seen as an earnest of greater things, and of victories to come.

The British made the most of it, immortalizing it in film and propaganda pamphlet, in stirring speech and kingly gesture, but still it was a genuine triumph. The five islands of Malta, with a total area of 125 square miles, lay some 60 miles south of Sicily, 180 miles north of Cape Bon on the North African coast, in so commanding a situation that Rommel himself saw them as the key to Mediter-

[1] Though his army was a failure, as Hitler had foreseen—'there are Indians', said the Führer sagely, 'who won't kill a louse, so they won't kill an Englishman either.' Bose himself is thought to have died in Taiwan after his Japanese aircraft had crashed there in 1945. Aung San, reconciled with the British after the war, became the first Prime Minister of independent Burma, but three months later, only thirty-three years old, was assassinated in the council chamber at Rangoon. The imperturbable Mufti settled after the war in Egypt—where I met him, I cannot resist recording, at the wedding of the King of Libya—and died in 1974 in Beirut.

ranean victory—in December, 1941, three-quarters of his supplies were sunk by ships and aircraft from Malta, and one Italian convoy, comprising seven merchant ships and ten destroyers, was completely annihilated. It was inevitable that the Germans and Italians should try first to neutralize, then to capture the islands: and between June 1940, when Italy entered the war, and November 1942, when the British won the Battle of El Alamein in Egypt, Malta was under siege. Some 16,000 tons of bombs were dropped on the islands: during fifteen days of April, 1942, there were 115 air raids.

There was never another imperial siege like this. Though the ships which supplied Malta, and the fighter aircraft which defended it, were all manned by Britons, the beleaguered population was notably *un*-British in everything but citizenship. Catholic, Latinate, speaking a language incomprehensible anywhere else, addled by political, religious and cultural rivalries, the Maltese were notoriously among the more difficult subjects of the Crown. Though since 1921 the colony had been ostensibly self-governing 'in all matters of purely local concern', the constitution had been suspended in 1936 because of this continual rumpus, and when the war came the Maltese had no responsible government at all. The British, as we have already seen, viewed the majority of the Maltese as less than absolutely pukkah, and generally looked upon the islands not as anybody's homeland exactly, but as a dockyard, a barracks and a sailors' tavern.

Yet the Maltese were to be the acknowledged heroes of the battle for Malta, and the prize examples of colonial loyalty. Though Mussolini had long laid claim to the islands, and though Valletta the capital was less than half an hour's flying time from Italian air bases, the Maltese never wavered in their faithfulness to the British Empire—or perhaps to the Royal Navy, whose ships had for a century provided a background for every facet of Maltese behaviour. The British got very sentimental about Malta as the war proceeded, and this was one wartime saga that was never discredited. Battered, half-starved, exhausted by noise and sleeplessness, living for the most part in subterranean galleries and dug-outs, the Maltese retained their morale from the very first trial black-out (a word which

Malta gave to the language) to the last fitful air raid from the retreating enemy. Baden-Powell would have been proud of them.

Actually it was a drama less in the British than in the Spanish style, very far from the tight-lipped heroics of Lucknow or Lady-smith. It was a siege in the romantic mode. Aesthetically it was thrilling. The splendid architecture of the islands looked more splendid still, manned for war. The searchlights played ritually around the great cathedral of St John in Floriana; the tracer bullets flew in coloured streams from the ramparts of the Knights; convoys laboured into the anchorage of Grand Harbour; black submarines slid shark-like from their pens beneath Fort Manoel. At night war-ships at sea reported Malta like an explosion on the horizon. By day Valletta was often shrouded in a pall of golden dust from the rubble of its bombed buildings.

Then the nature of the battle was stylish. Three elderly biplanes, at the start of it, represented Malta's entire air defence—*Faith*, *Hope* and *Charity*, whose indefatigable lumbering sorties against colossal odds were astutely publicized throughout the world and immor-talized in Maltese legend. Later fighter reinforcements were flown in from carriers offshore, and this too was a spectacle full of excite-ment—once reinforcements arrived actually in the middle of an air raid, so that within minutes of their arrival on the island they were refuelled and in the air again in action against the Luftwaffe. Once there was a suicidal Italian torpedo-boat attack upon Grand Harbour itself, when seventeen motor-boats, hurling themselves against the nets, chains and buoys of the harbour defence, were caught in the searchlights and blasted one by one out of the water by the rampart guns, while the Maltese lining the battlements cheered in the darkness.

There was the spectacle too, salutary to the British, of a Mediter-ranean people behaving night after night with a truly imperial stoicism, spiced perhaps with a trace of Cockney bravado picked up from so many generations of contact with the Navy. The Knights of St John had burrowed labyrinths of chambers and galleries in the soft limestone of Valletta, and there the Maltese slept each night, while the city was pounded into ruin above them. For a small fee they had the right to dig extra chambers for themselves, and sometimes they moved all their possessions down there too, put their own street number above the entrance, and made their home in some alcove of an ancient catacomb, or in the deep Fosse which, 300 years before, the Knights had dug on the landward side of Valletta. Priests moved about them at night; great ladies of Malta, in the English manner, brought them food and comfort; cameramen from the Ministry of Information frequently photographed them, and once or twice official war artists set up their easels in the half-light.

The Governor of the island, General Sir William Dobbie, was an epic figure himself. A Cromwellian figure, Churchill called him, and in fact he was a member of the Plymouth Brethren, an evangelical directly in the tradition of the God-fearing stalwarts of pre-Mutiny India. Dobbie believed implicitly in God's pro-British inclinations, and matched his faith with his bearing, so that despite his fervently protestant views he appealed to the fervently religious Maltese, and even had a pub named after him—the 'Everyone's Friend'. 'I call on all officers and other ranks humbly to seek God's help,' said Dobbie's first Order of the Day, 'and then in reliance in Him to do their duty unflinchingly.' Almost every morning he repeated the message in a broadcast to the islanders: he had fought in the Boer War and the Great War, both victories in the end, and in his sixties regarded his call to the command of Malta as a summons directly from the Almighty.

6

Most thrilling of all were the days when, like wildfire around the battered streets, there ran a rumour that a supply convoy was approaching the islands. At the worst moments of the siege Malta very nearly starved, and very nearly ran out of fuel, water and ammunition too—in August 1942 only two weeks' supply of petrol remained, and the populace was rationed to half a bucket of water a week. The Malta convoys, by which the Royal Navy tried to bring provisions direct from Britain, were among the most hazardous operations of the whole war, so ferocious were the air and submarine attacks upon them during their passage through the western Mediterranean. Each convoy was a major naval operation. Sometimes the greatest of the Royal Navy's warships, battleships, battle-cruisers, carriers, escorted the merchant ships, and hundreds of aircraft were deployed. Even so, of the 30 merchant ships which set out for Malta in the first seven months of 1942, 10 were sunk at sea, 10 were so badly damaged that they turned back, 3 were sunk in Malta harbour after their arrival, and only 7 survived to unload all their cargoes on the quay.

Single ships sometimes ran in and out of Malta—submarines, destroyers, or the 40-knot three-funnelled minelayers, *Welshman*, *Manxman* and *Ariadne*, which maintained a desperate shuttle service from Gibraltar. The arrival of a convoy, though, was a dramatic occasion, and the Maltese crowded to their ramparts, as they had thronged in happier times to watch the resplendent Fleet return from exercises, to see the rusted, battered and sometimes sinking blockade-runners limp in from the battle. The most momentous of them was Convoy S 26, Operation Pedestal, which sailed to Malta from the Clyde in August 1942. No convoy had reached the islands since June, and even the indomitable Dobbie had reported that the battle was nearly lost. Surrender plans had already been drawn up; some women and children had been secretly flown out in bombers to Cairo. Pedestal was accordingly the most powerful convoy of them all. The battleships *Nelson* and *Rodney* were accompanied by 3 carriers, 7 cruisers and 32 destroyers, and this great fleet was escorting no more than 14 fast merchantmen. Among them was the tanker *Ohio*, an American ship with a British crew, carrying 11,000 tons of kerosine and diesel oil, without which air operations from Malta would almost immediately collapse.

The voyage of this convoy through the Mediterranean was one of the great naval exploits of the war. The freighters and their escort left Gibraltar on August 11, and from dawn the following day, all the way to Malta, they were under continual attack by aircraft, submarines and torpedo boats. The losses were terrible. First the carrier *Eagle* was sunk, then the carrier *Indomitable* crippled. Two cruisers and 8 merchant ships were lost, 2 more cruisers, 3 merchants ships badly damaged. The *Ohio*, the one tanker in the convoy, was attacked more frequently and more heavily than any other ship, and on the 13th, 100 miles from Malta, her engines were put out of action. Twice her crew abandoned her, only to reboard her later. Sometimes she seemed to be enveloped in a sheet of fire. Once she was taken in tow by a destroyer, but she was holed so badly that she could not be pulled, and in the end, within sight of Malta, she was lashed to a destroyer and a minesweeper, one each side, and so carried rather than towed into the safety of Grand Harbour.

Only five merchant ships got through, but they saved Malta. As they sailed one by one into harbour, huge cheering crowds waved them in, crossing themselves in thanksgiving, while the bands played 'Rule Britannia' as in the old days, and the warships, all spick-and-span suspended, sailed to their familiar moorings beneath the walls. Such was the allegorical lifting of Malta's siege—the arrival, through three days of screaming dive-bombers and tor-pedo explosions, of a handful of damaged merchantmen and their exhausted escorts. There is a completeness to the Malta story that was rare in the imperial history, and its combination of unbreakable defence and snarling offensive spirit powerfully appealed to the British. 'The loss of Malta', Churchill had said, 'would be a disaster of the first magnitude to the British Empire', and when the worst was over the Empire recognized Malta uniquely. Just as Spanish cities, in the high days of Spanish chivalry, had been awarded corporate honorifics—the Very Noble, Very Loyal, Very Heroic and Invincible City of Seville, for instance—so now King George VI awarded his Colony of Malta the newly instituted decoration called the George Cross, 'for heroism and devotion that will long be famous in history'.

The King went to Malta in person to bestow it, sailing into Grand Harbour in 1943 on the bridge of the cruiser *Aurora*, and the medal, after being displayed like an ikon in every village of the archipelago, was laid up like a battle honour in the Governor's Palace.[1]

7

No such inspiriting epic sustained the imperial legend in the Far East. There it seemed less likely that the Empire would last Church-

[1] Though the islands are now officially simply the Republic of Malta, to this day they are still often called Malta GC, especially by the British.

Last Viceroy of India: Mountbatten in New Delhi

hill's Thousand Years. India was in a state of spasmodic turmoil, Congress supporters boycotting the war effort in a campaign called 'Quit India'. There were horrible riots, and mass arrests, and to complete the suggestion of impending catastrophe, the worst famine for nearly a century. Hong Kong fell to the Japanese after a battle that lasted seventeen days, Malaya did not survive much longer, and in 1942 there occurred the most humiliating single disaster in British imperial history, the fall of Singapore, an event which presaged unmistakably the end of Empire itself.

The island colony of Singapore, 225 miles square, was a British creation from scratch. Stamford Raffles had founded it, 120 years before, and it was specifically the alchemy of Empire which had made it, by the 1930s, the fourth port of the world. Flat, steamy, thickly humid, the island lay at the southern tip of the British Malay peninsula, linked to it by a causeway across the narrow Johore Strait. It was traditionally one of the main pivots of imperial power. The base the British had built there between the wars, their greatest military work of those decades, was intended to assure the security of the whole Empire east of Suez. It cost £60 million, it covered 21 square miles, and its complex of docks, warehouses and barracks lay massively along the Strait, well away from the open sea.

The base was specifically designed to service and support battle-ships, the chief weapons with which the British hoped to maintain their presence in the east. The plan was that a battle fleet would be sent there as soon as war in the Far East impended. The Australians and New Zealanders looked to Singapore as the main guarantor of their security, and it had been frequently represented to the world as impregnable. Some 700 miles of British territory lay to the north of it, while the sea-approaches to the island were commanded by batteries of 15-inch, 9.6-inch and 6-inch guns, hidden away in scrubby offshore islets, or sunk deep among the mangroves. There was an air base too, and a garrison of some 7,000 men, and though in normal times there were few warships at Singapore, and still fewer aircraft, it was assumed that in time of trouble powerful forces could get there in no time at all to impose the will of Empire on the east.

It was a handsome city, with its white cathedral above the green Padang, its stately offices of law and government, the spruce bunga-

Last echoes of the Raj: the British Consulate-General, Muscat

lows of its British rulers and the spiced haphazard quarters which housed the Chinese, the Malays, the Javanese, the Arabs, the Indians and all the other thousands who had flocked from half Asia to this profitable emporium. Nothing very terrible had ever happened in Singapore. Nobody had ever bombarded the place, still less invaded it. There was no political dissent, the colony being governed absolutely by the Governor and his Council: the large British commercial colony lived in Singapore complacently, enjoying the rewards of Empire without too many of its discomforts.

Its members believed explicitly, from experience, in the power of prestige, in maintaining above all the *respect* of their Asian subjects. Like the masses adopting some pastime when it is slightly passé among the fashionable, the Singapore British behaved rather as Anglo-Indians did fifty years before. 'Face' was all-important to them, and the maintenance of an invulnerable image. If the brown and yellow peoples thought them invincible, they reasoned, invincible they would remain: and so assiduously did they propagate this self-image that they had long come to believe in it themselves. It was inconceivable to them that British troops might actually be beaten by Asians, just as it was almost impossible to imagine an Englishman ordered about by Asians, or obliged to live Asian-style. It was not done, so it would not happen.

In this as in much else they were pitiably out of touch with developments at home. There, by 1940, the strategists had virtually dismissed the Far Eastern Empire, so overwhelming were the dangers nearer home. Smuts assured the Prime Minister that diplomacy was the only weapon available to the Empire in the east, and in fact the Americans had already been warned that the defence of the region was beyond British capability. As for Singapore itself, a defence conference had secretly reported that far from being impregnable, if Malaya were ever lost the island would be untenable —almost all its defences were on the seaward side, and there was almost nothing to stop an enemy crossing the Johore Strait from the north. The contingency plans for the colony, presented so reassuringly to the Australians and New Zealanders, had been shattered. The fighter aircraft which were to fly in relays to the east were at home defending Britain. The battleships supposed to rush to the base were engaged in the Mediterranean, left fatally vulnerable by the fall of France—in the summer of 1941 the Far Eastern Fleet consisted of three old cruisers and five destroyers.[1] There was not a single tank in the Malay Peninsula, and not a single modern fighter east of Egypt.

The Australians and New Zealanders, always afraid of Japanese intentions, pressed repeatedly for the reinforcement of Singapore, but British priorities were elsewhere. Churchill hoped in any case that the Japanese could be deterred from attacking the Empire by old-school imperial means—by a 'show of force'. In October 1941, when Tokyo's posture seemed particularly threatening, he ordered to Singapore two capital ships, the battleship *Prince of Wales*, the battlecruiser *Repulse*, the Prince of Wales's pleasure-ship twenty years before. They sailed there without air cover, their escorting carrier having been delayed by an accident in Jamaica, to intimidate by their presence a Japanese battle-fleet eight times the size and with a formidable air arm. 'Thus', said Churchill at a Lord Mayor's banquet in London, 'we stretch out the long arm of brotherhood and motherhood to the Australian and New Zealand peoples. . . .'[2]

Far away in Singapore the arrival of the great ships seemed no more than proper. The colony's sense of security remained absolute. Malaya was strongly garrisoned, an assault from the sea was inconceivable, and when Churchill sent out Duff Cooper as his special representative in Singapore, the colonists much resented his presence, and called him 'Tough Snooper'. No state of war was declared. There was no black-out. Life went on exactly as before, down to the daily tea-dance at Raffles. It accordingly came as a terrible psychological shock when in December 1941 the Japanese, the most Asian of all the Asians, who looked like monkeys, who could not see in the dark, who rode about on bicycles, who carried

[1] And its Admiral commanded it from London.

[2] The *Repulse* was twenty-five years old, the *Prince of Wales*, thanks to treaty commitments, was armed with 14-inch guns as against the 16-inch of Japanese battleships. Churchill reassured the New Zealanders, as he despatched the squadron, that 'nothing is so good as having something that can catch and kill anything', but only two months before he had privately deplored the inadequate armament of the *Prince of Wales* class—sorrow rose in the heart when one thought about it, 'or ought to'.

umbrellas, who talked in absurd high-pitched voices and were always comically bowing to one another, fell with a hideous efficiency upon the British possessions in Asia.[1] It was only to be expected perhaps that Hong Kong, isolated off the Chinese mainland, should soon be captured—though two battalions of Canadian infantry were rushed there when the attack began, arriving just in time to surrender. It was inevitable no doubt that the long line of island possessions, stretching away from the East Indies towards Australia, should be overrun. But it was an almost unimaginable blow when the two great capital ships, sailing from Singapore with four destroyers to prevent Japanese landings on the Malayan coast, were both sunk by Japanese bombers in the China Sea.

Even so, Singapore itself would surely stand. 'I trust you'll chase the little men off', said the Governor of the island, Sir Shenton Thomas, when the general in command told him that the Japanese had landed in Malaya. Even as the enemy divisions stormed southwards through the Malay jungle, thirty miles a day towards the straits of Johore, life in Singapore proceeded placidly. They still had their Sunday sing-songs at the Seaview Hotel. They still played their cricket on the Padang. When a young army officer suggested to his superiors the organization of an underground resistance movement in Malaya, he was dismissed as alarmist. Even when, on January 28, 1942, General Tomoyuki Yamashita arrived with his command post on the north bank of the Johore Strait, and looked through his binoculars at the green mass of Singapore beyond, etched with the nets and stilted shelters of the fishermen, at the cranes of the naval base away to the left, at the road that led away from the causeway to the great port itself—even then the British colonists of Singapore could hardly believe what was about to happen. When it was decided to make a strong point on the Singapore Golf Course, the soldiers were told nothing could be done until the club committee had met.

There were more than 100,000 soldiers on the island. The Malayan garrison, what was left of it, had retreated across the

---

[1] And simultaneously upon the American naval base at Pearl Harbor in Hawaii—thus ensuring a British victory in the end, for the Germans promptly declared war on the Americans too.

causeway, ineffectually blowing it up behind them, and British, Australian and Indian reinforcements had arrived by sea. But the muddle was terrible. Military clashed with civilian, commander with commander, outdated orders were scrupulously obeyed, initiatives were smothered in protocol and convention. The very word 'siege' was banned, in case it weakened British prestige among the Asians. General Arthur Ernest Percival, the commanding general, was a brave but uninspiring man, and the morale of the British in Singapore, falsely sustained for so long by myth and shibboleth, abruptly and disastrously collapsed. 'The city of Singapore', Churchill cabled, 'must be converted into a citadel and defended to the death. . . . Our whole fighting reputation is at stake and the honour of the British Empire. . . . There must be no question or thought of surrender.' But the battle was lost almost before it started.

Presently the Japanese were shelling and bombing the port incessantly, while queues of women and children waited at the dockside for passage on the ships about to leave. On the night of February 8th the Japanese crossed the Johore Strait; on the 12th a note from Yamashita was dropped by air, paying tribute to 'the honour of British warriorship', and threatening 'annihilating attacks' on Singapore unless the garrison surrendered; on the evening of Sunday, February 15th, General Percival sued for peace. The convictions of two centuries were knocked topsy-turvy by this event, and Asians were never to look upon Englishmen in quite the same way again. The Royal Navy had failed; the British armies had been outclassed; white men had been seen in states of panic and humility; the legend had collapsed in pathos—or worse still, bathos, for the generals were second-rate, the songs were banal, the policies were ineffective and even the courage was less than universal.

8

It was not all dishonour. The British surrendered Singapore partly because they had lost the imperial touch, but partly because they did not want innocent inhabitants to be killed, and partly because, lacking now the missionary zeal or the romantic passion, they saw no point in trying to hold it. Even Churchill soon abandoned his heroic exhortations, when he realized that defending the fortress to the death would be a useless sacrifice. Though the Japanese themselves, as it happened, were almost out of ammunition, Singapore was almost out of water. Surrender was an abject dénouement indeed, to the brilliant tale of this most successful of colonies, but it was the realistic and humane course of action, and if it hastened the demise of Empire at least it saved a few thousand human lives.

Nevertheless it was the worst reverse in the history of British arms, a disgrace and an ignominy. The surrender was signed in a drab place, the assembly plant of Ford Motors, three miles out of Singapore city on the Johore road. This was a dingy, brownish, single-storey building, fitted out in teak against woodworm, and its offices were dim-lit and glass-partitioned, and smelt slightly of india-rubber and typewriter ribbons. Here, soon after seven in the evening of February 15, 1942, the opposing generals met. Percival and his three staff officers, wan and exhausted, looked more like curates or schoolmasters than fighting soldiers: Yamashita looked bullish and aggressive in open-necked shirt and medal ribbons. Japanese war correspondents and military photographers jostled all around the table, Yamashita's generals sat impassive behind him, the tired bloodshot eyes of the British flinched in the flare of the flash-bulbs.

The proceedings were brief, for Yamashita simply demanded unconditioned surrender that evening. 'All I want to know is, are our terms acceptable or not? Do you or do you not surrender unconditionally? Yes or no?' Faintly, his head bowed, Percival agreed, and signed the document there and then, in a cramped schoolboyish hand. More than 120,000 soldiers and civilians were sent into captivity, and many ladies of the Seaview sing-song or Raffle's tea-dances were last seen, clutching a few bundles of their possessions, trudging along the coast road towards Changi Jail, whose gates they entered, for three years of terrible imprisonment, singing a tawdry but beloved patriotic song of the day, 'There'll Always Be an England'.[1]

[1] Singapore is transformed today as an independent City-State, but the Ford plant still stands and offers in its brown teak-furnished offices melancholy

9

The next edition of Whitaker's *Almanack* recorded Singapore as being 'temporarily in hostile Japanese occupation', and though before long the whole of the eastern Empire was taken by the Japanese, and even India was seriously threatened, still the British thought of these as 'temporary' setbacks. Though their war aims did not include any expansion of the Empire, they certainly did not include its abandonment either, and never for a moment were any of their territorial losses considered unredeemable. Throughout the war the Burmese Government solemnly sat in office in Simla, still listed in Whitaker's (*Governor, HE The Right Hon Sir Reginald Hugh Dorman-Smith, PC, CBE, Rupees 10,000*). Throughout the war colonial officials interned in Stanley Prison, Hong Kong, maintained that the British Government of Hong Kong was still in being, even if its writ ran no further than the prison walls, its officers still holding authority 'in virtue of their appointments and their oaths and duty to His Majesty the King'. A powerful stream of propaganda was directed towards the occupied colonies, assuring their inhabitants that the British Empire would soon be returning;[1] in most of them resistance movements were fostered or clandestine operations mounted. For many Britons these adventures, mostly organized by the Special Operations Executive, offered a fulfilment inadequately

---

evocations of the surrender. I asked my hosts there in 1974 if many Britons came to see the place. Not many, they said, very few in fact: but seldom a day went by without a coachload of Japanese tourists stopping at the factory gate.

General Percival, remembered by Yamashita later as being 'so pale and thin and ill', spent the next three years as a prisoner in Manchuria, where the Japanese kept their most important captives, but was released in time to be present at the Japanese surrender in Tokyo Bay in 1945, and lived until 1966. Yamashita, on the other hand, was hanged as a war criminal in 1946, the last victory being the one that counts.

[1] For example, in vernacular leaflets dropped into New Guinea: 'Japan bugger up finish. Altogether Japan man fraid too much, now all e like run away. By-n-by Govment makim good fashion along you fella, now you fella happy too much. Wait liklik time das all.'

supplied by expeditions and colonial skirmishes before the war, and more akin to the imperial satisfactions of their grandfathers' day.

Orde Wingate, for example, threw himself into adventures far more tremendous than his night patrols in Palestine, for with his Chindit guerillas he penetrated behind the enemy lines in occupied Burma for months at a time, until he was killed in an air crash in 1944. Other British forces in Burma, operating far beyond the Irrawaddy, got in touch with Aung San's traitorous Burma Defence Army and helped to turn it against its Japanese sponsors—Aung San himself being transformed from a 'traitor rebel leader', as Churchill had called him, into a gallant ally. The imperial administrators of the Solomon Islands, when the Japanese landed there, took to the bush with their radio transmitters and stayed there for the rest of the war, protected by their former subjects, sending information to the American forces. In occupied Sarawak the anthropologist Tom Harrisson, dropped by parachute with two companions, mustered the inland tribes into guerilla units, and when the time came, turned himself into Officer Administering Interior, Borneo. Even within occupied Hong Kong, isolated far to the north, a resistance movement was active. When the colony capitulated some seventy-five men, mostly Australians, escaped by motor torpedo-boat from Hong Kong island to the mainland, and for the rest of the war maintained a link between the Chinese capital of Chungking and the occupied colony: they arranged escape routes, they extracted intelligence, and they were even able to smuggle drugs and radios to the prisoners at Stanley Gaol.[1]

These were great adventures, terrifying and fulfilling, and conducted often with a revived panache. For one cameo of their spirit, let us imagine ourselves on an island beach off western Malaya on a summer evening in 1945, for there one of the archetypal imperial guerillas, Captain Freddy Spencer-Chapman, has an appointment with the Royal Navy. He has been in occupied Malaya for more than three years, organizing resistance teams and killing Japanese, and now he is to be withdrawn to Ceylon. The beach lies

[1] Twelve of whom were executed by the Japanese, in 1944, for the possession of a radio. They were shot on a pleasant sandy beach outside the prison, and a memorial marks the spot.

silver-sanded in the late morning, the surf rolls gently out of the
bay: when Spencer-Chapman and a brother-officer appear, emerging
cautiously from the thick jungle foliage behind, they look less like
British soldiers than a couple of beggars—dressed in ragged Chinese
clothes, heavily bearded, filthy, emaciated. They have with them a
sheet of white parachute silk, attached to a wooden frame, and every
quarter of an hour they hoist it above their heads for a few minutes,
so that it can be seen between the headlands of the bay. Between
times they swim, lie in the shade of the trees, or, propping themselves
against palm trunks, search the horizon with their binoculars.

Far away in England Spencer-Chapman is something of a
celebrity, a mountaineer, an explorer, a writer, but he is really a
compulsive adventurer of the neurotic kind, always seeking himself
in danger and in discomfort. He was one of Watkins' companions in
Greenland, where he fathered a child by an Eskimo girl, and he has
climbed in the Alps and the Himalaya, and even been to Lhasa with
the British diplomatic mission. It is only now, though, in this
desperate situation, three years hunted and disguised in enemy-held
territory, that he has truly found himself. Now he is an authentic
hero, and the ruthlessness he has so long been obliged to sublimate
in exploration and alpinism has found its true outlet. In one recent
fortnight he and his fighting patrol of Chinese and Britons have
killed 1,000 Japanese and destroyed thirty bridges. He has kept
perilous contact with the Chinese guerillas fighting the Japanese
in northern Malaya; he has been captured and escaped; he has tried
himself to the utmost, in exploits of astonishing courage and
endurance, and stripped and scoured by the war, allowed to kill in a
legal cause, he has, like many an imperialist of previous generations,
found his metier.

The afternoon draws on, and suddenly the Englishmen notice the
periscope of a submarine in the bay, cautiously twirling, and swimming this way and that. As dusk falls, to a whistling, clicking and
hooting of the jungle fauna, and rendezvous time approaches, they
flash their red torch into the gathering darkness, first every quarter
of an hour, then every five minutes, until at last, soon after nine
o'clock, when the moon has set and the beach is all in darkness, they
dimly see the hulk of the submarine rising from the sea. An English

voice sounds out of the night. 'How are your feet?' is the password. 'We are thirsty', shouts Spencer-Chapman in reply: and so the two adventurers, pulling their packs on their backs, flounder into the warm water, swim the fifty yards to the submarine, and are hoisted up its wet and slippery flank, bumpy with barnacles, and into the conning-tower hatch.

The engines start with a bellow, and with a spout of water HMS *Statesman* disappears once more into the night. The Royal Navy greets its jungly guests merrily, with whisky, for the ship's officers are all very lively, very young, very large, very bearded, and dressed only in sarongs.[1]

10

One by one, as the war progressed, the imperial territories were repossessed. The British 14th Army, with 700,000 Indians and two African divisions, slogged its way back into Burma, finally defeating the Japanese in the greatest of all land battles against them —one of the fiercest hand-to-hand engagements was fought around the Deputy Commissioner's bungalow at Kohima in the Naga Hills.[2] With a somewhat obvious symbolism the British returned to Singapore. Now it was the Japanese who awaited their conquerors, not in a gloomy factory, but outside the proud Government buildings in Empress Place; the Union Jack was the very flag beneath which Percival had marched to his surrender, among the watching crowds were thousands of emaciated soldiers released from Japanese prison camps, and when the imperial plenipotentiary arrived to accept the submission of the Japanese command, he turned out to be no less than a royal admiral, Lord Louis Mountbatten, Supreme Allied Commander in South-East Asia and a cousin of King George VI.

[1] I have adapted this episode from Spencer-Chapman's bestselling war memoir *The Jungle Is Neutral* (London 1949). Its author never refound himself after the war, and ended up as warden of a residential hall at Reading University—where, on a summer day in 1971, he took a gun and shot himself.

[2] The Kohima war memorial marks the site. The Deputy Commissioner's bungalow has disappeared, but the white lines of his tennis court have been marked out in concrete at the foot of the monument.

It was too late anyway. The return to the conquered territories might provide a transient satisfaction to the British, but it convinced nobody, least of all the subject inhabitants. The Empire could never be the same again. As the war proceeded the British had become progressively weaker than their gigantic allies, and more dependent upon American supplies and equipment. We see Churchill still sitting as an equal with Roosevelt and Stalin at the Yalta Conference, deciding the future of the world; we find Lord Wavell, Viceroy of India, deciding in 1944 that the British were 'a very great nation, greater than the American, and would remain so'; but it was more force of habit, or hope, than political truth. There was pathos to Churchill's insistence, after the collapse of Germany in 1945, that the American fleets in the Pacific should be joined by a British battle-fleet—so slow, so ill-equipped, so inexperienced in Pacific fighting, that it was an embarrassment to its American commanders. When it all ended the British were unmistakably junior partners in the Grand Alliance, and the grandeur of the reassembled Empire, as its Governors and District Commissioners resumed their plumes, privileges and responsibilities, was a somewhat hollow kind of consequence.

Yet as the world generously recognized, it had been, all in all, a fine conclusion to all the struggles, honourable and iniquitous, victorious or disastrous, by which the British Empire had established its presence across the world. It had been a last glimpse of greatness for the British—through the muddle and the miseries, as Churchill said, 'weary and worn, impoverished but undaunted, we had a moment that was sublime.' For a year or two the British peoples everywhere, whatever their colour, shared in the triumph and the honour. Who could ever forget the dark days of 1940, the Colonial Chaplain of British Honduras asked his congregation at a victory thanksgiving service? Not, it seemed, the British Hondurans, several thousand miles though they were from the nearest bomb or shell. 'It was with firm eyes and dry lips,' the clergymen reminded his listeners, 'and with a pregnant silence which surprised the world that we met those very dark days. . . .'

The full meaning of the Last War escaped the British at the time. They thought they were destroying a truly bestial enemy, but they were also destroying themselves and their heritage. Even their leaders, it seemed, seldom perceived this truth, and read in the story of the war only its heroic texts. Harold Macmillan, a future Prime Minister of Great Britain, was present at the victory parade in Tunis which marked the end of the campaigns in North Africa—the first in which American arms participated. The omens he read into the occasion were inspiring, but altogether false. First in the parade came the French and their colonial troops, brisk and colourful, then the Americans, young, tall and rather callow. A long pause followed, and Macmillan began to wonder if there had been a hitch in the arrangements. But no: faint, strange and magnificent over the crest of the road came the skirl of bagpipes, and then in a slow and steady pace the massed pipers of the British armies swung tremendously past the reviewing stand, followed by 14,000 bronzed and cocky British veterans of the desert war, each division led by its general. 'These men seemed on that day', Macmillan recorded, 'masters of the world and heirs of the future.'

In one sense they were. Even when the war ended, when Russian and American power was vastly greater than British, they controlled more territory than they ever had before. Not only was the whole of their Empire restored to them, not only did they share with their allies the governance of Germany, Austria and Italy, but to an unprecedented degree the Mediterranean was a British lake. It was an imperialist's dream. The whole of the North African littoral, the whole of the Levant was held by British arms, southern Persia was occupied and even Greece was more or less a British sphere of influence. With imperial armies deployed across the world, with a Royal Navy of 3,500 fighting ships and a Royal Air Force of unparalleled prestige, in theory the British Empire was a Power as never before, and the ageing Churchill, intoxicated by the honour of it all, was determined to keep it so. Things had worked out pretty well, he told an exuberant London audience. 'The British Commonwealth and Empire stands more united and more effectively powerful than at any time in its long romantic history.'

And if it was illusory, if the British victory was a defeat disguised, if the propagandists had given the people some false ideas about themselves and their prospects, if the Empire was not so united and powerful as Churchill suggested, if the grand euphoria of 1945 was all too soon to dissolve, still there was much to be proud of. They had stayed the terrible course, start to finish. They had lived up to the best in themselves and their civilization. For a few years the British Empire had achieved the condition, nothing to do with imperialism, which Burke had desired for England long before—not an England 'amusing itself with the puppet-show of power', but 'sympathetic with the adversity or with the happiness of mankind, feeling that nothing in human affairs was foreign to her'.

CHAPTER TWENTY-TWO

The Heirs Assemble

IT did not last long. The Churchills and the Macmillans might respond to glory still, the blood-bonds of Empire were momentarily strengthened by the experience of victory, British prestige in Europe stood higher than ever before, but still the mass of the British people were not interested in power or influence. A tremendous home-sickness seized the exhausted armies, and a sense of duty completed, chapters closed, lay upon the now shabby homeland. The people wanted only to live quietly and comfortably, and disillusioned as they had been by the torments of slump and unemployment in the 1930s, they saw in the aftermath of war the chance to make a fresh start—not in the distant fields of Empire, but in their own familiar island. They would not be gulled again by illusions of splendour. Churchill was dismissed from office at the first chance, and went smouldering off to write his war memoirs and to complete his *History of the English-Speaking Peoples*.[1] The Labour Government which took office under Clement Attlee in 1945 dedicated itself to social reform at home, internationalism abroad. The British Empire was held momentarily in abeyance, so to speak, while the new rulers of Great Britain decided what best to do with it.

Foreigners had firmer ideas, and the general opinion was, in the years after the Second World War, that it should now be abolished. Hitler indeed, before his death in 1945, had announced it to be doomed already, and called Churchill its grave-digger. 'It is at an end', he wrote venomously in the last weeks of his life. 'It has been mortally wounded. The future of the British people is to die of hunger and tuberculosis in their cursed island.'

[1] Which began with Julius Caesar and ended with Cecil Rhodes.

2

The Americans viewed the British Empire confusedly. There was still a powerful class of Anglophiles in the United States, especially in the Protestant Establishment of the east, and there was an inherited respect for the presence and the experience of the English, and for the common origins of the two nations—the imperial troops under his command, said a victory message from the American General Douglas MacArthur to Mr Attlee, had proved themselves 'fully worthy of the immortal tradition of our race'. British influence was still strong in Washington, for all the new disparities of power: through personal friendships and family connections, through the Common Law and analogous institutions, through the presence, always around the seats of power, of American Rhodes Scholars from Oxford. The Special Relationship, the Anglo-American axis forged during the war, survived into the peace, and made Britain, even in her straitened circumstances, more than a client State, but rather like a family adviser past the peak of his career.

This esteem and understanding did not generally extend to the British Empire. Nearly all Americans had been brought up to distrust the very notion of imperialism, whether they were descendants of American revolutionaries or refugees from the old tyrannies of the European continent. The Americans had experienced an imperialist spasm themselves, in the 1890s, when Kipling was moved to urge them to 'take up the White Man's burden' by occupying the Philippines. They had soon grown out of it, though, and indeed considered themselves the divinely appointed executors of universal self-determination, rather as the British had considered themselves entitled to interfere with slavery anywhere in the previous century. They had neglected the sacred task between the wars, by abandoning the League of Nations and withdrawing into isolationism, but the end of imperialism in the abstract, which meant British imperialism in the palpable, was certainly among their aims in the Second World War. They were astonishingly ignorant about the nature of the British Empire—it was commonly supposed even in Detroit and Chicago that Canada was ruled from London:[1] but they were

[1] On September 5, 1939, the Secretary of State himself, Cordell Hull, had

generally convinced that whatever its nature was, it was reprehensible.

In July 1941 Churchill went to Newfoundland to draw up with President Roosevelt the Atlantic Charter, a pietistic declaration of war aims. Churchill was determined to do it in the imperial style. He sailed on board the battleship *Prince of Wales*, fresh from the battle in which the greatest of German warships, *Bismarck*, had been sunk in the north Atlantic, six months from her own end in the China Sea: and it was beneath her 14-inch gun turrets, as the ship lay in Placentia Bay, off the southern coast of the island, that the two leaders were photographed at the end of their talks, surrounded by all the ancient circumstance of the Royal Navy. The Charter, however, which the Prime Minister found himself obliged to sign was anything but an imperial document, for it referred specifically to 'the right of all people to choose the form of Government under which they live'. Churchill tried vainly to write in a clause excluding the British Empire from this awkward philosophy, and lamely explained, when he returned to London, that it was 'primarily intended to apply to Europe'. It was not, though. It was intended to apply to Empires everywhere, not least Churchill's own.

The Americans certainly did not propose to support the British Empire after the war, however special the Special Relationship. Roosevelt himself was a vehement critic of the Empire—'the British would take land anywhere in the world,' he once remarked, 'even if it were only a rock or a sandbar.' 'Quit fighting a war to hold the Empire together', *Life* magazine demanded in 1942, and wherever in the course of the fighting the Americans came into contact with the imperial presence, they nibbled away at its authority, by diplomacy, by economic strategy, by example. As early as 1942 Roosevelt was suggesting to Churchill that India should be given immediate Dominion status, with freedom to secede from the Empire altogether—but of course, he disingenuously added, 'all you good people know far more about it than I do.' In the following year Lord Linlithgow, the Viceroy, had to ask the India Office to stop the

---

found it necessary to telephone the Canadian Prime Minister to ask if Britain's declaration of war meant that Canada was at war too.

flow of 'well-meaning sentimentalists from the United States to India, so that we may mind here what is still, I suppose, our own business'. In the last years of the war Colonel Louis Johnson, Roosevelt's personal representative in New Delhi, frequently maddened the British with his interfering ways, though Churchill, recognizing the priorities of the day, always restrained his resentments—'Anything like a serious difference between you and me', he wrote to Roosevelt after one particularly irritating intervention, 'would break my heart.'

In the Far East the Americans were especially determined that Hong Kong, when it was liberated from the Japanese, should be handed over to their Chinese ally Chiang Kai-shek. They considered themselves to have special interests in China, and the end of the British Empire on the Chinese coast became almost an *idée fixe* of the State Department. In 1941 it was suggested that Britain should sell the colony to the Chinese, the cost being met by the U.S. Treasury. In 1943 Roosevelt urged that Britain should give it up 'as a gesture of good will'. At Yalta he suggested privately to Stalin that it might be internationalized as a free port. When Hong Kong was liberated in 1945 it was only the immediate action of one Briton, the former Colonial Secretary Franklin Gimson, that kept it British—the moment he emerged from Stanley Prison, after three years of imprisonment, he ran up the Union Jack over Government House and resumed the colonial administration on his own initiative.

Nor were the American anti-imperialists much appeased by the disappearance of the die-hard Churchill and the advent of a pacific and reformist Socialist Government in Britain. On the contrary, now they feared the development of a *socialist* Empire—tea, rubber, tin and cocoa, reported the *New York Times* gloomily in August 1945, would probably all be nationalized. For economically the Americans were determined that in the reformed post-war world they would be dominant. They had always resented the imperial tariff systems, and they saw in the melee of the war their chance to break into vast new imperial markets. By 1946 their commerce was irresistible. The dollar had succeeded the pound as the chief world currency, and the American industrial machine, far from being debilitated by the war, like the British, was buoyant and productive as never before. Britain had lost enormous overseas assets, a loss calculated by John Maynard Keynes as thirty-five times that of the United States, while Lend-Lease, the American programme of war aid for Britain, had ensured that the entire British industrial machine was devoted to war production, so that when peace came again British technology could not compete with American. By 1946 Great Britain, for so long the greatest exporter in the world, was outclassed by the United States: American products of every kind, backed often by American loans, flooded the markets of the world, satisfying demands frustrated by four years of war, and elbowing out of the way many an old British trader, trying to recover customers and profits after the catastrophe.

By their example too, their style and their irresistible panache, the Americans irrevocably damaged the Empire. Their fleets and armies, pouring into the imperial territories, were like Hollywood come to life, offering dazzling new substitutes for the imperial way. Wherever they went the Americans presented an overwhelming image of opulence, vigour and generosity, more compelling by far than the dry modes of Empire. They cheerfully disregarded the old imperial taboos and superstitions, like the vital necessity of wearing topees in the tropic sun.[1] They lacked the proper sahibs' aloofness, they laughed at the carefully devised orthodoxies of the imperial system, passed so reverently from generation to generation of imperialist. And they had quite patently inherited from the British the magnetism of power. It was the Pax Americana now. 'Australia looks to America,' declared the Australian Prime Minister, John Curtin, 'free of any pangs as to our traditional link or kinship with the United Kingdom,' and young men throughout the Empire, of all races, turned to the American example as successor to, or relief from, the British manner of things.

All this was not immediately plain. It was like a shifting of transparent screens, the one set subtly, almost imperceptibly, replacing another. The existence of the Empire made the two Powers feel more like equals, and few Englishmen realized that the popula-

[1] A delusion long before exposed by T. E. Lawrence, who discarded both pith helmet and spine pad as an RAF aircraftsman on the North-West Frontier of India in 1928, and was said to be the first British serviceman to do so.

tion of Great Britain was now no more than a third that of the United States (in 1897, Diamond Jubilee year, it had been almost a half). On most conscious levels Britons and Americans remained friends and allies. The Pilgrim Society still avidly toasted Anglo-American amity,[1] the Rhodes Foundation still welcomed its scholars from Yale or Idaho, American servicemen paid sentimental visits to their wartime airfields in East Anglia, the Daughters of the American Revolution devotedly recalled the unvarying dukedoms of their English pedigree. Churchill was far more devoutly honoured in America than he was in England.

But it was happening all the same, and as the British presence began to fade beneath palm and pine, so the Americans stepped in with a jazzier hymn and a bourbon on the rocks—making a welcome change, so even some Englishmen thought, from the eternal gin and tonic.

3

A more baleful view of the Empire prevailed in Moscow. This was only to be expected. Even in Tsarist days, as we have seen, Russia was considered the most insidious of the Empire's enemies, and the Bolshevik Revolution had done nothing to ease the mutual suspicion. Just the opposite, for while the Tsarist threat had been purely strategic, the massive pressure of one expanding empire upon another, the Communist threat was profoundly ideological. Lenin had defined imperialism as the last stage of capitalism, and Ireland had always been the Marxist archetype of oppression: now the anathema was extended to the whole colonial empire. Over the Himalaya in Tsarist times there had been an enemy indeed, but an enemy not so different in kind from the British themselves, whose Mongolian agents and Cossacks in disguise talked much the same language as the spies and adventurers sent from British India. Now there brooded beyond the steppes a different kind of enemy altogether, altogether alien and virtually beyond communication.

Now there was no frontier with Russia—or rather, the frontier was everywhere, and the Great Game was played in every continent.

[1] Or even occasionally, as it once did by a pleasant slip of the tongue in my presence, 'Anglo-British friendship'.

The very name of the Union of Soviet Socialist Republics indeed, containing as it did no territorial reference at all, implied that frontiers were artificial anyway, and eventually disposable. Churchill's alliance with Soviet Russia, concluded in the desperation of war, was purely opportunist. He detested Communism, and had consistently and frankly opposed it since he had first sent troops to fight the Bolshevik armies in the winter of 1918. All his fine words about Russian heroism and brotherhood, the propaganda adulation of the Red Army, the jewelled Sword of Stalingrad sent to Russia in tribute by the British people, the perilous convoys taking supplies to Murmansk, the Hurricane fighters sent to fly with the Red Air Force—all this meant nothing, beyond the immediate necessities of war. There was hardly a true point of sympathy between the two Powers. They were altogether antipathetic, and *au fond* they wanted nothing better than the extinction of the other.

It was 'ridiculous', Stalin had once observed to Ribbentrop, Hitler's Foreign Minister, 'that a few hundred Englishmen should dominate India'. It was also heretical, for it did not fit the Communist dogma. The Marxist view of history as a purely economic process meant that Empire too was simply an instrument of capitalist profit. No allowance was made for quirk or anomaly, sentiment or habit. 'Imperialism' was something much more cruel, much more sinister than a handful of elderly Englishmen filling in forms in Simla. It was racialism, exploitation, capitalism, all mixed up, and was presently to become the favourite pejorative of the entire Marxist glossary.

For some years after the war the Russians seemed to let the Empire be. They were as exhausted as the British by the conflict, and they were inhibited perhaps by the fact that American nuclear power dictated the state of the world. Unlike the Americans, too, they seemed to miss the truth of the British decline, and far from depicting the Empire as a decaying and impotent tyrant, greatly overstressed its power. It was a decade before they achieved any obvious success in the imperial territories, and generally they behaved towards the British presence with no more than a timid kind of disrespect, like schoolboys trying to mock the headmaster.

Economically they posed no threat. Their export industries were vestigial, their currency more or less private. But as the spectacle of America dimmed the allure of Empire, so in a subtler way the message of Communist Russia nagged away at the Empire's composure. The intelligent young of the subject peoples, denied real political expression, were inevitably attracted to an ideology which specifically declared itself to be race-less and class-less, at once anti-imperialist and social-revolutionary. Communist books, papers and magazines had circulated tentatively in India and Africa even before the war, and many of the early nationalist leaders had been inspired by Marxist ideas. Now a new generation of African and Asian leaders were waiting to learn their lessons.

Soon most of the infant nationalist movements of the Empire had their Marxist cadres, and their leaders found their way by one means or another to founts of Communist teaching, generally in London, sometimes in Moscow itself. In England a West Indian Communist agent, George Padmore, instructed generations of young Africans, who went to England to study medicine or the principles of Common Law in the Mother Country, and went home politically animated from Moscow. The young Kenyan Jomo Kenyatta, to be the nationalist leader of his country, was one of his protégés, and spent some time at a revolutionary school in the 1930s: so was the Gold Coast leader Kwame who aspired to the leadership of a West African Soviet. Most of the guerilla leaders of Malaya and Burma and having fought against the Japanese during work against the British in the peace.

knew what was happening, and seldom beneath the surface of colonial life, move-intensity were gathering strength. They cal activity in the subject colonies, and the instinct of white supremacy had settlers of Kenya were as racialist as forcible separation of the races, basis of social life; in Southern ped into the gutter when they street. The young activists political concept) naturally turned to the force most unlike their rulers, and realized for the first time how false was their subjection. It was bluff! They were not after all, as they had been led to believe, immature races incapable of political action. They need feel no gratitude to the Empire, no sense of duty or even loyalty. They became at once more confident and more inhibited than their fathers had been, frustrated by their own growing knowledge, wounded by slights which earlier generations had hardly noticed.

'What *is* Communism?' a visiting inspector from the Colonial Office was asked by a simple farmer in Nyasaland in 1949, and instantly he knew that change was on the way. Communism was the *deus ex machina*. Communism was the party of universal brotherhood, sworn to rid the world of racialism and imperialism, and cut the lordly British down to size. The remarkable thing was not the 1... of Communist appeal, nor the scale of Communist activity in the decade after the Second World War, but its restraint and even its ineffectiveness: but perhaps that was because the dictator Joseph Stalin, the living embodiment of the creed, was a born imperialist himself.

<h1 style="text-align:center">4</h1>

Anyway, the groundswell of discontent among the subject peopl was essentially patriotic. The British preferred to call it 'nationali a word which had an obscurely disreputable ring to it, and leaders were always said to be 'agitators', but it was really just of country, of culture, of one's own people and one's own his Social reform played little part in these early stirrings of reb 'Give us chaos!' Gandhi demanded, for he knew very we British administration, in India as everywhere else in the was likely to be fairer, stabler and more efficient than its in successors. That was not the point. When a nationali harangued a crowd in Karachi, or started a strike in Lagos not generally for ideological reasons at all, but simply wanted his country to rule itself. Independence was all

The intensity of national feeling varied greatly. S like the Ashanti, had never lost their profound an savage sense of nationhood. Some, like the Jamaicans, one, so that their leaders had to summon it into existence. India had been welded into nationhood by the British themselves, while Uganda contained a separate kingdom within itself, and was in a perpetual state of tribal schizophrenia. Yet almost everywhere in the Empire, as the 1940s moved into the 1950s, patriotic movements were awake, from the experienced political associations of India to the first ill-informed nationalist cells of British Guiana or Somaliland. In 1945 a Pan-African Congress was held in the town hall of Chorlton-on-Medlock, a suburb of Manchester dominated by the University. Delegates came from most African territories, and they passed a resolution demanding autonomy and independence for the whole of black Africa. The London newspapers took little notice. The citizens of Chorlton-on-Medlock were not much impressed by the shabby black men assembling in the drizzle. But the delegates at that meeting, the plans they decided upon, the new resolves they took home with them to Africa, were presently to change the world.

Among the emergent patriots were some remarkable men. Some were familiar and comprehensible figures, like the brilliant Nehru, Harrow and Cambridge, who wrote an exquisite English prose, behaved like a gentleman, and spoke a language of politics that needed no interpreters. Others were much stranger. Nobody knew quite what to make, for instance, of Jomo Kenyatta the Kenyan, *né* Kamau wa Ngengi, the grandson of a Kikuyu magician whose education had been part Church of Scotland evangelicism, part tribal ritualism, part London School of Economics. This powerful and clever man baffled the British first to last. He had been a Communist in his time, he had spent many years in England, he had written an anthropological study of his tribe, he had been an active Kenyan nationalist since the 1920s and returned to Nairobi in 1946 as president of the Kenya Africa Union: and as his thick-set calculating figure rose from obscurity to fame, petty agitator to celebrated national leader, they never did master how many of his motives were explicable impulses, urges towards power that westerners could understand, and how many were mysteries of Africa, from another morality and another sensibility.

Another riddle was Alexander Bustamante of Jamaica, founder of the Bustamante Industrial Trade Union, and twice a political prisoner of the British. He was an exotic of a very different kind, but scarcely less disconcerting to the imperial tradition, for nobody really knew the truth about him. Why had he changed his name from Clarke to Bustamante? Was it or was it not true that he had served with the Spanish Army in Africa? What was the thread that bound his extraordinary career—waiter, salesman, dietician, money-lender, police inspector, New York tramway company official, labour leader and finally royalist, anti-socialist, anti-atheist nationalist politician? Or what about Gamal Abdel Nasser, the Egyptian revolutionary leader who was presently to oust both Farouk and the British themselves? The British had never heard of the fellow, until he thrust himself to power, for he was a life-long conspirator, full of charm and good sense, but recognizing no accepted channel of dissent, publishing no manifestos, addressing no meetings, simply working quietly and anonymously until the moment of revolution came.

Such men were baffling to the Empire. Fair play did not allow for them, and the rules did not apply. The Arab leaders who now set about destroying British suzerainty were men of a kind the British hardly knew—middle-class townspeople or country bourgeoisie, who had never played a hand of bezique with the Ambassador up at the palace, nor even been invited ('well it *would* be kind, dear, useful too') to take tea with the Manager's wife on Saturday afternoon. In Singapore the easy-going Malays, whom the British liked and patronized, were being supplanted by the formidable and secretive Chinese, whom the British feared. In Cyprus a learned Archbishop, Makarios III, theological graduate of Boston University, was allying himself with a murderous guerilla, Georgios Grivas, in a campaign to achieve *enosis*. In South Africa the unshakeably alien Afrikaners rid themselves of Smuts for the last time and entrenched themselves it seemed indefinitely in power.

It was not simply a new world, as it had been after the First World War. Now it was *several* new worlds, surrounding the perimeters of the British Empire, and erupting within. This great movement of change and discontent was more organic than deliberate. No universal conspiracy linked the scattered patriots. Some were Communists, some tribalists, some men of religion, some cultural irredentists. Some were pursuing old grudges, or evolving new ambitions. Many were only doing what their masters had taught them, pursuing those very principles of political freedom which were, they had so long been told, the glory of English history. All, though, were encouraged by one common instinct: the British Empire was dying, and the heirs must prepare themselves.

5

The blindest observers of the process, except of course those myriad primitives who had not yet learnt what an Empire was, were the British themselves. They continued to believe in the power of their prestige, and sometimes it seemed indeed as though prestige *per se* was the object of their foreign policies. Prestige had been the basis of their imperial power for so long that they clung to it faithfully now—it had been, as Elspeth Huxley had written of its manifestations in Kenya, 'like an invisible coat of mail, or a form of magic', and even after Singapore the British believed in it. Sometimes indeed it still worked, and the safest man in a nationalist riot, Cairo to Calcutta, was often the man who proclaimed his Britishness frankly and proudly. But it was a deception too, and prevented the British from coming to terms with history, and accepting their reduced circumstances in the world.

For even now, at the end of the 1940s, most of them saw that world through Victorian glasses. They believed in their hearts that things British were necessarily things best. They believed that they, above all their Allies, had won the war. They saw themselves still, like their grandfathers, as a senior and superior race. Not only within their Empire, but across half the world, they expected to be treated with deference, to use the best hotels, to get the best table, to be waited upon, the Australian writer Alan Moorehead observed, 'like the children of rich families'.

The islanders were as chauvinist as ever, for all their experiences on the battlefields, and attitudes at home had not much changed since Victoria's day. Foreigners were still inferior. Coloured peoples were still of a different class. Abroad was still comic.[1] The popular

---

[1] *Englishman:* Do you have frogs' legs?
French waiter: Mais oui, monsieur.
Englishman: Well hop off and get me a cheese sandwich.

Press remained the strident instrument that Northcliffe had made it, in the first excitement of popular literacy, and it spoke even now in the raucous voice of the New Imperialism, braggard and insensitive. 'Mightier yet!' sang the audiences at the Promenade Concerts in Queen's Hall, swinging into 'Land of Hope and Glory' to a tumult of balloons and paper caps, and one would hardly guess that Great Britain was monumentally in debt, industrially crippled and strategically strained almost to breaking-point.

But it was not an *imperial* braggadocio. Except in the vague and satisfying abstract, except in matters of 'prestige', the British cared little about their Empire, and knew still less. A poll in 1947 revealed that three-quarters of the population did not know the difference between a Dominion and a colony, that half could not name a single British possession, and that 3 per cent thought the United States was still a British colony. Their pride was in themselves, not in their Empire, and in this if nothing else their instinct was right. The days of the European Empires were over, and if Britain was to be prosperous and influential in the future it must be as an island Power off the coast of Europe. Now as always, it had not been the British Empire that the world really respected. It had not even been, as a matter of fact, Great Britain. It had been England, the heart of it all, England of Shakespeare and the Common Law, England of the poets and the liberators, Churchill's England of the white cliffs and the Cockney courage:

> *May you be saved, Shakespeare's island, by your sons and daughters*
> *and glorious ghosts.*
> *Here, from far-off shores*
> *I summon them and they respond,*
> *thronging out of their numberless past,*
> *mitred and iron-crowned,*
> *with Bibles, swords, and oars,*
> *with anchors and bows. . . .*[1]

[1] By Jorge Luis Borges, 1940, translated from the Spanish by Norman Thomas di Giovanni. Borges was of partly English ancestry, and lost his job as a librarian in Buenos Aires because of his British sympathies in the war.

CHAPTER TWENTY-THREE

1947

ON March 22, 1947, a new Viceroy arrived at New Delhi to take up office, and he and his wife were met by their predecessors on the steps of Lutyens' palace. They had all known each other for years, but never was there such a contrast in styles between the old incumbents and the new. Down the steps came Lord and Lady Wavell of Cyrenaica, elderly and benevolent: he dressed in his Field Marshal's uniform, his face grey and haggard, his one eye heavy-lidded (he had lost the other in the Great War), she smiling in the background in a low-waisted dress of floral silk, a tea-gown perhaps, wearing a silver ornament around her neck and looking for all the world like an evangelical bishop's wife welcoming a new curate.

Up the steps came Lord and Lady Mountbatten of Burma, in the prime of worldly life, dashing, good-looking, confident: he in the uniform of an Admiral, slashed with the medal ribbons of a triumphant wartime career, she svelte in green cotton, as though she might be going on later to cocktails in Knightsbridge, or a theatrical party on 51st Street. The Field Marshal was a cultivated, gentlemanly but not very demonstrative soldier, aged sixty-three, the Admiral a pushing, rather conceited and brilliantly enterprising sailor, aged forty-six. One man bowed to the other, the younger woman curtseyed to the older, but later that day the Viceroyalty passed from Wavell to Mountbatten, and the courtesies were reversed. Their meaning was ironic, for the Admiral's sole purpose in assuming this, the greatest office the British Empire had to offer, was to end the Raj in India, and conclude the long line of the Viceroys once and for all.

2

Since the constitution of 1935, India had muddled on, and she was no nearer independence ten years later. This was not for lack of trying. The nationalists had maintained their pressure throughout the war, the British, nagged by their allies, had gone so far in 1942 as to offer immediate Dominion status after victory, with the right to secede from the Empire too. Every initiative, though, had ended in deadlock. The most forceful of the Indian Hindus wanted instant independence—'Why accept a post-dated cheque', Gandhi is supposed to have asked, 'on a bank that is obviously crashing?' The most intransigent of the Indian Muslims wanted autonomy for themselves. The British, and especially Churchill, did not really want independence to happen at all.

Most Englishmen still doubted if Indians were ready for self-government: for one thing the populace was now apparently irrevocably divided on religious lines, for another there was the problem of the myriad Princely States, not part of British India at all, but direct feudatories of the imperial Crown. Most Indians still doubted if the British were sincere, suspected that the Muslim-Hindu rivalry was encouraged for imperial purposes, and had no patience with the pettifogging and sycophantic princes. The Hindus believed themselves to be the natural successors to the British as rulers of all India: the Muslims believed themselves to be natural rulers *per se*, never subject to Hindu rule and never likely to be: the Sikhs believed themselves to be separate from, superior to and irrepressible by any other parties in the dispute: the British thought themselves, *au fond*, indispensable.

The war had sharpened these multiple antagonisms. The British were disillusioned by Indian behaviour during the war. Though Indian soldiers had fought on nearly every front, they had really fought more as mercenaries than loyalists—'India's soldiers', Gandhi wrote in 1942, 'are not a national army, but professionals who will as soon fight under the Japanese or any other if paid for fighting.' The British were dismayed to find the defectors of the Indian National Army greeted as heroes by the populace, and their leader Subhas Chandra Bose hailed after his death as a martyr and a liberator. None of the chief Hindu leaders had helped in the war—they were all imprisoned for subversion in 1942—and the 'Quit India' movement had brought the country to the brink of revolution at the most vicious moment of the conflict: large areas had been altogether out of Government control, communications between Delhi and Calcutta were cut and more than 100,000 people arrested.

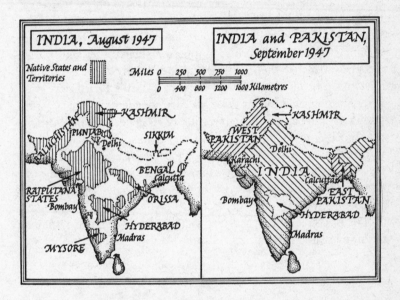

Conversely British prestige had been irrevocably eroded by the war. The Viceroy's unilateral declaration of war had been bitterly resented. 'There was something rotten', Nehru thought, 'when one man, and a foreigner and representative of a hated system, could plunge 400 million human beings into war without a slightest reference to them.' The fall of Singapore fatally weakened the British military reputation, and as more and more Indians succeeded to senior jobs in the Government at home, so the Raj itself lost its power of aloof command. Every Club had its Indian members now, and the mystery had gone. In wartime the British might be as ruthless and resolute as ever, but in peacetime, as many a percipient Indian understood, they would never hold on to India by sheer force.

It was not mere loss of will—to many members of the ICS the liberal tradition of their service was now reaching its fulfilment, and the transfer of power would be the honouring of an old purpose: but whether it was weakness or high principle that the imperialists displayed, cautiously the Indian opportunists edged away from them, looking ahead to new patrons.

Anyway, ever since the Salt March Indian leaders had felt themselves to be masters of their own destinies. It was only a matter of time, as Churchill had foreseen. The Indians, Hindu and Muslim, were perfectly conscious of their power—all over the world public opinion, ignorant or informed, supported their cause, and it was inconceivable that the British would indefinitely defy it. Gandhi, now the *guru* of the Congress Party to Nehru's Presidency, was already recognized by many of his correspondents as the *de facto* President of India, and the universality of his appeal, the implication that India represented, now as always, deeper spiritual values, gave to British actions a sadly parochial, almost suburban air. It seemed to the British only their duty, to arrange matters with order and dignity: it seemed to the rest of the world only sophistry and procrastination.

The British aimed, as usual, at compromise. The Indians wanted nothing less than absolutes.

3

Many of the British, even now, failed to grasp their true relationship with India. The habit of sahibdom was too ingrained, the attitude of condescension, even mockery, still natural to them. It was through a veil of false constructions that they groped their way towards an Indian settlement, falling back often upon legalities and constitutional pragmatisms. They were torn between themselves, and *in* themselves, for even the most sincerely liberal Englishman felt a pang of regret, when he considered the ending of the Indian Empire. Subconsciously, no doubt, they often hoped for failure, as the long negotiations intermittently continued, and frequently officials in the field felt London reluctantly lagging and prevaricating at the other end of the cable. The chief obstacles to progress, said the

Viceroy Lord Linlithgow, were Indian political stupidity and British political dishonesty, but the chief one really was lack of decision. Two Viceroys had presided between the 1935 Act and the arrival of Lord Mountbatten: both were men of honour, both sincerely worked for an Indian settlement, but neither had the decisive powers to achieve it.

Lord Linlithgow was the first. 'Hopie', as his friends called him, was a Tory aristocrat, like most of his predecessors—of the eight twentieth-century Viceroys, six had been the sons of peers and five had been Etonians. He was a very tall man, 6 foot 5 inches, an elder of the Church of Scotland, a devotee of English music-hall and a former deputy chairman of the Conservative Party. Doreen his wife was almost six foot herself, handsome, sociable, daughter to an antique baronetcy, and indefatigable in good works. They made a formidable couple, and their intentions were altogether admirable, but they seem in retrospect out of their historical depth.

Diligently and tactfully Lord Linlithgow grappled with the bickerings, sectarian rifts and snaky rivalries of Indian political life: like obelisks of British probity Doreen and her family stood at his side, dignified always, never daunted, and often breaking into the chorus of 'The Roast Beef of Old England' when the band played them into dinner at the Viceregal palace. Linlithgow was Viceroy for more than seven years, and his knowledge of Indian affairs was enormous. Yet he achieved, in effect, nothing. Though he assiduously implemented the Constitution of 1935, nothing was added to it in his time, and the reason perhaps was not lack of integrity, or lack of intellect, or even the frequently meddlesome interference of London, but incompatibility. Lord Linlithgow, who once admitted to his private secretary that he had never set eyes on an Indian rupee, was as absolutely removed from the roots of the Indian problem as it was possible to be—in temperament, in background, in experience.

Instinctively he played for time, and sought to preserve the past rather than hasten the future. It was Linlithgow who declared war on India's behalf without consulting a single Indian, and his promises of self-government to the Indian people were hedged all about with qualifications and reservations—'entering into consultation', 'in the light of the then circumstances', 'such modifications as may seem desirable', 'subject to the due fulfilment of obligations'. It was not always the Viceroy's fault—every sentence of his declarations had to be argued out with London—but still it was hardly the stuff of generosity. Lord Linlithgow had never, as the Indian conservative leader Tej Bahadur Sapru said, 'touched the heart of India'. If he was a great man to his Scottish tenants, or even to his adoring staff at the Viceroy's palace, set against the scale of the Indian future he was a man out of his class—a good, cautious man, promoted, like so many a pro-consul of Empire, beyond his genius.

His successor was Wavell. A kinder person seldom lived, and there had never been a Viceroy more earnest or benevolent than he, or one more truly anxious to give Indians their rights. He seemed in many ways the ideal Englishman. He had written a scholarly biography of Allenby, and lectured at Cambridge on the arts of generalship. He had memorized so wide a repertoire of English verse that he later turned it without addition into a book. His subordinates loved him. His colleagues found him modest and helpful. Yet he was a loser, often petulant and frequently depressed. Churchill recognized it, and never had much faith in him, which is why in June 1943 Wavell was appointed Viceroy of India instead of getting some great command in the assault on Europe. His war record, generally through no fault of his own, had been disastrous. His only victories, which had made him the idol of the British middle classes, had been won against the inept Italian armies in Africa, and he had never won anything again: he had lost Greece and Crete, would have abandoned Iraq but for Churchill's prodding, and as Commander-in-Chief in south-east Asia, failed to save Malaya, Singapore or Burma. He was condemned, he said himself, 'to conduct withdrawals and mitigate defeats'. He lacked panache. He felt a fool when the Indians, in their effusive way, garlanded him with flowers. He was so taciturn as to be disconcerting in company. He was reluctant to take initiatives, and had no confidence in his own powers as Viceroy. 'I very much doubt', he wrote in his diary on New Year's Day, 1946, 'whether my brain-power or personality are up to it.'

He was right. His personality was *not* up to it. Churchill, having considered Eden and Miles Lampson of Cairo as possible Viceroys,

had perhaps chosen Wavell precisely because he did not want a solution in India—he hoped still that the problem could be indefinitely postponed. Wavell, himself, though, genuinely believed in Indian independence, and was ashamed of the often specious policies which emanated from London—'we were proposing a policy of freedom for India, and in practice opposing every suggestion for a step forward.' He was embarrassed by the insincerity of British attitudes, and he responded generously to Indian feelings—'If India is not to be ruled by force, it must be ruled by the heart rather than the head.'

Yet even so he never achieved a rapport with the Indian leaders. They rather looked down on him, as the simple soldier he often claimed himself to be, and he quite failed to perceive their stature. He called Nehru's approach to Indian independence 'sentimental', and wilfully refused to recognize the greatness of Gandhi, whom he described at one time or another as obstinate, domineering, double-tongued, unscrupulous, impertinent, malevolent and hypocritical. Just as Lampson had habitually called Farouk 'the boy', so Wavell, with the same defensive contempt, liked to speak of Gandhi as 'the old man'.

These were the two statesmen, helped as often as they were hindered by their instructions and advices, who for the twelve years after 1935 tried to cut the Indian knot. They were hamstrung by circumstance, historical and political, and distracted by the exigencies of a terrible war: but some of their disadvantages lay in themselves.[1]

4

The knot was worse than the Gordian, and everyone was tightening it. '*Beware the Gandhiji*', wrote Wavell one evening at the end of a long day:

> *Beware the Gandhiji, my son,*
> *The satyagraha, the bogy fast,*
> *Beware the Djinnarit, and shun*
> *The frustrious scheduled caste.*

[1] Linlithgow became chairman of the Midland Bank, dying in 1952: Wavell became president of the Royal Society of Literature, and died in 1950.

The long withdrawal: Independence Day, Ghana, 1957

The year was 1946, and he had been grappling with a Cabinet mission sent out by the new Labour Government to make yet another fresh start. By then the prospect of a peaceable transfer of power to a united independent India seemed remote, for every Indian aspiration was subdivided into lesser hopes, and complicated by deceptions, illusions and contradictions—as Gandhi said, every case had seven points of view, 'all correct by themselves, but not correct at the same time and in the same circumstances'.

All the parties were at odds. Congress claimed to be the natural successor Government still, but the Muslim League, under the inflexible M. A. Jinnah, was now demanding a separate State for the country's 90 million Muslims—Pakistan, 'Land of the Pure'.[1] But there were many lesser disputes and anomalies. The Sikhs wanted a Sikhistan, the 584 Indian States mostly wanted to remain within the Empire, the Kashmiris were a Muslim community under a Hindu prince, the Hyderabadis a Hindu community under a Muslim prince. Even the British themselves pursued varying interests. The Indian Political Service was deeply committed to the cause of the Princely States, the British commercial community was anxious about its future profits, the Indian Civil Service was planning another competitive examination for British entrants, and an Indian Army Commission, comprising three British officers and one Indian, had lately recommended that half the Indian Army's officers should continue to be British.

And gradually, as they argued, the Raj was cracking. Government was running down, the ICS was now half Indian, and Wavell himself admitted that the British had lost nearly all power to control events. Riots and strikes swept the country. Illegal organizations proliferated. The Indian Navy mutinied. There were rumours that the Afghans were about to invade the North-West Frontier Province, that the Sikhs were about to rebel in the Punjab. The integrity of the police became ever more suspect, and even the civil service became for the first time politically tainted, as some of its Indian officers tacked understandably to the wind. Above all the

[1] A conception first devised by Indian students at Cambridge in 1932. P stood for Punjab, A for the Afghan areas of the North-West Frontier, K for Kashmir and S for Sind, while PAK handily meant 'pure', in a religious sense.

spectre of communal war, Muslim versus Hindu, now stalked the country. In August 1946 the two religions clashed so violently in Calcutta that in a single day 5,000 people were said to have died. Demagogues of both sides gained eager audiences everywhere, and from Dacca to Peshawar people prepared to kill or be killed, in the cause of Kali or at the bidding of Allah. 'We shall have India divided,' wrote Jinnah, 'or we shall have India destroyed!' 'I tell the British,' cried Gandhi, 'give us chaos!'

Still Wavell laboured on, studying his interminable instructions from London, cabling endless memoranda, reasoning with Jinnah, debating with Nehru, hammering away at constitutional niceties or communal discrepancies. He blamed it mostly on Churchill's Government. 'What I want', he wrote, 'is some definite policy, and not to go on making promises to India with no really sincere intention of trying to fulfil them.' Once he seemed almost to succeed, when the Indian leaders, assembled in conference at Simla, appeared ready to accept a constitutional settlement: but when that hope collapsed too, the exhausted, dispirited and now embittered Viceroy gave up. Perhaps the only way, he thought, was simply to leave, without devising a solution at all—giving them chaos, if that was what they wanted. Perhaps they should withdraw province by province, women and children first, then civilians, then the army, leaving India to murder, burn and loot itself as it wished. He sent the plan home to London, and called it 'Operation Ebb-Tide'.

> *Twas grillig* [he wrote]; *and the Congreelites*
> *Did harge and shobble in the swope,*
> *All jinsy were the Pakstanites,*
> *And the spruft Sikhs outstrope.*[1]

5

Presently he was sacked. Attlee, the Prime Minister of the new Labour Government, had firm views about India. He had gone there on a Parliamentary mission as long before as 1928, and had concerned

[1] His Indian Jabberwocky is printed in *The Viceroy's Journal*, edited by Penderel Moon (London 1973).

himself with the subject ever since. He had long ago reached the conclusion that only the Indians themselves could solve their own problems, freed of all British constitutional restraints, and even before the war he had argued that India should be given Dominion status within a fixed period of years. By 1946 he was sure that the transfer of power must be made as soon as possible—peaceably if possible, with the rights of minorities protected if they could be, but above all quickly, and absolutely. The first necessity, he thought, was to be rid of poor Wavell, whom he considered defeatist and ineffectual, 'a curious silent bird'. It did not take him long to find a successor. 'I thought very hard,' he wrote, 'and looked all around. And suddenly I had what I now think was an inspiration. I thought of Mountbatten.'

Mountbatten! The perfect, the allegorical last Viceroy! Royal himself, great-grandson of the original Queen-Empress, second cousin of George VI, though by blood he was almost as German as he was English he seemed nevertheless the last epitome of the English aristocrat. He was a world figure in his own right, too, for as Supreme Commander in South-East Asia he had commanded forces of all the allied nations—one of the four supremos who, in the last year of the war, had disposed the vast fleets and armies of the western alliance. Moreover he was a recognized progressive, sympathetic to the ideals of Labour, anything but a reactionary on the meaning of Empire, and with a cosmopolitan contempt for the petty prejudices of race and class.

'What is different about you from your predecessors?' Nehru asked Mountbatten soon after his arrival in India. 'Can it be that you have been given plenipotentiary powers? In that case you will succeed where all others have failed.' The Viceroy had in fact *demanded* such powers, enabling him to reach swift decisions on the spot. He had also committed Attlee to a date for the end of British rule in India, with no escape clauses. The Raj was to end not later than June, 1948, when complete power would be handed to Indian successors.

This renunciation meant that Britain had no bargaining power any more: she was genuinely disinterested at last, and was concerned only to see that India was left a workable State, preferably a member of the Commonwealth, at least friendly to Great Britain. She had nothing much to offer in return, now that liberty was so firmly pledged, but nevertheless Mountbatten was marvellously, some thought overweeningly, self-confident. As he said himself, he thought he could do anything, and sure enough the combination of prestige, assurance and clear intention made him a much more formidable negotiator than the aloof Linlithgow or the despondent Wavell. It meant that he was arguing, if not from strength, at least from style.

The Mountbattens brought to the viceregal office an element of *brio* absent since the days of Curzon. They sustained the swagger of it all, the thousands of servants, the white viceregal train, the bodyguards, the curtseying and the royal emblems, but they made it contemporary. Gone were the ancient shibboleths of the court. The only royal Viceroy was the least grandiose of them all. At viceregal dinners now half the guests were always Indian, and earlier incumbents might have been horrified to observe how frankly the Mountbattens talked to natives of all ranks. It was an abdication in itself, for it was the very negation of imperial technique, but it was proper for the times and the purpose.

Mountbatten hoped to leave behind a federal united India, Hindus, Muslims and Princely States constitutionally linked. As second-best, he aimed at a peacefully divided one. He was adamant from the start that there would be no reservations or hidden clauses. 'All this is yours', he said to Gandhi one day, when the Mahatma asked if he might walk around the viceregal gardens. 'We are only trustees. We have come to make it over to you.' No Viceroy had ever talked like that before, and no Viceroy had ever ventured into such intimate political relationships. During his first two months in India Mountbatten had 133 recorded interviews with Indian political leaders, conducted always in an atmosphere of candid urgency—if the Indians wished to inherit a peaceful India, they must decide fast how to arrange it. He talked to scores of politicians, but the fate of the country was really decided by four men: the Viceroy himself, Gandhi, Nehru and Mohammed Ali Jinnah.

Mountbatten recognized the force of these men. Day after day he received them, usually together, sometimes separately, in the sunny and fresh-painted study at his palace. Times had greatly changed, since the half-naked fakir had first penetrated this imperial sanctum to negotiate with the austere Lord Irwin, while the palace servants gaped to see a political agitator exchanging badinage with the Viceroy. Now the negotiators met on an equal footing, like distant relatives assembling to divide an inheritance. The talks were seldom easy, for the issues were colossal and Indian passions ran high, but they were not generally rancorous, for at last it was patent that the British interest was not in keeping India, but in honourably getting rid of it.

Mountbatten's relations with the three leaders greatly differed. Gandhi, past the peak of his career, he recognized as a kind of constitutional monarch: he was baffled by him, charmed by him, often, like all Englishmen, irritated by him—'judge of my delight', he reported once, 'when Gandhi arrived for a crucial meeting holding his finger to his lips—it was his day of silence!' Alone among the senior British officials of India, though, he became a friend of the Mahatma, and Gandhi in return gave him his affection. He was, as always, free with his advice—'Have the courage to see the truth, and act by it!'—but he was attracted by the soigné youthfulness of the Mountbattens, their combination of the simple and the very urbane. Though he was never reconciled to the idea of a divided India, still by making his friendship publicly clear, by appearing often in happy companionship with the Viceroy and his wife, Gandhi gave his *imprimatur* to the course of events.

Jinnah was a very different negotiator. He was dying of cancer, but nobody knew it: he was as decisive as Mountbatten himself, and as confident too—since the date of independence had already been decreed, he knew that he had only to keep arguing to ensure that Pakistan came into being. His lawyer's brain was sharper than the Viceroy's, his purpose more dogmatic, and as the months passed towards independence day he became ever more adamant that the only solution was the partition of India and the establishment of Pakistan under the government of the Muslim League. Though a Muslim only in theory—he was the grandson of converts, and could speak no Urdu, the language of Islam in India—rather than submit to Hindu rule, he said, he would have a Pakistan consisting only of the

Sind desert. A gaunt, wintry, rather alarming-looking man, wearing a monocle said to have been inspired by Joe Chamberlain's, and suits of irreproachable cut, Jinnah was very Anglicized: he had a house in London, and had spent much of his life in England. He was, though, impervious to the Mountbatten charm, and noticeably resistant to logic or sweet reason. Mountbatten thought him the evil genius of the drama, the wrecker, and called him a haughty megalomaniac.

It was Nehru who became closest to the Mountbattens, and this was not surprising. Nehru was an agnostic intellectual, but of a sensual, emotional kind, a patrician like Mountbatten himself, a charmer and a lover of women—the Nehrus were famous philanderers. He was a Kashmiri Brahmin, with much of the Kashmiri melancholy and introspection, and though he had spent his life fighting in the patriotic cause, he was highly susceptible to personal magnetisms. He needed a cause, a love, a leader. He was the devoted subject of Gandhi, and in a subtle, tacit way he was the passive collaborator of Mountbatten. The two men were of an age and of a taste, in many ways complementary to one another: through all their tortuous talks an understanding ran, an acceptance perhaps that they had more in common as men than they were at variance as statesmen.

Getting to know these three men, consulting many others, weighing the opinions of his administrators, nevertheless Mountbatten soon made up his mind about the fate of India—too soon, his critics were to say, and there was something impetuous to his solution, something inherited perhaps from his experiences of war. As the months passed so the British grip on security weakened, until the army could no longer guarantee order, the police were helplessly over-strained, the intelligence services had disastrously decayed and even the last Britons of the administration were demoralized. The conviction had gone, and as the country slid towards anarchy few Britons in India would now argue the case for Empire—Field Marshal Claude Auchinleck, the Commander-in-Chief, said of his own army that any Indian officer worth his salt was now a nationalist, 'and so he should be'.

Mountbatten's press attaché, Alan Campbell-Johnson, likened

India in 1947 to a ship on fire in mid-ocean with a hold full of ammunition. The Viceroy himself, finding the parties irreconcilable, decided that partition was inevitable, and the sooner the better. 'There was in fact no option', Campbell-Johnson thought. Two Dominions would be created at once, with immediate independence. There would be no interim Government of any kind, no gradual transfer of power. Punjab and Bengal, with almost equal numbers of Muslims and Hindus, would be bisected. The Princely States would be urged to join one Dominion or the other. Everything would be partitioned, the Indian Army, the National Debt, the railway system, down to the stocks of stationery at the New Delhi secretariat, and the staff cars of GHQ.

Mountbatten flew to London to get the Government's approval, and persuaded Churchill, in Opposition, not to delay the process. Then, in an enormous hurry to prevent the whole administration falling apart in communal violence, he set the plan in motion. Congress and the Muslim League both accepted the proposals, and together their leaders announced it to the nation on All-India Radio —Jinnah in English, his speech being then translated into Urdu. Mountbatten himself gave a press conference to announce that the British could not wait until the following year after all—independence had to come before the end of 1947. Pressed by reporters to name a date, he decided there and then upon the anniversary of the surrender of the Japanese in 1945. They would withdraw from India, he said, completely and irrevocably, by August 15, 1947, and sovereignty would then be surrendered. After 250 years on Indian soil, the British had given themselves seventy-three days to retire.

6

'Plan Balkan', the precipitate partitioning of India, was hardly a dignified process, but it was decisive. It was, like Dunkirk, a failure dashingly achieved, with a touch of sleight of hand. The unity of India, the proudest achievement of British rule, was to be deliberately sundered. The great institutions of British India were to be split down the middle. The Indian Princes, who had lost their own independence in return for the protection of the Crown, were to be betrayed. A sub-continent on the brink of civil war was to be subjected to an enormous social and governmental upheaval, millions of people to assume new nationalities, Ministries to be shuffled here and there, loyalties to be suddenly jettisoned or invented. Mountbatten himself called it 'sheer madness, fantastic communal madness'.

But it was final. Gone were the old fumblings and second thoughts which had, for forty years, alternately delayed, hastened and obfuscated the transfer of power in India. Seventy-three days! In London the India Independence Bill ran through all its Parliamentary stages in a single week, ending at a stroke all British claims to sovereignty in India, and abrogating all the hundreds of treaties concluded between the Crown and the Princely States. Mountbatten kept a large calendar on his desk, to mark off the days, like the count-down of a space launching, and with hectic resolution the British in India prepared the obsequies of their paramountcy. It had taken them centuries to pacify and survey the immense expanses of their Indian territories, father succeeding son in the great task: now in a matter of weeks a boundary commission sliced the edifice into parts, laying new frontiers like string on a building-site, under the dispassionate instructions of an English barrister, Sir Cyril Radcliffe, who had never set foot in India before. The Indian Army was bisected, regiments split by squadrons, companies that had served together for a century suddenly distributed among alien battalions. The white Viceregal train chuffed away from Delhi for the last time, for it was allotted to Pakistan: the officials of the Kennel Club were relieved to be told that its assets would remain in India.

All went! All the dreams of the Empire-builders long before, all the Curzonian splendour of imperial function, and as the British relentlessly cleared their office desks, India subsided into anarchy. Eleven million people abandoned their homes and moved in hordes across the countryside, hastening to the right side of the new communal frontiers. The roads were crammed with refugees, people clung to the steps of trains, or crowded upon their roofs, and the old gypsy confusion of India, the crowding and the clutter, the always familiar scenes of exhaustion, bewilderment and deprivation, were multiplied a thousand times. Violence erupted on a scale never known in British India before, even in the Mutiny. It was like a gigantic boil bursting, an enormous eruption of frustrations and resentments suppressed for so long by the authority of Empire. Whole communities were massacred. Entire train-loads of refugees died on the tracks, to the last child in arms. In the Punjab gangs of armed men roamed the countryside, slaughtering columns of refugees, and thousands of people died unremarked in the streets of Amritsar, where the death of 400 had horrified the world twenty-five years before.

The Viceroy was not deterred. Working day and night the last British officials of the Indian Civil Service established administrations for the two new Dominions, one with its capital at New Delhi, the other at Karachi. A Boundary Force, commanded by a Welshman with Hindu and Muslim advisers, was hastily put together to try to keep the peace, and the British Army was steadily and unostentatiously withdrawn: one by one the regiments left the soil of India, embarking on their troopships at Bombay while the bands on the Apollo Bunder played 'Auld Lang Syne', and the old hands of the Royal Bombay Yacht Club looked on sadly from their verandah.

The last days passed. The last of the wavering Princely States opted for one Dominion or the other, leaving only Kashmir, Hyderabad and Junagadh with their futures unresolved. The last of the British imperialists desperately worked against the clock, creating at least rudimentary Governments to succeed themselves. As the bloodshed and the turmoil continued, as the migratory masses laboured this way and that across the Indian plains, in the midst of it all the British Empire came to an end in India. At 8.30 a.m. on the appointed day, August 15, 1947, the Union Jack was hauled down all over the sub-continent, from frontier fort and Governor's palace, from law court and town hall, from the water-front at Surat where it all began, from the Viceroy's crowning palace, from the ruined Residency at Lucknow where the flag had flown night and day since the Indian Mutiny—

> *Shot thro' the staff or the halyard, but ever we raised thee anew,*
> *And ever upon the topmost roof our banner of England blew.*[1]

[1] Said to have been the 200th in the line of succession from the original

In Karachi Jinnah was sworn in as Governor-General of the Dominion of Pakistan. In Delhi Lord Mountbatten gave up the Viceroyalty, with its almost despotic power, and became the Governor-General of the Dominion of India, with no power at all.

Incalculable crowds celebrated the event in Delhi—the largest crowds anybody could remember seeing, in a country of multitudes. The grand military parade planned for that evening was cancelled, the military being so engulfed in the crowd that only the coloured tips of their puggarees were visible from the grandstands. When the Mountbattens returned in their State coach to the palace, thronged all about by hysterically happy crowds, Nehru sat on top of the hood, 'like a schoolboy', Mountbatten reported, four Indian ladies with their children clambered up the sides, a Polish woman and an Indian newspaperman hung on behind. Thus queerly loaded, and accompanied by several thousand running, cheering people, the last of the Viceroys clattered the long mile up Kingsway to his great house, while fireworks sprayed the evening sky above him, and the thunder of a million people echoed across Delhi.

It was done. The flags of the new Dominions, the Indian decorated with the Wheel of Ashoka, the Pakistani with Islam's star and crescent, flew all over the sub-continent, and all the paraphernalia of authority was handed over to the successor States. King George VI, Emperor no longer, sent his somewhat stilted greetings. Mountbatten proclaimed it 'a parting between friends, who have learned to honour and respect one another, even in disagreement'. Attlee said it was not the abdication, but the fulfilment of Britain's mission to India. The British Press was self-congratulatory, the British public, casting a cursory eye over the reportage, thought that on the whole they were well out of it. Some 200,000 Indians had died.

7

'The wheels of fate', Rabindrinath Tagore had written on his

---

mutiny flag, the Lucknow Union Jack was sent to King George, at his own request, to be kept at Windsor Castle.

death-bed, 'will one day compel the British to give up their Empire.
. . . What a waste of mud and filth they will leave behind!'

Still, it was fulfilment of a kind. The British left India euphorically
surprised and touched by the goodwill they were shown, retaining
most of their commercial advantages, and leaving behind at least
cadres of government. Since both the new Dominions chose to
remain within the Commonwealth, they persuaded themselves that
the Empire had been preserved anyway, and they persuaded the
world that their withdrawal had not only been generous, but had
been planned all along, flourishing many a hoary quotation, from
Macaulay, Montagu or Ramsay MacDonald, to prove it. Few of
them felt bitter: many felt regret. Soon after the withdrawal Queen
Mary, widow of George V, had a letter from her son the King. 'The
first time Bertie wrote me a letter with the I for Emperor of India
left out,' she noted in her journal. 'Very sad.'

It was left to Gandhi to enact by sacrifice the bitter breaking of
his own long dream. He had always recognized the merits of the
British Empire. He stood himself for the best of its values—
tolerance, duty, service. His life-long fight against the Raj, his fasts
and imprisonments, his frail self-exhibitions high above the adulating crowds—all were full of contradictions, for he was not at heart
a politician at all: his real concern was not the immediate cause of
political emancipation, but the far greater one of human dignity. He
was a philosopher and a moralist, not an ideologue.

It seemed to him now that India's freedom was being gained
without Godliness, against an ironic *mise en scène* of bigotry. He had
been unassailably opposed to partition, which was a surrender to
religious prejudice, and thought the British would have done better
to allow Jinnah to preside over a united India for the remaining
years of his life, or even to adopt Wavell's policies of despair. As the
events of 1947 approached their climax, he withdrew from political
life into prayer and fasting, emerging into public only to preach to
the crowds that assembled wherever he appeared, or to calm by his
very presence some communal disturbance. He did not believe his
powers to be waning. He thought he would live to be 125—'or
according to learned opinion', as he used to say, possibly 133—and
he knew that his influence upon events remained immense. He

exerted it now, though, only in conciliation, to still the consequences of his own victory.

Time and again he appealed for reconciliation between the fanatic religionists; twice he entered into protracted fasts, to shame them into peace; he personally quelled riots in Delhi and Calcutta, so effectively that Mountbatten called him 'a one-man boundary force'. When independence came, and partition became a fact, Gandhi accepted it regretfully and fairly—too fairly, many of the Hindu extremists considered. They thought him altogether too conciliatory to the Pakistanis, just as they had always disliked his liberal ideas on caste; and so the chief architect and ornament of Indian independence, the most beloved and admired character to be given to the world by the entire imperial process, was killed by his own people for his opinions.

Every week Gandhi held a public prayer meeting in the garden of Birla House, the home of a rich sympathizer only half a mile from the Viceroy's palace in New Delhi. There, on January 30, 1948, he was martyred, the only apposite death perhaps for a man whose life was never ordinary. It was a crisp winter evening, and the usual throng of people had walked through the dusk to Birla House, and were sitting on the big lawn behind the house. The Mahatma emerged from a side door, spindly as ever and bowed now over his staff, for he was in his seventy-ninth year, and made his way to his prayer platform at the far end of the garden. As he climbed the shallow steps, and turned to greet the people with his *namaste*, the joined hands and bow of Indian courtesy, a young man in khaki clothing stepped out of the crowd and fired three shots at him with a revolver. Gandhi died at once, and they took his body upstairs in Birla House, and placed it on a balcony for all to see, illuminated by a searchlight from the road below.

As soon as he heard the news Mountbatten hastened down to Birla House. A huge crowd was still milling about the place, and as the Governor-General shouldered his tall way to the door a voice cried, 'It was a Muslim who did it!' Mountbatten did not know who had done it, but he reacted instantly in the tradition of Empire. 'You fool,' he cried, so that the crowd could hear, 'you fool! Don't you know it was a Hindu?'—and so the imperial instinct, for so long the keeper of the Indian peace, saved in the moment of its extinction a few last Indian lives.[1]

[1] Mountbatten returned to the Royal Navy in 1948, and became successively First Sea Lord and Chief of the Defence Staff before retiring in 1965. By 1976 his honours included KG, PC, GCB, OM, GCSI, GCIE, GCVO, DSO, FRS, Hon DCL (Oxford), Hon LLD (Cambridge, Leeds, Edinburgh, Southampton, London, Sussex), Hon DSc (Delhi and Patna), Legion of Merit, DSM (USA), Greek Military Cross, Grand Cordon of Cloud and Banner (China), Croix de Guerre, Order of White Elephant of Siam, Order of Lion of Netherlands, Knight of St John, and Grand Crosses of the Orders of George I (Greece), Legion of Honour (France), Star of Nepal, Isabella Catolica (Spain), Crown of Rumania, Avis (Portugal), Seraphim (Sweden), Agga Maha Thiri Thudhamma (Burma), Dannebrog (Denmark) and Seal of Solomon (Ethiopia). He was a member of seventeen clubs, if you include the Magic Circle.

My account of his Viceroyalty is subject to my own preferences for colour, panache and decision, but I should add that his policy has been much criticized. He is accused of being precipitate and partial to the Hindu cause, and it is argued that a more gradual approach, perhaps as Gandhi suggested arranging for Jinnah to head a united administration, might have obviated partition altogether. But Mountbatten did not know that Jinnah was dying: Gandhi possibly divined it.

CHAPTER TWENTY-FOUR

The Last Rally

THE end of the Raj in India made the end of Empire certain. 'As long as we rule India', Curzon had said, 'we are the greatest power in the world. If we lose it we shall drop straight away to a third rate power. . . . Your ports and your coaling stations, your fortresses and your dockyards, your Crown colonies and protectorates will go too. For either they will be unnecessary, or the tollgates and barbicans of an Empire that has vanished.' So it proved. Half the structure of Empire was mere scaffolding for the possession of India. Many a possession now lost its point, and the whole British attitude to the world, governed so long by the great possessions of the east, slowly and painfully shifted. 'If India becomes free,' Gandhi told Roosevelt in 1942, 'the rest will follow.' So it was, and after 1947 the British Empire was in a constant condition of dismantlement.

It was hardly a trauma for the British, because few of them recognized what was happening, or realized that with the loss of India three-quarters of the imperial population had gone anyway. Besides, the new Labour Government held confused convictions about imperialism. On the one hand they rightly saw the Empire as a phenomenon of class, tended to agree with Marx that it was an unwarrantable extension of capitalism, and believed in the essential equality of human kind. On the other hand they were by no means Little Englanders. They were patriots, proud as anyone of their part in the Second World War, and convinced that Great Britain should still be a power in the world. Ernest Bevin, the Foreign Secretary, believed that the day of empires was over, but he did not argue for any helter-skelter withdrawal from the frontiers, only for the establishment under British guidance of some collective partnership among all the imperial peoples. Hugh Dalton, the Chancellor, accepted the need to retreat, but not apparently on any high moral grounds: 'If you are in a place where you are not wanted', he said, 'and where you have not got the force to squash those who don't want you, the only thing to do is to come out.' The *habit* of Empire was almost as strong among Labour men as it was among Conservatives, and Attlee's Ministers knew all too well how heavily Britain's economy now depended upon the imperial territories—more than half British exports went there, nearly half British imports came from there—82 per cent of oil from the Middle East, 81 per cent of tea from India, 50 per cent of grain from Canada. Besides, many Socialists hoped that the colonies, especially the white settlement colonies, might be proving-grounds for their own social theories.[1]

But once they had surrendered India, everything changed. Almost at once Burma and Ceylon fell away from the Empire, without opposition from their imperial masters—'I only want to see Burma happy', said Governor Dorman-Smith, benignly surrendering his seals of office. Burma chose independence outside the Commonwealth, Ceylon became a Dominion like India and Pakistan. All at once the Empire became more fluid, more complex, more ungraspable than ever, as it became apparent to the policy-makers in London that the whole immense deposit was likely to start sliding, like a melting ice-flow, year by year towards independence. By 1948 it was an even stranger assemblage than Disraeli had known, when he called it 'this most peculiar Empire' a century before. It was no longer held together by the will, the force, or even the prestige of Great Britain, for as the war-clouds cleared it became clear that the British, though they might still enjoy some sensations of imperial power, had lost the ability to retain it. The idea of imperialism itself was almost universally discredited, only a handful of unpersuasive

[1] Upon which hangs a footnote. In 1836 five of the Tolpuddle Martyrs, the agricultural trade unionists of Dorset, returned from transportation to Australia and decided to try their luck in Canada. They made a pact among themselves that they would never tell their children the story of their imprisonment, to ensure them an absolutely fresh start, and it was only after they and their children had all died that their grandchildren discovered the truth, from a British Labour Party delegation to Canada in 1912.

diehards still arguing the moral right of one people to rule another against its will: and anyway the British could no longer afford, even if they wished it, to keep the huge fleets, expensive bases and scattered armies needed to keep half the globe in order. The British were no longer 'the world's policemen', as they had loved to call themselves before the war, and most of the old justifications of Empire were now invalid.

What to do with it all? How best to adapt it to the times, and employ the remaining imperial years? The British Commonwealth consisted by now of some eighty-five territories, in every stage of development or esteem. There were the four old white Dominions, in all respects independent nations. There were the three new brown ones. There was the anomaly of Southern Rhodesia, which was not quite a Dominion but not quite a colony either, being self-governing in some things but not in others. There was Newfoundland, which had been a Dominion once, had gone bankrupt and was now a colony again. There were the mandated territories like Palestine and Tanganyika, which were in the Empire but not quite of it, and equivocal territories like the Suez Canal Zone or the Sultanate of Muscat and Oman, which were of the Empire but not strictly in it. There were vast rich countries like Nigeria which possessed no political rights at all, and infinitesimal ones like Mauritius which were half-way to sovereignty.

All this survived, and infinitely more, the withdrawal of British power from India: to some an assurance of greatness still, to many more a bore or a deadweight, to the majority only a vague irrelevance. It fell to Attlee's Government to decide its future: and like so many of its predecessors, that Ministry decided that what the British Empire needed, even in epilogue or afterthought, was logic.

2

They rationalized the Commonwealth. This brainchild of Rosebery, Smuts and Balfour had been an oddity from the start, and was now a benevolent enigma. The white members, whose combined population was about 75 million, were now joined by the brown, with a combined population of 400 million, overwhelming the bonds of blood and common culture which had been the truest meaning of the association. The Statute of Westminster had been a sufficiently flexible conception, but even it could not easily embrace these new realities, especially as before long other Afro-Asian newcomers would certainly be joining too. Determined nevertheless that the Commonwealth should not dissolve in bewilderment, the British decided to make it more elastic still, blurring it into a kind of half-imaginary federation, rather like the Holy Roman Empire, whose purpose would be as vague as its allegiance, but whose very existence would keep alive, so it was hoped, some residual majesty of the old order.

In 1953 Queen Elizabeth II, Victoria's great-great-granddaughter, was crowned Queen of England on a drizzly day in May. The occasion was greeted optimistically as an omen of renewed greatness, a new Elizabethan age. The Coronation ceremony was still recognizably imperial, and all the Commonwealth Prime Ministers were there: Nehru svelte in his silken jacket, bluff Robert Menzies from Australia ('British to my boot-straps'), D. S. Senanayake the Ceylonese tea-planter, even the dour Dr Malan of South Africa, who had been a frank pro-Nazi, who was an outspoken republican, and who viewed all things British, not to speak of all things black and brown, with a properly Krugerian mixture of suspicion and disdain. The imperial symbols were paraded as always, the flags and the bearskins, the battle-honours and the horse-drummers: on the night before the ceremony the Queen was given the news, rushed just in time by runner and diplomatic radio from the Himalaya, that Mount Everest, the last objective of imperial adventure, had been climbed by a British expedition.

But when, amid the arcane splendours of Westminster Abbey, before the hushed peers and the silent trumpeters, the ninety-ninth Archbishop of Canterbury proclaimed Elizabeth Queen, he did so in evasive terms. Her father had been 'by the Grace of God of the United Kingdom of Great Britain and Northern Ireland and of the British Dominions beyond the Seas, King, defender of the Faith, Emperor of India'. His daughter was 'of the United Kingdom of Great Britain and Northern Ireland and of Her other Realms and Territories, Queen, Head of the Commonwealth, Defender of the

Faith'. The fanfares blared, the congregation stood, *Vivat! Vivat!* rang out across the fane: but there was no denying the bathos of this grey title, or hiding the process of retreat that had given birth to it. No longer was the Queen an Empress, and in an association of nations in which Hindus and Muslims outnumbered Christians by three to one, she was only debatably a Defender of the Faith.

So the fantasy of the Crown faded with the reality of the Empire. An immense sentimental web of loyalty had extended from Windsor Castle to the far corners of the British world, and even when nationalists were rioting, angry politicians were hurling calumnies at Governors, colonial newspapers fulminating about imperial exploitation, the Crown had been immune to insult. To have met, or even to have seen, a King, a Queen or a Prince of Wales remained, for millions of the old imperial subjects, one of the great experiences of life: some virtue was attached to the royal legend, and people of every kind aspired to its munificence.

Now this magic was following into oblivion that other, cruder sorcery, the spell of Empire itself. During the next decade the Commonwealth was to be expanded beyond the reach of the Crown, as old imperial territories were promoted from subjection to nationhood, its members being of varying races, colour, religion and culture, linked only, *au fond*, by the fact that the British had once ruled them. Now a man was a citizen of India, or Canada, or Ceylon, while the only British subjects were the islanders themselves, and the residents of the colonies they still governed direct from Whitehall. Gone was the Roman aspiration of a common citizenship spanning the world. One by one the Commonwealth countries deserted even the imperial law—each year the Legal Committee of the Privy Council, the Empire's ultimate tribunal, heard fewer cases from Yukon or the Outback, and esoteric litigations of the tropics seldom enlivened the law reports of *The Times*. Never again would there be an Imperial Cabinet, as there had briefly been during the Great War, and never again would a man like Jan Smuts rise through the agency of Empire to the summit of the world.

The word British disappeared from the title of the Commonwealth. The old Dominions and India Offices were merged into a Commonwealth Relations Office. Each member country was free to give the Queen its own title, so that 'Defender of the Faith' soon dropped from the royal honorifics when local circumstance, such as a universal devotion to Mohammed, or 99 per cent papistry, made it inappropriate. Logic in short was given to the association, but it was a logic shrouded more than ever in imprecision, so that in the end people were less clear than ever what the Commonwealth was all about. It had no constitution. Its members were bound by no obligations. They could leave when they pleased, devise their own status, decide whether they would have a Queen or not, adjust almost anything to suit themselves. It was a very obliging club.

The Indians had decided they wanted to be a Republic, so they became one. The Pakistanis decided they could do without a Governor-General, so they did. The Australians and New Zealanders concluded a defence alliance with the United States, excluding Britain. The Singalese preferred to regulate their affairs by the Buddhist calendar.[1] The South Africans pressed ahead with racial arrangements which were repugnant to all their fellow-members.[2] In the United Nations the Commonwealth members voted against each other whenever they felt like it. 'We want no unwilling partners,' Attlee declared, so rather than have rules broken, the creators of the new Commonwealth did without rules.

But at least the development of this cloudy new brotherhood helped to muffle the breakdown of Empire. The British deluded themselves that it *was* the Empire, more or less, reconstituted in contemporary form—was not the Queen to be seen gracefully presiding over its assemblies, or sweetly chatting with the black and brown men who, more numerous with each year as the decades passed, soon came to sit in the front row of the group photographs, instead of smirking self-consciously in the background? Ennobling epithets were attached to it—Family of Nations, Brotherhood of

[1] Or the lunar phases. 'Bank Closed', said a notice starkly when I went to cash a cheque in Kandy once, 'On a/c Full Moon Day'.

[2] Until at last the Afrikaners achieved their old dream, and declared the country a republic. The institution of *apartheid* as formal Government policy was too much even for the pragmatic Commonwealth to stomach: in 1961 South Africa's application to remain a member was rejected at a tense Commonwealth Conference, and the imperial connection with the country came to an end.

Heritage—and the veteran Milnerite Leo Amery, hopeful to the end, thought it might prove to be the nucleus of a new world order.

As for the other members, most of them approached it with an ambivalent mixture of cynicism, respect and affection. Old habits die hard, even among militant nationalists, and it was still splendid after all to be honoured among the grave monuments of imperial London, familiar to every Commonwealth citizen through generations of indoctrination, and to be received by the Queen of England herself, in a diamond tiara and the Order of the Garter. Besides, there was still something to be gained from membership of the sterling area, the financial structure which underlay it, and which was still one of the great economic blocs of the world. There were many fringe benefits too, technical aid and economic perquisites, access to professional bodies or athletic competitions, even a sort of half-nostalgic bonhomie, like the comradeship of an old school tie, which gave the Commonwealth delegates, even those who had lately fought, imprisoned or hurled insults at each other, an inner membership at international gatherings.

Nobody aspired to much more. The purpose of the Commonwealth, as it was officially defined, was nothing more than to 'remain united as free and equal members . . . freely cooperating in the pursuit of peace, liberty and progress'. It did look comforting on the map, though: for even if the cartographers only striped the Commonwealth countries in red, or merely speckled them with pink dots, at a cursory glance it looked as though a third of the world was still, in a manner of speaking, British. The Commonwealth was the last opaque reflection of the grand illusion.

3

But we must narrow our focus, for the British Empire was now something else. The countries of the Commonwealth became members of the wider world, and so slip away from our story: but in the 1950s there were still several million people, in all four continents, directly subject to the rule of London. Towards them the Labour Government looked in a very different spirit, avuncular and improving. It had always been a Fabian view, *pace* Marx, that a good colonial system could usefully reform and educate, and those few Labour politicians who interested themselves in the Empire clung to the vision of an imperial trusteeship. Generally less worldly than their Conservative counterparts, less experienced in foreign affairs, they were if anything more paternalist towards the coloured peoples, and believed them to need, however shamefully they had been exploited in the past, several generations of kindly British socialist supervision. Giving the African colonies independence, said Herbert Morrison the Home Secretary, would be 'like giving a child of ten a latch-key, a bank account and a shotgun'. Whatever had happened to the Dominions and India Offices, nobody suggested winding up the Colonial Office, and young men went out to Africa in the early 1950s looking forward with perfect confidence to a lifetime's useful career.

The nonconformist strain of Labour thinking was easily translated into an imperial earnestness, not so far from the idealism of Exeter Hall in the previous century, and so in the late 1940s the British colonial empire found itself not hastily jettisoned, as Tories had prophesied, but temporarily rejuvenated. Not since the heady days of the New Imperialism had an administration approached colonial affairs with such positive intentions, and Arthur Creech-Jones was the most enthusiastic Colonial Secretary since Joe Chamberlain himself. He believed the Empire had a purpose still— to prepare the subject peoples economically for political favours to come: to abandon the task would be 'to betray the peoples and our trust'. There was nothing transcendental to this view. It was a purely materialist, Benthamite approach, and little was done to influence the native peoples politically. Socialists though they were, hearty singers of 'The Red Flag' at party conferences, the Ministers of the Labour Government looked on the colonial Empire, as Chamberlain did, as a property in need of development.

Much of it was derelict. It had been British policy always that each Crown Colony must pay its own way. The passing of the wartime Colonial Development Act, as much a propaganda gesture as a change of attitude, had been the very first time metropolitan Britain had agreed to pay for the economic and social progress of the colonies. By now most of the tropical possessions were in a sorry state, many of them never having recovered their prosperity since the abolition of slavery, others never having achieved any prosperity to recover. The sagging verandah, the peeling façade, the beggars touting on the quayside, the slow cycle of malnutrition and disease, the scourges of crown-worm or tsetse fly—these were the images which, in every traveller's mind, were summoned by the British tropical colonies then. Bathurst, in the Gambia, was likened by a visiting academic in 1937 to 'a water-logged sponge floating in a sea of its own excreta'. 'General impression of Trinidad', noted the novelist Evelyn Waugh during a Caribbean journey, 'that I don't want to see it again. General impression of Georgetown that I don't mind how soon I leave it.' The war had brought a specious revival to some of these unlovely dependencies, but they had soon relapsed into indigence, while many of their citizens, returning to their humid huts or dilapidated shanties from war service abroad, for the first time looked around them with open eyes, and made their first dismayed comparisons.

The Labour Government, surveying this baffling preserve, resolved to give it order. It became an Empire of planners, developers and economic theorists. In 1947 the Colonial Development Corporation came into being, publicly funded and charged with the creation of public utilities in the colonies—the roads, power stations, water supplies and irrigation works which would be the foundation of their progress. The Colonial Office itself burgeoned as never before. Its staff tripled, its expenses were quintupled, and thousands of new young men, mostly wartime officers, were recruited to its ranks. For so long the fusty club of a few gentlemanly and detached administrators, the old institution had never seen anything like it, and the economic satirist Northcote Parkinson was inspired by its explosive expansion to devise his own theoretical interpretation, Parkinson's law, which postulated that a bureacracy would grow as fast as there were desks to accommodate it.

Every month, it seemed, a new commission was established, a new inquiry was instituted, a new committee was assembled to discuss health services, legislative reform, educational priorities, the possibility of growing sugar-beet on Ascension Island or the incidence of bilharzia in Dongola. Team after team went out to Africa, the

Caribbean or the Far East, and each territory was invited to produce a comprehensive development plan for its own development. Many an old imperial anomaly was weeded out—never again would a Brooke rule in Sarawak, a chartered company govern an African tribe, or a wandering English adventurer raise an eponymous regiment: down on the Embankment the Crown Agents for the Colonies, who had been in existence since 1833, were busier than ever concluding contracts, engaging staff, ordering postage stamps or investing funds on behalf of all the separate colonial governments, wherever they were.

Never had the colonies been so elaborately governed. Everywhere local establishments were vastly increased, sometimes tripled. In five years 6,500 new men were sent out to the colonies, six times as many as had administered the whole of British India in the heyday of Empire. Even the smallest and most neglected colony, a St Helena or a St Lucia, found itself suddenly invaded by technical experts and advisers, nearly all expatriate Britons—in Mauritius Mr Kenneth Baker, former president of the English Fire Brigade Union, was attached to the Governor's staff as Adviser on Trade Unions. It was the age of town-planning, and scarcely a colonial town escaped its inspection by disciples of Sir William Holford or Sir Patrick Geddes, nearly all of whom had grandiose ideas for plazas, ring roads and housing estates, and almost none of whom ever saw their plans implemented.

Money was poured into the estate out of Britain's denuded treasuries, and for the first time since Chamberlain's day the British tried to apply to their colonial empire the latest devices of technology. The Colonial Development Corporation, with its sibling the Overseas Food Corporation, poured cash, men, tractors, test-tubes, bulldozers, specialists and technicians into many a desolate tract of Empire, providing jobs for thousands of Britons otherwise difficult to employ in the officer-like circumstances to which they had become accustomed, and plunging hosts of tribesmen bewilderingly into the deep end of western method. Some of these activities were successful, some were famous fiascos, like the Groundnut Scheme for Tanganyika which sold not a single groundnut, or the Egg Scheme for Gambia which exported not an egg.

Earnest but over-sanguine, too, were the several schemes by which the British hoped to group their colonies into more rational political entities, ready to graduate in time into the ranks of the Commonwealth. They all made sense in theory. The West Indian islands, for example, could surely be united in a federation as a prelude to independence—they shared a common history, a common language, common customs and common problems: and so they were, Trinidad and Jamaica, Barbados and the Windwards, all with a common Assembly in Trinidad, and a smiling Governor-General in Jamaica. In East Africa Lord Delamere's old federal dream was revived, and Kenya, Uganda and Tanganyika were placed under a single East African High Commission.[1] Further south the plan was that Southern Rhodesia, Northern Rhodesia and Nyasaland should federate: the whites of Southern Rhodesia would supply the skills and the capital, the blacks of the other territories the labour and the mineral resources—a partnership, breezily suggested the Prime Minister of Southern Rhodesia, Godfrey Huggins, like that between a rider and his horse. Another federation was planned for south-east Asia, coalescing the various States of Malaya with Singapore, Sarawak and north Borneo, while in south Arabia the British hoped that fusion with the reliable rural protectorates might temper the potentially subversive urban society of Aden colony. Even among the ostensibly independent Arab States to the north, the British tried their methodological best: for it was they who first dreamed up the Arab League, a last muffled attempt at an indirect Viceroyalty of the Middle East.

All this was order! More and more complex constitutions were devised for the colonies, reported *in extenso* in *The Times*, as they progressed towards responsible Government: unabated came the flow of pamphlets and encouraging statistics from the presses of the Colonial Office. The official policy towards the colonies was perfectly clear: 'It is', said Command Paper 7533, 'to guide the colonial

---

[1] By the end of the Empire in East Africa, the three territories shared a railway, a currency, a postal union, a customs union, a common market, a university, an airline, a court of appeal, a tourist association, a development bank, a harbour board and an income tax department. By 1977 it had almost all disintegrated.

territories to responsible self-government within the Commonwealth in conditions that ensure to the people concerned both a fair
standard of living and freedom from aggression from any quarter.'
Yet now as always, it seemed, the British, even the most liberal of
the British, did not really believe that the objective would ever be
reached. It was simply an article of faith, ritually repeated. The
young men still went out in the hope of a life's career; the theorists
presupposed at least a couple of generations for their projects to
mature; all those brave federal schemes had as their object some
continuance of the old order, some maintenance of British influence
or supervision, or at least the survival of Britishness. Even promotion to the Commonwealth seemed to many Britons no more than a
confirmation, enabling fledgling States to enter into a fuller state of
grace, for while intellectually the British might see the truth about
their colonial empire, emotionally they clung still to the old illusions, and hoped it might somehow be induced, by all these infu-
sions of energy, cash and calculation, to last more or less indefinitely.

Almost opposite Westminster Abbey, in one of the finest sites in
Europe, a German bomb had created a large empty space, where the
Westminster Hospital used to stand. There the British, in the last
decades of their Empire, decided to build a new Colonial Office. The
India Office had gone, absorbed so thoroughly into the Foreign
Office that already very few Londoners could say where it had
functioned—even its library, that incomparable repository of
oriental knowledge, had been shifted somewhere east of Temple
Bar. But in a last hallucination of imperial hopes, these late imperialists determined to erect themselves a headquarters worthy of
their history. It would be in the neo-Georgian mode, the predominant style of the declining Empire, eight floors high, with
the royal crest large above its symbolically decorated doors, twin
flagstaffs on its roof and a façade facing across the square to the
ancient purlieus of the Abbey.

Here, passing through the pillared entrance hall, one would find
the general staff of the rejuvenated Empire, its Under-Secretaries
and its Deputy Secretaries, its legal officers, its information officers,
the editor of *Corona Magazine*—the librarians and archivists, the
code and cipher clerks, the radio specialists with their transmitters

on the roof. High on the top floor would be the Colonial Officers' Club, conveniently next door to the Treatment Room, and the ground floor would contain comfortable guest quarters designated on the architectural drawings simply as 'Visiting Governor'. It is true that the plan allowed for a Tea-Making Room rather larger than the office of the Controller of Overseas Communications, but still, standing as it would directly opposite to Westminster Abbey itself, in rational antithesis to that mysterious shrine it would represent everything frank and enlightened in contemporary colonial government, urban sewage to female education, seal farming in the Falklands to East African Federation. As Sir Alan Burns, a former Governor of the Gold Coast, wrote in his book *In Defence of Colonies* in 1957, 'Many years ago Britain undertook the gigantic task of helping the people of various under-developed territories to overcome the handicaps imposed on them by nature and environment. . . . In many parts of the world the task has not yet been completed, and it is inconceivable that we should abandon it half-done.'

4

It never happened of course. 'It will clearly be some time', wrote Sir Charles Jeffries, KCMG, *apropos* of that new headquarters, 'it will clearly be some considerable time—though, one may hope, not forty years in the wilderness—before the Office can move into the promised land.' But the more time passed, the remoter the promise, until at last even Sir Charles, even Sir Alan, were obliged to admit that the day of the Empire was over, the task must be abandoned uncompleted after all, and the Office never would reach Abbey Square.[1]

For the last rally of colonial enthusiasm in the 1940s and 1950s was one of history's more endearing misjudgements. To almost everybody but the British activists themselves, all the signs and arguments were against it. Colonialism was excoriated in every corner of the world, and the growing power of African and Asian nationalism, openly backed now by the Soviet Union and certainly

[1] Though the site is vacant to this day, being used bathetically as a House of Commons car park.

not discouraged by the Americans, was apparent for all to see. One did not need to be a George Curzon to realize how insubstantial the imperial structure had become, now that the grand mass of India had been removed. The tide of history, irresistible and one would think unmistakable, was sweeping the imperial idea irrevocably out to sea.

So that brave new building was never even started, and the structural reforms of the colonial empire, too, were mostly aborted or abandoned. It made no difference that throughout the 1950s the Conservatives were back in power. The Empire crumbled nonetheless. The West Indians, being far less homogeneous than they looked on the map, and far more ambitious to be their own Prime Ministers, soon deserted their Federation and split into petty autonomies. The East Africans, being of many different tribes, several different religions and three different colours, never were persuaded into unity. The blacks of the Central African Federation, rightly surmising that it would mean their indefinite subjection to white minority rule, demolished it as soon as possible. Singapore was rejected by the rest of the Malaysian Federation, to set itself up in dudgeon as a City-State on its own. The subversive Adenis, far from being disciplined by the loyal sheikhs of south Arabia, subverted the sheikhs instead, while the Arab League, instead of being a bastion of British influence in the Middle East, soon became the principal organ of Arab anti-imperialism. The hopeful departments of colonial development were dropped by the reference books one by one, converted into organs of foreign aid, or silently dispersed. Soon the Colonial Office itself, still in Great Smith Street after all, would seem a quaint anachronism, and its functionaries would find their way to more contemporary offices of State, like the Ministry of Social Security, or the Totalisator Board.

Let us then, since it is almost our last chance, take a walk through Westminster, one morning in the later 1950s, in search of *homo imperialis*. It is a relatively prosperous moment of British post-war history, the nation standing for the moment between economic crises, and the civil servants hastening from the St James' tube station look, for the most part, well-dressed, well-fed, ordinary kind of people—not so very different from Belgians, say, or Norwegians. Up on the morning train from Beckenham or Guildford, they are

settled into the mould of western urban man, and want nothing much
more than a quiet life, a television set and an annual holiday, with
pension rights assured. Among them, it is true, are more colourful
figures to remind us, even now, that London is a world capital—
black men in white robes and curious hats, an Arab or two, Malays
or Chinese, a few huddled Indians: but then in a world now en-
circled by the jets one may just as easily meet a sheikh in Zurich, a
Jamaican waiter in Manhattan or a Nigerian doctor completing his
training in Düsseldorf. It is a cosmopolitan crowd, but in the middle
of the twentieth century any great city is cosmopolitan, and London
looks scarcely more imperial than Stockholm.

But there, look, swinging briskly around the corner from the
Abbey, courteously stepping into the gutter to overtake the pavement secretaries, oblivious it seems to the curses of taxi-drivers—
there is a figure you will not find in Copenhagen! He is not a young
man now, in his fifties perhaps, and he is slightly stooped, as
though a succession of fevers has warped his spine. But he is slim,
stringy, rather rangy, and his face is so heavily tanned, not simply
a sunburn but a deep, ingrained tincture of brown, that physically
he scarcely looks like an Englishman at all. Yet British he unquestionably is, the most British man in sight, his expression, his movement, his every gesture reflecting a Britishness that has almost
vanished from England. Even his clothes are yesterday's. He wears a
brown floppy trilby hat, looking as though it has been repeatedly
soaked in rainstorms and dried in the sun, and slightly scuffed suede
shoes. His overcoat looks like a reconstituted British warm. Tucked
under his arm to read in the bus (for one suspects he seldom uses the
underground, disliking the fug down there), he carries a book from
Harrod's Library—General Slim's new volume of memoirs, perhaps,
or Alan Moorehead's *The White Nile*—he doesn't go in for fiction
much. On his finger he wears a signet ring, and as he swings his arm
one can just see, beneath the sleeve of his tweed sports coat, the
glint of oval cuff-links. He wears braces, one wouldn't wonder.

Is there something wistful to his worn if still agile figure? There
is. He looks out of touch, out of time. He meets nobody he knows,
for he has few friends in London now; even at the Office it's all new
faces, and he's never bothered with any of those damned clubs. He

averts his eye from the passing crowd, for to be honest he doesn't much like the style of Londoners these days. He is not much looking forward to his interview with Sir What's It, who doesn't know a bloody thing about Totseland anyway. He doesn't like the climate. He doesn't like the traffic. He detests what they've done to the South Bank. The young men need a haircut. That play at the Royal Court was a load of old rubbish.

He is a foreigner in his own capital. He is a true exotic among the cosmopolitans. He is the last of the British Empire-builders, home on leave and hating it.

CHAPTER TWENTY-FIVE

The Last Retreat

QUITE suddenly it was to go, like the whisking away of an opera set on the revolving stage. The brief revival of purpose spluttered out, and even the sages of the Colonial Office acknowledged the truth. 'Mankind has struck its tents', pronounced Jan Smuts, 'and is on the march', and suddenly the imperial idea seemed not merely distasteful, but preposterous. It was like waking from a dream. Young men from England going out to rule the Ashanti, or preside over the courts of Sarawakis! English civil servants in plumed topees receiving the salaams of potentates! A huge department of State, in a middle-sized nation of western Europe, devoted to the governments of people thousands of miles away! What had seemed to the late Victorians romantically splendid seemed to mid-century Britons perfectly nonsensical. In the fantasies of the Groundnut Scheme and the New Town Plan for Totse City, the imperial conviction trailed away in absurdity.

It was inevitable, for by now the British Empire, for so long the backcloth of world events, had been replaced by newer sets, and players from other companies were in rehearsal. Even in Britain a generation was arising who had never experienced its stimulations, never thrilled to the red on the map, and as its elaborate old scenes were dismantled, one by one, only a few traditionalists in the stalls, English gentlemen, Indian princes, African Knights of the British Empire, sentimentally demanded curtain-calls.

2

It was in Palestine that the British imperialists, for the first time, frankly abandoned the imperial responsibilities, and there the last retreat began. 'No promotion', Storrs had written, 'after Jerusalem', and in a way the possession of the Holy City, and the establishment there of the first Christian Government for a thousand years, had marked the summation of the Empire itself. Jerusalem had set a seal upon the adventure, and the governance of the Holy Land had been the crowning privilege of Victorian imperialism. Yet there the Empire first admitted impotence. The withdrawal from India could be rationalized, even romanticized: the withdrawal from Palestine was without glory.

Exalted though the duty was, Britain's rule in Palestine had never been happy, for it was based upon equivocals. It was a Mandatory government, for one thing, so that in theory at least the British were not absolute masters. For another it was tinged with the suggestion of betrayal, since so many Arabs, and not a few Britons, believed that Palestine should properly have become part of an independent Arab kingdom. And it was embittered by the ambitions of the Zionists, who had professed to want only a National Home within a multi-racial Palestine, but who really aimed, it had long become apparent, at an independent Jewish State there. The little country, hardly 200 miles from north to south, sacred to three religions, was racked from the start by envy and suspicion, and the benign rule of the British Empire, which made it materially the most advanced country in the Middle East, degenerated over the years into a squalid regime of force and self-protection. Sometimes it was the Arabs who broke the peace, sometimes the Jews, and so inflammatory was the situation after the Second World War, when hundreds of thousands of European Jews desperately sought a new home, that the Holy Land became hardly more than an armed camp.

By now the Zionists, financed by Jews throughout the world, had rooted themselves in cities, farms and desert settlements all over Palestine. Though they were still only a third of the population, they were much better organized than the Arabs, and with their powerful supporters in America, far richer and more influential. The Arabs feared and loathed them, and the British by now, after fluctuations of sympathy, tended on the whole to agree. By 1947 the administration, though theoretically impartial, was virtually at war with the Jewish activists. Terrorists kidnapped and murdered British soldiers; Jewish settlements were repeatedly raided and searched by the Army; there were ambushes and explosions and reprisals and threats; the whole country was in a state of fear, racked by violence and conspiracy, meshed with barbed wire, and patrolled always by the armoured cars of the Empire.

Jerusalem was ravaged by these miseries. The British had governed it for only thirty years, but with their gift for balance and decorum, their sense of history, their love of things rooted and traditional, they had made it more truly Jerusalem the Golden than it had been for centuries. Never had the Colonial Service possessed such a city, and its officers had guarded it lovingly. The walled city they preserved intact in all its mediaeval intricacy, its cavernous bazaars and its dusty wrinkled alleys—the Muslims meditating in the Haram esh-Sherif, the black-capped Jews pushing their paper supplications into the crevices of the Wailing Wall, the Catholics, the Greeks and the Monophysites incessantly processing, with bells, censers and harsh canonicals, from one shrine to the other of the Church of the Holy Sepulchre.

Outside the walls New Jerusalem had arisen under the imperial aegis, in golden stone too, but flat-roofed and spacious. Here a new mixture of cultures had been fostered. Here one could see the new generation of westernized Arabs, British-educated and gentlemanly, dressed often in well-cut tweeds or cavalry twill, speaking an exquisite English, and constituting the most highly skilled and widely cultured elite of the Arab world. Beside them the urban Jews flourished, refugees often from Vienna or Berlin, running bookshops, serving scrumptious cakes at Viennese cafés, rehearsing with the Palestine Philharmonic or presiding over intellectual tea-parties behind the Hebrew University. The members of these two communities were not natural enemies: they had much in common, and were much alike: it had been the highest ambition of the British to bring them together, fuse them into a governing class, and so bring their government of Palestine to an honourable conclusion.

But by 1947 they had no such high hopes, the city they had cherished with such pride was all barbed wire and sandbags, and their own presence was defensive, even furtive. Since the war they had been repeatedly urged to admit more of the millions of Jewish refugees made homeless and destitute by Hitler's war, and by now the floodgates were almost bursting. The Americans were pressing them; the Jews themselves, desperate from the slums and concentration camps of Europe, were sailing to Palestine in their own rickety steamers, half-submerged with the weight of their passengers, only to be driven off the beaches by British troops, or turned away to internment camps in Cyprus and Mauritius.

The Arabs were no less passionate in opposition, and were supported by the Arab States which ringed Palestine, and so Jerusalem festered in a state of incipient tragedy. High barbed-wire barricades closed the streets to Government offices, armoured cars rumbled ungainly through the city, patrols of infantry laboured along the pavements. Not a generation had passed since Allenby entered Jerusalem in triumph in 1917: yet here were the sons of his soldiers, angry and cynical, keeping the Holy City precariously in order by a perpetual show of weaponry. One wing of the King David Hotel, behind its festoons of wire, lay in ruins, having been blown up by Zionist terrorists. Many of the shops kept their shutters down all day, in case of trouble, and the few people in the streets did not loiter, but did their business briskly and hurried back indoors. Nobody felt safe in Jerusalem now, and sometimes one heard a shot from the suburbs, or a sudden rattle of machine-gun fire.

The British had planned to keep Palestine as a Middle Eastern power base, to replace the Suez Canal Zone, and even now they were building a new military complex near Gaza, in the south: but by the middle of 1947 it was obvious that they could be no more than policemen there. All their energies went into keeping Arabs and Jews from each other's throats. Two divisions of British troops were committed to this unproductive task, and as the year proceeded, as Jewish pressures increased and Arab resentments mounted, it became more and more like war. Casualties were frequent, the cost was enormous, the public at home wanted nothing of it, the world at large watched the sordid drama without gratitude, blaming the British both for creating the problem, and for failing to solve it.

It had been the imperial intention to establish a self-governing

Arab-Jewish State in Palestine, but even before the war a commission of inquiry had declared the idea unworkable, and had suggested partitioning the country into three—an Arab State, a Jewish State, a British enclave around the Holy Places. After the war the United Nations came to a similar conclusion, and in November 1947 voted for the creation of Arab and Jewish States. The Arabs rejected the plan, the Jews accepted it, the British refused all responsibility for it. Nagged by the United Nations, pestered by the Americans, bewildered by the Zionists, insulted by the Arabs, excoriated by world opinion, exhausted by the strain of it, impoverished by the cost, disillusioned, embittered, in December 1947 the British Government announced that, like Pilate before them, they would have no more of it. They washed their hands of the Holy Land. On May 14, 1948 the last British soldiers embarked on their troopships at Haifa: and even as they sailed away, behind them the disputing peoples of the Holy Land, emerging from their fox-holes and secret arsenals, hurled themselves upon one another, and, splitting the country furiously between them, prepared to live savagely ever after.

3

Palestine was a declaration. The British would no longer fight to the finish. For old hands this was a bitter realization. Churchill, in opposition, foresaw 'a steady and remorseless process of divesting ourselves of what has been gained by so many generations of toil, administration and sacrifice'. Many of his supporters argued that parts of the Empire were essential to Britain's role in the world, however diminished her power. Ernest Bevin himself, Attlee's Foreign Secretary, had staked his reputation upon finding a solution for Palestine. Later a Conservative Minister was to say of Cyprus that the British Empire could 'never' abandon the island of Cyprus, while one of his colleagues was to declare that there were some imperial possessions, Malta for example, which could not for strategic reasons ever hope to rule themselves. They all had to eat their words. The world had overtaken them. When it came to the point the British would never again stick it out to the end, though partly in self-esteem, partly to ensure a stable succession, they often felt it necessary to offer a brief rearguard action.

So through the fifties and into the sixties, as people after people awoke to the realization of patriotism, or were goaded into it by politicians, the imperial retreat proceeded. The barbed wire and armoured cars of Palestine were duplicated across the world, as successive colonies flared into revolt, and the British Army whose power had so stirred Macmillan at Tunis a decade before was reduced to squalid duties of repression and withdrawal. It was the Easter Rising magnified a thousand times, and dispersed across the Empire: the same passions, the same ironies, the same waste, sometimes the same poetry, always, in the long run, the identical conclusion. For the rebels these eruptions of patriotic spirit were often splendid, and were to be commemorated for ever in street names, national holidays and heroic legend: for the British they were generally petty and often misguided, for it was apparent to nearly everyone that whatever else the subject peoples would get from independence, it would not be better government.

The British were spared, by their own common-sense, or perhaps lack of will, any such terrible conflicts as the French fought in Indo-China or Algeria. By now the nationalist rebels were generally abusing the converted, for most Britons felt, in the text of their own truest ideology, that it was no longer fair to coerce unwilling subjects. The British people would not have tolerated great wars of reaction: if the blacks wanted to rule themselves, all right then, good luck to them, let 'em get on with it. Nevertheless, in the twenty years after the Second World War they were seldom without a conflict somewhere in the old Empire—as they had seldom been without one, indeed, since the first days of Victoria's rule. Once these skirmishes of Empire had been stimulating, good practice for the soldiers, good sport for the officers. Now they were good for nobody, but merely served to embitter the rebels, and turn the British themselves more wanly against the profession of arms, the pretension of prestige or even the pursuit of power. Furse, watching it all sadly from his retirement, as the last of his young men packed their bags and handed over their files, thought it was like 'batsmen playing dangerously hurried strokes at hostile and unaccustomed bowling, on a tricky wicket, in a bad light, confused by contradictory advice yelled at them from the pavilion and by the spectators generally. . . .'

We see them in every climate and every landscape, always at their roadblocks, barricaded in their barracks, guarding post offices, escorting pale English children to school or squatting behind sandbags on the roofs of Government Houses. In Cyprus, which had never proved of the slightest use to the Empire, they struggle year after year against Greek guerillas; in Lord Delamere's White Highlands they fight the enigmatic and murderous Mau Mau; in Malaya they wearily stalk guerillas through Spencer-Chapman's jungle; in Egypt they run down young patriots across Wolseley's battlefield of Tel el Kebir; in the Shatt el Arab, where Townshend disembarked for Kut, the Royal Navy stands fruitless guard over the oil refineries of Abadan, soon to be nationalized by the Persians.

Often they succeeded, and curbing the impatient passions of the local patriots, managed to restore order in a colony before handing it over to their successors. In Malaya a patient and methodical campaign finally contained the Communist guerillas, returning them to their havens to await more propitious times, and allowing the Malaysian Federation to get off to a peaceful start. In Kenya a ruthless and sometimes brutal operation subdued the rebellious Kikuyu, enabling Jomo Kenyatta, the most famous of the tribe, to become Prime Minister. In British Guiana an inconvenient Communist coup was suppressed by British troops before independence was granted.

Elsewhere the withdrawal of British power, as in India, as in Palestine, left bloodshed behind. Hardly had the last aircraft withdrawn from the bases of Iraq than the young Harrovian king, with all his family, was murdered by the rebellious mob, and the Prime Minister Nuri es Said, a friend and ally for thirty years, was cut in pieces and dragged through the streets of Baghdad. And in Aden, the very first acquisition of Queen Victoria's Empire, the British left shooting to the last. Step by step they withdrew from the city to the harbour and the airfield, and while Royal Marines kept the indigenes at bay, a stream of aircraft flew off the last of the imperialists. Offshore two carriers, a depot ship and a submarine waited;

helicopters clanked heavily around the harbour; at the airfield transport planes arrived in a ceaseless flow from Cyprus, refuelled again and took off with their loads of refugees. Gradually the British perimeter contracted, closer and closer to the shore, while outside it rival groups of Arab guerillas sniped, looted and skirmished. The High Commissioner flew off in a helicopter to the carrier *Eagle*. The last commandos raced for their helicopters. The last flag was lowered. The last flotilla of the Royal Navy, its crews smartly lining their decks, its radars twirling, sailed away from Steamer Point into the Red Sea.

Behind them the guerillas fell upon the abandoned stores and barracks, swarmed up the steps to Government House, and shot at each other from rooftops.[1]

4

Far more often, though, the sequence of farewell was peaceful, and rather touching. The happiest of the imperial exits were stage-managed by Furse's protégés of the 1930s, now the Governors and Chief Secretaries of their colonies, and they were characterized by the same tolerance and guileless optimism that Santayana had admired in them in their youth.

The band that played out the Raj on the Bombay waterfront was to perform often again, as colony by colony the Empire was dismantled. Down came the flag, out rang the last bugle, and once again, until they got tired of the performance, the heart-strings of the British were momentarily tugged. The procedure became almost standard, like an investiture. The chief nationalist leader, lately released from detention and propelled into fame, wealth and power, found himself greeted by the retiring British Governor with a

---

[1] Next day the Southern Yemen People's Republic was proclaimed, and the country is now a Marxist State. At the height of the Aden troubles, when the port was in a state of open war, I went for a walk in the hills above the town and stumbled by chance into the garden of the Chief Justice of the colony. His wife, emerging at that moment from the back door, was not in the least perturbed to find me there, but greeted me with a classic imperial inquiry. 'Good evening', she said. 'Are you a visiting MP?'

comradely new bonhomie, and was saluted by white guards of honour as he arrived, dressed in his own ethnic fineries, at the Independence Day parade. Out from England had come some scion of royalty; and there was an Independence Day ball, at which the new Prime Minister danced enthusiastically with Her Royal Highness; and there were sundry ceremonies of goodwill and fraternity, a message from the Queen, a presentation of maces, or crests, or Speaker's Chairs, and an editorial from *The Times* quoted in the local paper, and lots of stamping of boots and quivering salutes by British military men determined to demonstrate their loyalty to the new regime (for if they often had doubts about the ability of coloured people to rule themselves, they were not generally averse to appointments as Chiefs of Staff or even Commanders-in-Chief of emergent armies).

These ceremonies impressed everyone, even in most cases the patriot leaders themselves, and they were quoted all over the world as examples of British liberal good sense. How civilized it all was! With what good grace Her Royal Highness went in to dinner on the arm of a tribal politician of Marxist leanings until recently imprisoned with hard labour in a desert penal camp for subversive activities against the Crown! How moving it was to see the rituals of Westminster and the Inns of Court translated so faithfully to that tropic setting—the Speaker of the new House, preceded by his brand-new mace, attended by his solemn Serjeant-at-Arms—the judges and barristers of the new Supreme Court, authentically wigged and tabbed—the black colonels in their gleaming Sam Brownes and red tabs—the black bishops fluttering in starched canonicals—even perhaps a black naval officer or two, caps tilted at the proper Beatty rake, from the harbour defence launch down at the harbour. Just for the moment they all meant it, and it seemed hardly more than a passing of tradition from one hand to another, or a coming-of-age.

5

One by one they went, all through the 1950s and into the 1960s, Nigeria, Kenya, Ghana, the Sudan, Uganda, even the never-to-be-abandoned Cyprus. Sometimes they changed their names with independence, confusing older imperialists and infuriating carto-graphers, but generally the moment of transition was smudged. Most of the new nations passed into the Commonwealth, that limbo of Empire, and anyway manners, methods and even people lingered from the old regimes, and gave an impression of continuity. Some-times the Governor himself remained, by request of the new Govern-ment, to represent the Head of the Commonwealth, and as his political responsibilities declined, so his geniality flourished, and he was often to be seen slapping the backs of former terrorists, or laughing at old tribal jokes.

In the House of Assembly, too, as likely as not, one or two Britons would sit, representing planters, or commercial interests, and generally looking, set in their pale tropical suits against that vivid polychromatic parliament, urbane but unmistakably impotent. Up at the regimental mess Major Carruthers still presided over his orderly room, clipped and expressionless, with crossed assegais replacing the lances upon his cap badge, and beside the swimming pool at the club Mrs T and Mrs Z agreed that, though one could not of course resist progress, and though neither of course was in the least colour conscious *as such*, still one could not help noticing that the ladies' room was distinctly messier since, well, since in-dependence and all that.

And for a year or two, in gently falling cadences, the systems of Empire survived. The British constitutionalists, for ever devising more perfect forms of Government to leave behind, seemed to sup-pose that their inventions were actually organic, more than mere artificial formulas, and for a time the new arrangements did seem to have self-generative powers. The Common Law was upheld in all its dignity. The Westminster rules were faithfully honoured. Captain Abdullah Khan's handbook for young officers of the Pakistan Army suggests that just as the proof of being a gentleman should always be discernible in an officer's moral standards and mannerisms, so he should not as a general rule carry an umbrella on parade. In the Sudan, when the black District Commissioner puts on his pith helmet with its gay feathered tuft, and strides out of his hut for morning inspection, one can see in his very walk the example of

Marlborough and Trinity, and hear in his voice—'Mark you it would all look a bit greener if we hadn't had such a rotten summer'—unmistakable echoes of the imperial castes.

But gradually the recent and the remoter past became curiously jumbled, as the alien authority dissolved, and loyalties long suppressed came to life again. For a few years all was mixed, tribal taboo with democratic shibboleth, Crown with immemorial fetish. Often the first of the new leaders expressed these paradoxes in their very persons. We have long been used to the spectacle of Gandhi and his disciples talking the most sophisticated language of western political theory, while dressed in loin-cloths and sitting at spinning-wheels. Now the anomalies were to be stranger still. Here the astonishing Ngwenyama Sobhuza of Swaziland, billowing with the plumes and skins of his regality, glaring with a kind of stylized fury all around him, and accompanied by the court functionary called the Eye of the King, passes through the lines of his devoted subjects, all kneeling, or even lying flat on their faces, to open the new session of the Swaziland Legislative Council. Here the Dinkas of the southern Sudan lope nakedly into Juba to cast their votes in the general election—each party being represented for convenience sake, since 98 per cent of the electorate is totally illiterate, by a party symbol, a bicycle, a butterfly, a crowing cock or a companionable pipe.[1]

One of the first Prime Ministers to take office in Africa was Kwame Nkrumah of the Gold Coast, a country which, as soon as he achieved authority, he renamed Ghana after an older empire. Nkrumah was to be seen any session day decorously at his place in the House of Assembly, below the Speaker's Chair, deftly parrying the opposition in the best Westminster manner, and he had become quite friendly with the last British Governor. People had high hopes for Ghana under this attractive leader, and Nkrumah figured largely in the pamphlets of the British Central Office of Information, and was made much of, with his delightful smile and engaging manners, when he appeared in London for Commonwealth conferences.

[1] Short-lived emblems—in 1969 all political parties were banned in the Democratic Republic of the Sudan. As for the bold Ngwenyama, he presently became King of Swaziland, and soon doing away with the democratic paraphernalia, assumed all power himself.

The transition, though, was not as straightforward as the propagandists implied. Educated partly in Britain, partly in America, Nkrumah was hardly your natural Parliamentarian. He was a Catholic, but a revolutionary Marxist too. He was a Bachelor of Law, but aspired to mystic brotherhoods, oaths of loyalty, blood-vows. Sometimes he dreamt of uniting all Africa under his presidency, sometimes he saw himself as a divinely appointed Messiah. In London once he was photographed leaning easily against a staircase in the garden of 10 Downing Street exchanging pleasantries with his colleague the Prime Minister of Great Britain: but when he was at home he often summoned magicians and soothsayers to his official residence, and sometimes he made the pilgrimage over the frontier to Kankan, in French Guinea, where the most famous of African oracles advised him how best to defeat the Opposition's amendments to the Municipal Housing (1954) Enactment Bill, or alternatively how best to obliterate the Opposition.[1]

6

Sometimes sceptically, sometimes indulgently, the British observed all this. Progressives were delighted at the course of events, conservatives were saddened, the mass of the public seemed indifferent. A nation does not watch its power shrivel away, though, without some moments of bitterness, and as the great Empire dissolved a strain of resentment and self-pity fitfully entered the British attitudes. The last retreat might be necessary, even honourable, but it was not much fun.

Only once did it flare into paranoia, as in a last impotent revival of the aggressive spirit the British tried to reverse the course of history. It was in 1956, and the retreat was already precipitate. Anyone, it seemed, could now cock a snook at the British. There was no respect for the Flag any more, no gratitude among the emanci-

[1] Which he effectively did in 1964 by declaring Ghana a one-party State, under his own life presidency. He was deposed by the army two years later and took refuge in the Republic of Guinea, whose president sympathetically appointed him a co-head of *his* State instead: and there he remained, writing revolutionary handbooks, until his death from cancer in 1972.

pated colonies; angry correspondents to the *Daily Telegraph* drew
bitter conclusions from the decline of Empire, reminded the Editor
about the fate of Rome, and reproachfully quoted poems—

> *We sailed wherever ships could sail,*
> *We founded many a mighty State.*
> *Pray God our greatness may not fail*
> *Through craven fear of being great.*

or

> *Only a dream, I know, and yet it means I must be ill.*
> *One thing a soldier said at last that I remember still.*
> *He said, 'We went to carry on the work begun by Clive;*
> *If you did not want an empire, we might have been alive'.*[1]

Most people were less coherent: but two decades of change, improvisation and finally withdrawal had left their mark upon the public
consciousness, and even those least chauvinist or raucous in their
patriotism felt, just the same, a sense of waste, unfairness and help-
lessness. Was this why they had won the war, simply to subside into
the ranks of the minor Powers? Was the whole imperial achievement
a deception after all?

Among those most bitterly affected was Anthony Eden, who was
born in the year of Victoria's Diamond Jubilee, and had succeeded
Churchill as Conservative Prime Minister in 1955. Eden had spent
his life close to the sources of British imperial power, and he thought
of Great Britain ineradicably as one of the arbiters of world affairs.
He had stood at the right hand of Churchill, he had experienced that
last triumphant exertion of British will which had defeated Nazi
Germany, and raised the nation in victory to the moral summit of
the world. The idea of a Britain to be defied with impunity by any
impertinent sheikh or corruptible politician was not simply repugnant to him, but almost inconceivable. He stood still, posed for ever
in his beautiful London suit, at Churchill's shoulder, next to President Roosevelt, at one of those conferences which had, only a few
years before, decided the future of the world.

[1] The first quotation, one of Curzon's favourites, is from Tennyson's
'Hands All Round', the second from Lord Dunsany's 'A Song in the Ruins'.

Eden developed a particular and peculiar antipathy towards one of the most persistent of all the Empire's opponents, Gamal Abdel Nasser of Egypt. Nasser's revolutionary movement of army officers had deposed King Farouk in 1952, setting up a republic, and had obliged the British to give up their vast military base in the Suez Canal Zone—60,000 men even in the 1950s. Then by intrigue, propaganda and force of example Nasser had inflamed almost the whole of the Arab world against the British connection, effectively ending British suzerainty in the Middle East. He had established Egypt as the anti-imperialist leader of the Arabs, and everywhere from Mosul to Oman he had turned men's minds against the British.

In 1956 this ambitious dictator, who had galvanized his own people into a new pride and confidence, nationalized the Compagnie Universelle du Canal Maritime de Suez, the French company which had run the canal since its construction, and in which the British Government held a substantial share. He then seized the canal, announcing that the Egyptians would henceforth run it for themselves, using the profits for their own national development. The company's concession was due to expire anyway in 1968, shareholders were promised full compensation, and in principle at least Nasser's action was not specifically directed against the British. There was no reason why Egyptians should not operate the canal quite capably, or pilot ships through it safely, and strategically it had obviously lost much of its importance for the British, now that their eastern empire was gone. But just as the assumption of British greatness was inherent to Eden's political thinking, so the Suez Canal remained an inescapable totem of it—'in some essential sense', said Lord Hinchingbrooke, MP, 'part of the United Kingdom'. Suez was 'the life-line of Empire', the 'Imperial jugular'. 'East of Suez' was a synonym for British world power. A Suez Canal in unfriendly hands would be, so British traditionalists cried, a Britain that had forfeited not merely her Empire, but her very freedom of action—her independence, in fact!

So Eden launched the last and most forlorn of all the imperial initiatives, a new and much more disastrous Jameson Raid, the action known euphemistically as the Suez Adventure of 1956. It was an operation clouded in secrecy, duplicity and irrationality—for as the crisis developed Eden came near to nervous breakdown. It was also a cruel parody of the British imperial style. Eden cast himself as an elegant younger Churchill, saving the world by his exertions. Nasser he portrayed as a Muslim Hitler—'I want him destroyed!' cried the Prime Minister to one of his Ministers. The ultimatum that was presented to the Egyptians, requiring them to restore the Canal to its rightful owners, rang with the righteous zeal of 1939. The invasion force that was assembled was like a punitive expedition of old. The Royal Navy mustered its ships and landing craft at Malta, jet bombers of the Royal Air Force were concentrated upon Cyprus, and in the streets of southern England convoys of Army trucks, painted a desert yellow, hastened to the southern ports as in greater days before. Buller at sea in the *Dunnotar Castle* would have recognized the temper of the time. So would Hamilton leaving Mudros for Gallipoli. Even General Gordon, perhaps, catching his train at Charing Cross for his martyrdom at Khartoum, might have responded to the 1956 theme of self-righteous retribution. The old tag *casus belli* was knowledgeably quoted in London clubs, and among the ageing imperialists of England the general view was that they had a just one.

But just or no, they were deluding themselves. Britain could no longer punish trouble-makers as she pleased, with a resounding statement in the House of Commons and a brisk expeditionary force. Nothing was so simple now, in the complex world of the 1950s. A plan must be concocted with the French, who were in a similar mood of national frustration, and with the Zionists, who had now established their own State in Palestine, and considered the Egyptians their most threatening enemies. There were the Americans to consider—could they be trusted to help, or would they intervene to hinder? There was the now amorphous Commonwealth, some of its members reliable enough to put in the picture, some more safely left in the dark. There was the United Nations, vociferously anti-imperialist still. There was world opinion in general, as unsympathetic to British imperial causes as ever it was in the Boer War. There was opinion at home, split furiously on the issue. And finally there was Russia, now the greatest imperial power, which at this fatal moment found itself confronted by a rebellion of its own, among the subject patriots of Hungary.

Through this maze Eden and his advisers moved as in a dream, cunningly. They held secret meetings with the French and the Israelis. They told less than the truth to the Commonwealth, and actually lied to the Americans. While they pretended that they would occupy the Canal Zone only in order to pass the Canal into United Nations care, they really planned the overthrow of Gamal Abdel Nasser himself, the wildly popular leader of the Egyptian people. All was shame-faced and underhand. As the world watched aghast and unbelieving the ultimatum was delivered to President Nasser, requiring him to withdraw his forces from the Canal: almost at the same time the Russians, invading Hungary with overwhelming force, and putting down the rising with infinite cruelty, let it be known that if the Anglo-French invasion of Egypt were not aborted, Russian rocket missiles might soon be falling upon London.

This was another world, beyond the capacity of the imperialists. Of course the invasion was aborted—they had no choice. The invasion force, laboriously assembled and poorly equipped, did indeed invade Egypt, capturing Port Said and advancing down the canal. The Egyptian air force was virtually destroyed by attacks on its airfields, and scurrilous leaflets, ludicrously inciting the Egyptians to rebel against their leader, were dropped in the streets of Cairo. The Israelis occupied the east bank of the Canal, the British and French pushed southwards to occupy the west bank. But in no time the British lost their resolution, as the terrible truth dawned upon them that they could no longer behave imperially. The whole world was against them, even their oldest friends, and even in Egypt it seemed, the most despised of all their dependencies, they could no longer honour their own convictions. Even a Wog had a voice at the United Nations now, and all the splendours of the past, assembled in such pitiful pastiche in the familiar waters of the eastern Mediterranean, could not save the British from ignominy. The invasion force was withdrawn, and the imperial ghosts turned uneasily in their graves.

The British were numbed by this unnecessary disaster, even

those who had most passionately opposed the invasion. They did not like to talk about it, and a veiled reticence fell upon the subject. Eden himself, ill and distraught, retired from public life for ever, handing over to Harold Macmillan. It was as though a developing neurosis, erupting into a moment of schizophrenia, had subsided once again, this time for ever, leaving behind some shattered nerve or atrophy, and never again did the British stand up for their imperial privileges. Before we leave the spectacle of the last retreat, to which this provided a sad, misguided and untypical climax, let us visit Port Said, at the northern end of the Suez Canal, during the brief British occupation, for never again shall we see a distant seaport seized by the forces of the British Empire, or observe the White Ensign dominating the sea routes to the east.

In a way it was like a home-coming. Everyone knew Port Said. Everyone had soldiered there, or sailed by. Everyone had smelt that special smell, that blend of dust, dirt and oil which, reaching the approaching ship far out at sea, before even the first flicker of the lighthouse, told the imperialist that he was nearing the east once more. Generations of memsahibs had wandered around the scents, silks and brass-studded camel-saddles of Simon Artz, taken coffee on its balcony while the ships sailed by, or laughed at the gully-gully men conjuring chickens from their sleeves on the street outside. Almost every British regiment had taken its pleasures here, at one time or another, and hardly a British warship had not refuelled in the roads. The arcaded offices of Eastern Telegraph Company had been there almost as long as the canal itself, and the whole tempo of life at the Casino Palace Hotel was geared to the schedule of the P and O boats, passing so majestically to and from Bombay.

Port Said had never been beautiful, but it was familiar, and in their rough way the British had been fond of it. Awful though it was, pimps, touts, slums and all, still it was part of their heritage. It is like a nightmare to find them back in these familiar streets as enemies. The hush that hangs over the town is a hush of shock. Nobody can quite believe it, but it is true. Offshore lies the invasion fleet, overhead the helicopters and the bombers fly, and sprawled across Port Said is the British Army. A squadron of Centurion tanks lies in the churned-up mud by the airport causeway, their crews drinking tea to a crackle of static from their radios, and all along the beach soldiers have bivouacked among the beach-houses, sweeping the sands with mine-detectors, hanging out their washing, and sometimes, in a bitter memory of Gallipoli, bathing in the Mediterranean against the background of the silent ships. Infantry patrols wander dustily among the back streets, officers drive about in requisitioned Citroëns, British sailors stand sentry at the dock gates and officers can be seen moving importantly through the domed offices of the Suez Canal Company. Simon Artz is shuttered like the grave; the Casino Palace has been turned into a field hospital, and there is no sign of its courtly tarbooshed manager, who used to ask so fondly after General Hindlesham or Miss Packer, and wonder how the weather had been in Poona. Here and there a shop has been looted, and there is a litter of broken glass and empty boxes on the pavement. The streets are deserted, but for the soldiery, and an occasional scuttling scavenger, and one or two merchants sitting listlessly on kitchen chairs outside their shops.

'Can it be *real*?' they say when they recognize you, raising their hands helplessly, palms up, to embrace the whole hideous scene. 'Never, never would we have expected it of the British. . . .' And the British too seem to find it unnatural. They talk in the idiom of all the British wars, but self-consciously, as though they know this is somehow fraudulent: the old jokes ring false—'Elephant and Castle!' say the soldiers, as they pile into the three-ton truck, but the quip has no savour to it. The ethos of Empire, as of war, was acceptable to the British when it was backed by convictions of honour—by the belief, false or misguided, that the British were acting rightly, for the good of themselves and the world. Fair play! In most of their wars the British had been so convinced, and there was dignity to the cocky good humour of the soldiers, and true beauty to the unwavering patriotism of the British people.

Now, in Port Said, 1956, there was only pretence—a sham virility, a dubious cause, a nation divided, an army with little verve to its campaigning. Port Said, shattered and appalled, stood as a bitter memorial to the last display of the imperial *machismo*: and beyond the quays the funnels and masts of sunken ships, blocking all passage through the Lifeline of Empire, ironically illustrated the point.[1]

[1] This is an eyewitness account, and is coloured by the fact that I went to Port Said from Sinai, where I had been watching the Israeli army in the field: the contrast in spirit between the two forces, the ruthless Israeli so brilliantly aggressive, the genial British apparently so half-hearted, powerfully influenced me in the writing of this book.

CHAPTER TWENTY-SIX

On the Beach

IT was nearly over now. Future historians may well say the British Empire ended at Suez, for there it was finally made plain that the imperial potency was lost. In the 1960s it became clear to the staunchest of the British imperialists that their Empire was gone, and in a frame of mind more bewildered than resentful their leaders half-heartedly set out to find a new role in the world— as mediators between east and west, as Athenians to America's Rome, as the ageing chatelaine of the increasingly skittish Commonwealth, or, as a last resort, as offshore islanders of a new Europe. Nostalgia set in, and while novelists and playwrights still made fun of the imperial blimps and postures, many other Englishmen looked back with a wistful if puzzled affection at the spectacle of their grandeur, fast dissolving into memory.

After Suez they never resisted again. They recognized the tide of history, and bowed to it—or more pertinently, perhaps, they remembered that politics was only the art of the possible. In January 1960 Harold Macmillan, the Conservative Prime Minister, set out on a tour of Africa. He was a man of the imperial age. The grandson of a Scottish crofter, the son-in-law of a Duke, the son of an American mother, he was a product of Eton, Balliol and the First World War, and for him the true Britain was still the Britain that had basked so expansively, so genially, in the flowered days of the Edwardian era. It was less than twenty years since he had represented the British Empire, in the last display of its power, as Minister of State in Cairo during the Second World War, or since, standing on the dais at Tunis while the Highlanders appeared over the crest of the road, he had believed the Empire to have the world at its feet.

He was a politician first, though; he had lived through the trauma of Suez to pick up the pieces of Eden's policy afterwards: and in Cape Town in January 1960 it was he, the twenty-first Prime Minister of Great Britain since the accession of Queen Victoria, who formally recognized, as he might have recognized a new State or a new alliance, the end of the imperial idea. Ever since the end of the Roman Empire, he said, one of the most constant facts of political life in Europe had been the emergence of independent nations. 'In the twentieth century, and especially since the end of the war, the processes which gave birth to the nation-States of Europe have been repeated all over the world. We have seen the awakening of a national consciousness in peoples who had lived for centuries in dependence upon some other Power.'

This was, Macmillan implied, something inevitable, something true. It was not the work of agitators or false prophets. It was, he said, in the last of all the truisms, euphemisms, hyperboles and *obiter dicta* that had enriched the vocabulary of imperialism, only the wind of change.

2

Let us end the story gently, on a loyal note, for not everybody saw the Empire as wicked—people all over the world admired it still, for all its weaknesses and excesses, as a force for good, a kindly force despite it all, and a shield against those 'scientific blackguards, conspirators, churls and fanatics' whom Santayana foresaw as its supplanters. Even in the 1960s many a possession and dependency preferred to stay within the old fold, remote, dreamy, contented or simply ill-informed, governed still by English gentlemen, and visited sometimes by the spick-and-span frigates of the shrunken Royal Navy. By then Hong Kong, once among the least promising of British possessions, was as heavily populated as all the rest of the Empire put together, and the Colonial Office no longer existed: but Sir Ralph Grey was still Governor and Commander-in-Chief of the Bahamas, £2,800 was the going salary for the Governor of the Falkland Islands, Mr Gribble was still the Government Printer of Fiji, and if you happened to be going to the Turks and Caicos

Islands Geoffrey Guy the administrator there, though it is true his salary was only £2,150, would be sure to look after you.

These scattered settlements did not look much on the map, set beside the sweep of the imperial crimson not so long before, but still it was pleasant for the wandering Briton to stumble upon such half-forgotten relics and anachronisms, flotsam on the beach of history, still retaining some air of the imperial reassurance, some promise of relative good manners, some suggestion of punctuality or prospect of fish and chips. Let us then, having inspected so many fortresses of Empire, visited so many great cities, witnessed such scenes of splendour or of tragedy, end our own journey by dropping in upon some of the places which, while the world shifted all around them, seemed to be governed still by Victoria's presence.

3

Muscat was one such place, far away at the south-eastern tip of Arabia. It had never actually been an imperial possession, only an ill-defined protectorate, but by the 1960s nowhere was more redolent of the lost empire of the Raj. For a century and more the Sultans of Muscat and Oman had been feudatories of the Indian Empire. A treaty in 1891 had given British subjects ('and their servants') extra-territorial rights in Muscat, and gave Muscatis in return the right to enter with their vessels 'all ports, creeks and rivers in the British Empire', and to live, trade, travel, possess houses and shops in any imperial territory, except self-governing colonies.[1] By 1903 the Sultan was swearing 'eternal devotion and fidelity' to Lord Curzon, beneath the awnings of HMS *Hardinge* in the harbour, and his son was appearing as a loyal feudatory at the Delhi Coronation Durbar for King Edward VII, taking with him a presentation model of a camel and a palm tree, fashioned in Muscati silver. Ever since then the Sultanate of Muscat and Oman had been a British ward—or puppet, if that was the way you saw it. It was subsidized by Britain, its foreign relations were controlled by Britain, and the British had deliberately kept it insulated there

[1] In the early 1800s the Muscatis had a fleet of seventy-five warships, but it has been unkindly suggested that only one could be manned at a time.

against foreign examples or predations, autocratically governed now by the Sultan Said bin Taimur, and supervised as always by the British Consul-General.

Here British India survived, and the ghost of Curzon still seemed to lean from the rail of the British India steamships, on their weekly visit to the capital. The ships indeed remained Muscat's only public contact with the outside world. There was no civil airfield, and even the desert hinterland behind, linking Muscat proper with its southern dependency of Dhufar, had only recently been crossed from coast to coast by its first European.[1] No nationalist party had yet arisen to disturb the equanimity of these arrangements—the only subversives were of a thoroughly traditional kind, fractious sheikhs and tribesmen of the interior who were, so to speak, essential to the *genre*. Sailing into the little capital of this almost unknown country really was like sailing back into the eastern Empire.

Two great fortresses, built by the Portuguese, stood sentinel over Muscat harbour, but they had no guns in them, and on the rocks above the anchorage the Royal Navy had, over many generations, left the graffiti of its supremacy in the Gulf. Scores of ships' names were inscribed there, some faded, some fairly fresh, some of forgotten sloops and gunboats, *Teazer* or *Surprise*, one or two of old familiars like *Hardinge* herself. It was in this harbour, during the Napoleonic wars, that the British frigate *Concord* had captured the French *Vigilante*; Nelson came here as a midshipman on the *Seahorse*; here, during the Second World War, a Japanese submarine sank a British freighter, adroitly aiming its torpedo through a gap in the harbour rocks.

Once inside the anchorage, there was the long water-front of the capital's twin towns, Muscat and Mattrah, gleaming white buildings in front, a jumble of *suks*, lanes and high-walled houses stretching away to the grey hills behind. Two flags flew bravely over this suggestive scene. Over the palace of the Sultan, gleaming and massive at the water's edge, there flew the red flag of Muscat: but over the pleasant white residence of the British Consul-General, larger, grander and rather better laundered, there flew the Union Jack. It had an air of indolent arrogance, and the posture of the

[1] Me!

533

house itself, which stood at the end of the water-front, slightly separate from the town, was ineluctably prefectorial. 'Unquestionably', wrote the explorer Theodore Bent when he visited Muscat in 1895, 'our own Political Agent may be said to be the ruler in Muscat'; Curzon observed that Percy Cox, the incumbent in 1903, virtually ran the place; and though the Agency had now been tactfully metamorphosed into a Consulate-Generalship, still the old house remained the source of ultimate decision in Muscat.

It was a lovely house—more than a house, for it had a compound in the truest Anglo-Indian manner, grouped around a gravel courtyard with a huge flagpole in it—upon which to that very day runaway slaves still sometimes threw themselves, to clasp it with their brawny arms and claim emancipation. There was a wide verandah flagged with stone, on which Englishmen in white ducks could still be observed being provided with long cool drinks by silent Indian servants, and there were old prints of Empire in all its corridors, portraits of former Agents, weapons from imperial skirmishes, carpets from the marts of India, relics of Lord Curzon's visitation or souvenirs of naval occasions.[1]

Elsewhere in town, too, Englishmen were still living the imperial style. The Sultan had an English Wazir, who had formerly been in charge of the prisons of the Sudan, and was now one of the great men of Muscat. He lived in a magnificent old Arab house in the heart of the capital, and he appeared on ceremonial occasions in a long black Arab robe, wearing a beret and carrying a ciné-camera. The commander of the Sultan's forces was an Englishman, and around the bay lived the English mercenary officers of his army. Many of them had gravitated into the Sultan's service from the Indian Army, and had brought with them all the attitudes of Anglo-India, here in its last incarnation. Their messes smelt of metal polish, pipe tobacco, whisky and dogs. English magazines and soldierly books lay all about, there were fat cats and baskets of flowers, and one major in the Sultan's employ, a former Royal Marine, had taught the mess servants to obey British boatswain's

[1] 'Good drawing-room,' noted Storrs characteristically when he visited the house in 1917, 'with a new Collard and Collard and two large China rice vases. . . .'

calls, whistled between the teeth—'Do you hear there? Do you hear there? Cooks to the galley! Hands to muster on the quarter-deck!'

It was all a quaint echo of lost times, and Muscat sheltered these late imperialists kindly. The Sultan preferred to live in the past too, for more than sentimental reasons, and he saw to it that his little capital changed as minimally as possible. This was hardly beneficial to the ordinary Muscatis, who were by now the most backward and deprived of all the Arabs, but it was certainly agreeable for the wandering Briton. How comforting, to withdraw through the harbour mouth into this little Shangri-La of sahib, sultan and respectful servant! How pleasant to find that here, if nowhere else, the Union Jack still commanded the deference of the natives! At night, as if to exclude the fantasy from the real world outside, a big gun was fired from the harbour fort, and the gates of the old city were slammed. Arab levies took their muskets to guard the exits, each gate being the responsibility of a particular tribe, and Muscat went to its beds in the old way—guarded at the gates, beneath the protection of the British Empire, to the lap of the peaceful sea. Perpetual curfew was the rule, and nobody might venture in those streets without a lantern, so that only an occasional flitting of robes, a twinkle of moving lights, the stir of the warriors at the walls, disturbed the shuttered silence of the town.[1]

4

Mauritius was another relic. Here the British authority was more direct, for the island had been a Crown Colony since 1810, but it was no less easy-going. Mauritius had never much mattered to the Empire anyway. Ceded by the French after the Napoleonic wars, its only imperial purpose had been as a convenient and not too disagreeable place of exile—Boers, Jews and the Shah of Persia had all been

---

[1] Much more disturbs it now, for though in 1977 the British still play an equivocal role in the affairs of the Sultanate, Muscat has been transformed by the accession of great oil royalties: a busy corniche runs along the waterfront today, air-conditioned hotels welcome the visiting executive, and you can travel to that Byronical seaport by direct flight from London.

sent there at one time or another. Very few Britons knew where
Mauritius was, and it was famous in England only as the home of
the Dodo, *dodus ineptus*, who had waddled his last long before the
British arrived, and of the Mauritian 2d Blue, issued in 1847 to
cover the expenses of a fancy-dress ball at Government House.
Because there was a sizeable and cultivated French community
there, few Britons were ever needed to run the island, and fewer
still chose Mauritius for the making of their fortunes: only the top
echelon of Government had ever been British, and even its members
were regarded by the local French gentry with a certain condescension.

But it had been a happy enough association, and by the 1960s
there was a fragrant, almost festive feel to the British presence on
Mauritius. It might not be a very important appointment for a
Governor or a Colonial Secretary, but it was very agreeable. The
island was beautiful—Darwin described it as 'elegantly constructed',
and it did possess a quality of graceful disorder that seemed almost
contrived, like a landscape garden. Most of it was fine open country,
such as the British always loved, with wide reaches of down and
moorland, where deer roamed for the hunting, and small lakes lay
darkly in the sunshine like Scottish tarns. Most of it was high, too,
so that gusts of fresh winds often blew exuberantly off the sea, and
the British could build their villas far above the sunburnt coast.
Most of the Empire's tropic islands were essentially ordinary, and
the imperialist who had served in one often felt he had governed
them all, but Mauritius was an endearingly odd place. Its fauna was
odd—unique lizards and otherwise extinct birds—and until the
seventeenth century no humans had lived on it at all. Also it had
been periodically ravaged by hurricanes, so that all in all there was
nothing very old on the island, no aboriginal artifacts or last descendants of forest dynasties, and a sense of inescapable transience
gave to exile on the island an air of holiday.

Most of the population was Indian, and the British presence was
unobtrusive. There was an old fort on a hill, looking rather
Khyberesque, and an Anglican cathedral with a white spire, and a
few mild emblems of Empire like red pillar boxes and statues of
Queen Victoria or Sir John Pope-Hennessy. There was a Royal

Navy communications base, and a pleasant Gymkhana Club popular among the Mauritian bourgeoisie. But Port Louis, the capital, still looked much more French than British, with its balconied Provençal houses behind walled gardens in bleached quiet lanes, and its high-collared savants dissecting rare reptiles in the Natural History Museum, or strolling towards that evening's meeting of Le Cercle Littéraire. The exquisite little Théâtre Royal was in fact built by the British, but if you sneaked into the wings during a Saturday rehearsal, so resonantly did the French theatrical voices echo through the empty stalls, so determined was the talk of Sartre, Camus or Molière, that you might well think yourself back-stage in some Théâtre Municipal of Seine or Languedoc.

Even the Governor's mansion, Le Réduit, was more French than British. It was perhaps the *nicest* of all the Empire's Government Houses, but it had nothing in common with Dacca or Lagos, or even with Jamaica. It was built by the French as a refuge to which women and children could flee when pirates or Englishmen raided the Mauritian coast, and it still looked like a gentlemanly French country house. Governors' wives loved it—it was so unpompous, so natural, so home-like—and visiting bigwigs from England, too, found it a pleasant change from the usual gubernatorial barracks. The British had set their stamp upon it, of course, but it was a tentative, gentle stamp, an imprint of croquet, charades and early morning tea. They had planted heaps of trees, and *re*planted them after successive hurricanes: they had laid tennis-courts, and built gardens, and enlarged the mirrored ballroom, and littered the terraces with comfortable seats.

Here the Governor and his lady pottered through their generally uneventful terms of office. They were left much to themselves, for generally speaking there was not a great deal to do—Governors of Mauritius had not usually been in the front rank of public servants, being men who deserved pleasurable late postings in an undemanding colony. The British had never been assimilated into Mauritian life, but there were always delightful French guests to enliven their dinner-parties, and nobody was going to be rude to them, or throw bottles at their cars, or accuse them of capitalist exploitation. The worst hazard, hurricanes apart, was nothing more perilous than a lecture on the superiority of French civilization from the president of the Historical Society.

It was not, one must admit, a very energetic colonial regime. Progress had been sluggish in matters like education, health, housing or political advance. On the other hand the usual excellent roads had been built, telegraphs and cables happened, new techniques of growing tea and sugar were introduced. Most visitors to the island, in fact, remembered as vividly as anything the Mauritian Railway, a most beguiling little prodigy of Empire, which rambled all around the island puffing endearingly and frequently whistling, stopping at stations with names like Sans Souci or Circonstance, delicately pausing at the Governor's private waiting-room outside Le Réduit, and faithfully taking to their offices each morning, behind the slatted windows of its khaki four-wheeled coaches, the sun-helmeted merchants and lawyers of Port Louis.[1]

If the regime was regarded patronizingly by educated Frenchmen of Mauritius, it was never actually unpopular. In both world wars Mauritians of all races had freely gone to war for the British Empire, and turned up unexpectedly on fighting fronts all over the world. The Empire had come, one day perhaps it would go, in the meantime it was there. Nobody fussed about it much. Above Port Louis there was a mountain, Pieter Both, upon whose spire-like summit there resided a large boulder, precariously balanced. It was an Old Mauritian Belief, so the guide-books and folklorists claimed, that so long as that boulder stayed in place, the British would rule the island. As the apes were to Gibraltar, that rock was to Mauritius. In most dependencies of the British Empire, by the 1960s, some patriot would have climbed up there and pushed it off, daubed all over with vituperative slogans. In Mauritius nobody cared, the boulder remained placidly upon its mountain-top, the savants talked regretfully about lost French supremacies in the Indian Ocean, and the croquet continued after tea on the lawn at Le Réduit.[2]

---

[1] They closed it down in 1962. During a general election the following year one ballot paper was scrawled accusingly: *'You have taken away the railway jobs that Queen Victoria gave us!'*

[2] Even in 1975, though Mauritius was by then altogether independent

5

Far, far away was British Honduras, on the steamy shore of Central America. Muscat remained imperial out of ignorance, Mauritius out of good nature, but British Honduras stayed within the Empire out of self-protection. For thirty years the neighbouring republic of Guatemala had laid claim to the place, even including the old colony on its own national maps, and contumaciously calling it the Province of Belice—and indeed the boundary was, as the *Encyclopaedia Britannica* had observed in 1898, 'of a purely conventional character', being an imaginary line drawn by surveyors through virtually untrodden jungle. The possession of British Honduras was scarcely vital to the welfare of Great Britain, but for good reasons nobody in the colony wanted to be annexed by the Guatemalans: so anomalously the flag still flew above the little capital of Belize, and the duty guards from the permanent British garrison, one of the very last in the Empire, stamped up and down outside Government House.

It had been more or less British for 200 years, the first settlers being rapscallion communities of woodcutters, mostly of mixed Scottish and Negro blood, but it was declared a Crown Colony only in 1884. Since then nothing much had changed. Floods and hurricanes ravaged the capital sometimes, but it was rebuilt much as before: there was a Legislative Council, but the Governor was its president and held the deciding vote. Most of British Honduras was wild country, mountain, forest and savannah, and nearly half its people lived in Belize, the capital. Here the sugar, the teak and the citrus fruit came down to the sea: here all the varied peoples of the colony, the Negroes, the Mayans, the Indians, the Europeans, mingled on the foreshore, and here, in its modest premises around the bay, the imperial authority still resided.

Belize was everyone's idea of a tropical port, fretted, woody, shabby, jolly, cheek-by-jowl, smelling of rum and fermenting fruit,

---

within the Commonwealth, I found that the chief of police, the Cabinet secretary and the comptroller of Le Réduit were all Britons; the rock was still on the mountain-top, too.

loud with car-horns and market cries. It was surrounded by mangrove swamps, and drained by a series of gaseous canals, and it lay unreformed by town planners after all on both sides of an inlet called Haulover Creek. Here the teak and mahogany logs came floating down the Belize River—Old River to the locals—and Belize still felt rather like a lumber-camp, makeshift and temporary. Its houses were mostly of shabby clapboard, stilted against heat, floods and rats, and they all seemed to look down to the river mouth, where rafts of logs were towed out to sea by ancient tugs, where fishing boats bobbed at the quay, and long outboard motorboats, with quaint names like *Passenger Lady or Nigger Gal*, passed to and fro beneath the iron bridge, all their passengers sitting bolt upright and facing forward, like Indians in log canoes. Since the people of Belize were overwhelmingly half-caste, a tangled mixture of white, black and Indian, and since everyone seemed to know everyone else, visitors got the impression that everyone was related, and this heightened the sensation of a community encamped there, waiting to move on to some more settled country, but in no hurry to go.

Yet its loyalties were old and rooted. The thirteen original quarters of town maintained their identities—Cinderella Town, Lake Independence, Queen Charlotte Town—and though Front and Back Streets, the original trading streets, had been officially renamed after Queen Victoria and Prince Albert, everyone called them by their old names. East and west across town the street names commemorated the imperial heirarchy—Bishop Street, King Street, Prince Street, and pointedly supervising both the St Ignatius Catholic Church and the Wesleyan Chapel, Dean Street. Nearby Basra Street, Allenby Street and Euphrates Avenue honoured the Belize men who had fought for the Crown in the Middle East during the Great War.

The imperial establishment was small, but absolute. Nobody but Britons had ever ruled this colony, and Britons distinctly ruled it now. The Governor was Uppingham and Cambridge, the Development Commissioner had spent most of his life in Basutoland, the Chief Justice was the author of *The Law of Compulsory Motor Vehicle Insurance*, the Conservator of Forests had published *A Working*

Plan for Settlement Forests in Lango District of Uganda. The garrison, a battalion strong, was quartered along Barracks Road, north of the town. The Dean of Belize lived in a trim clapboard Deanery, with flowers on its verandah and a lawn of buffalo grass. Prominent in the Legislative Council were old Belize families who had made their fortunes in timber—the 'mahogany kings' who dominated local society, the Bradleys, the Stuarts, the Gabourels. Sometimes a Royal Navy ship put in, on a Caribbean cruise, and sometimes an official from the Colonial Development Corporation flew down from Miami (for the only way to fly in or out of British Honduras was by way of the United States).

The climate was awful but they lived cosily enough. Local politics could be rip-roaring, but their animosities were seldom directed against the colonial government, and there were no racial tensions, religious rivalries or tribal vendettas.[1] Alone in the British Empire the Legislative Council had evolved directly from the Public Meeting, the forum of the early settlers, and even now it had some of the family quality of a New England Town Meeting. In British Honduras even the relationship between owners and slaves had been relaxed, since both sides were utterly vulnerable in that remote environment, and equally subject to the rapine of Spanish buccaneers, so that society in the colony had escaped the blood bitterness of Africa or Jamaica. There was a recognizable camaraderie to the place, most happily apparent at the race meetings which were held on public holidays at the polo ground along Barracks Road in Belize. Everyone went, and had a grand time. Soldiers marched here and there to the thump of a military band, the administrators' wives assembled in their best cottons on the club verandah, the races were started gallantly by bugle calls, and at the corner of the pavilion the merry Belize children queued for free lemonade at His Excellency's expense.

It was a little living relic, an enclave of the past. It had been, on the whole, a successful imperial enterprise, and though by the

---

[1] The only British Governor to die violently in Belize was Hart Bennett, in 1918: observing the Court House flagpole smouldering during one of the capital's not infrequent fires, he promptly ordered it to be cut down, and it fell on his head.

1960s they were preparing a new constitution for the colony, the first step towards independence, it was still cheerfully loyal to the Crown—when, in 1962, twenty filibustering Guatemalans invaded the colony from the south, announcing the liberation of 'Belice', what they did to symbolize the great day was to burn in a village market-place photographs of Queen Elizabeth and her husband, together with a Union Jack (they were tried at Stann Creek Town Assizes, but within the year were all safely home in Guatemala). The Union Jack indeed flew all over Belize. There was one outside the Court House, and one on the Customs House, and one on the Anglican Cathedral, and one on the Fort George Hotel, and one on the Bliss Institute.[1]

And at the end of the foreshore, fluttering over Yarnborough Lagoon as the Consul-General's flag at Muscat commanded the harbour, or the Governor's at Le Réduit rose above the camphor trees, the biggest of them all flew over everybody's idea of a colonial governor's residence, basking among its lawns and flamboyants between the cathedral and the sea. This house was enough to make a Nehru or a Kenyatta nostalgic for Empire. It was a square building, not very beautiful, painted a dazzling white and mounted on stilts, and it breathed a mingled suggestion of Virginia, Queensland, Jamaica, Nova Scotia and the Carnatic—an anthology in itself of the imperial yearnings. It was not air-conditioned, its plumbing was erratic, its attics and boxrooms smelt a little musty and were frequented by tropical weevils. But upon its rickety verandahs were placed the chaise-longues and shabby sofas of the imperial afternoon, in its shrubberies cats licked themselves and spaniels bounded, around its gravelled paths the white-helmeted policemen dutifully patrolled, and through its dining-room windows one could sometimes see, beneath the not very skilful portraits of his

[1] A library and cultural centre named after an engaging imperial philanthropist, Henry Victor Bliss, JP, of Marlow, Bucks, 4th Baron Bliss in the Portugese kingdom, who sailed into Belize in 1926 upon his yacht *Sea King*. Before he had time to go ashore he died, but he left his fortune to the colony, and his body to be buried in a granite tomb at the harbour point. His bequest is still being put to good purposes—what other library in the tropics has a copy of Charlotte S. Morris's *Favourite Recipes of Famous Musicians* (1941)?

predecessors, His Excellency the Governor and Commander-in-Chief of British Honduras, smiling agreeably at the Archdeacon's wife across a less than epicurean bean stew.[1]

6

Across the old Empire many another community looked back wistfully to its heyday, and felt itself abandoned or betrayed by the course of history. Millions of half-castes, especially in India, were left to fight their own battles in a world where it was a handicap rather than advantage to be able to claim descent from a corporal of artillery or a planter's assistant. The Coptic gentry of Egypt, for so long the acolytes of Empire, were now left defenceless and reproachful in the flaking grandeur of their Assiut palaces, hung with Pharaonic devices and portraits of Lord Allenby. The Malays of Singapore were overwhelmed at last by the ambition and acumen of the Chinese, the Arabs of Israel festered in the occupied villages of their homeland or rotted down the generations in sordid refugee camps. Loyal servants and grateful deputies everywhere remembered lost friendships and comradeships—kind Mrs Weatherby who loved the baby so—Colonel Repton Sahib, a gentleman through and through—my dear old friend Judge Torrington, to whom I owe so much and to whom I affectionately dedicate this little memoir—Mr Glover of Public Works, who would *never* have allowed this kind of thing—Holden Bey, who still writes every Christmas—or dear Annie Lyttleton, the Governor's wife, you know, who was one of my *very* dearest friends, and who gave me that particular embroidered cushion you're sitting on, as a matter of fact, embroidered with her own dear hands . . .

Here and there they tried to stem the tide. The Maltese and the Seychelloise unsuccessfully proposed integration with the United Kingdom itself, while the Falkland Islanders steadfastly preferred the rule of London, personified by a genial Governor whose official car was a London taxicab, to the rule of the Argentine,

[1] To this day (1978), though a new capital has been built inland from Belize, the entire colonial structure almost uniquely survives—Governor, garrison and all.

personified as often as not by dictatorial criminals and military thugs. The Sultan of Brunei clung to his British protectorate when all about him were losing theirs. The Protestants of Northern Ireland were as ready now as they had been in 1914 to defy the Catholics by force of arms—a favourite banner of the Orange Order in the 1960s showed Queen Victoria presenting a Bible to two kneeling black men, above the motto 'The Secret of England's Greatness'. While the white leaders of Southern Rhodesia plotted once more to break away from the Empire and maintain their own supremacies for every, the black leaders looked to Britain still to impose a fair solution. When the Gibraltarians held a referendum to decide whether to stay British or join Spain, 44 voted for Spain, 12,138 for the Empire: the Governor of the Rock was still a serving officer of the British Army, the apes flourished, and every night in the Ceremony of the Keys, at the Main Guard, the sentries bawled out their imperial catechism:

> *Halt, who goes there?*
> *The Keys.*
> *Whose Keys?*
> *Queen Elizabeth's Keys.*
> *Pass, Queen Elizabeth's Keys. All's well.*

7

But the passion was spent, for or against Victoria's Empire, and so, except for these quaint or adamant anachronisms of loyalty, it came to an end calmly and almost apathetically, like an old soldier pacified at last by age, pain and experience. The last garrisons were withdrawn from the distant fortresses. The great fleets were no more. The British, turning their backs upon the great adventure, made themselves once more a European nation. Abroad the emancipated peoples soon adopted new styles and philosophies of government, or even acquired new overlords: at home a generation came of age which had never heard the trumpets.

CHAPTER TWENTY-SEVEN

Home!

IN the winter of 1965 Sir Winston Churchill, aged, beloved, hazed by brandy and long campaigning, died at his home in London. He had by then passed beyond the bickerings of party politics, and had become the living examplar of British glory. Loathed and reviled in earlier life, he was to be calumniated again after his death, in the way of legends: but for the moment, as he lay massive on his bed in death, ninety-one years old and the most universally honoured man on earth, he was beyond criticism. He was a dead spirit of grandeur, and for a day or two not only his own nation, but half the world paused wondering and reverent to mourn him. It was like that moment of antiquity when, the wild god Pan having died, strange music sounded and spirits moved from one end to the other of the classical world. Churchill had gone, and a sigh, part regretful, part wry, part sentimental, went around the nations.[1]

In him the lost Empire of the British, bad and good, had found its fallible embodiment—brave, blustering, kind, arrogant, blind in many things, visionary in others, splendid but often wrong, lovable but frequently infuriating. For seventy years Churchill had lived the experience of Empire more intensely than any other man, through six reigns and many wars, from the brazen climax of the old Queen's jubilee to the melancholy disillusionments of Suez. He had forced the Malakand Pass in '98. He had argued with

[1] Reaching me at the newly rebuilt Shepheard's Hotel, Cairo, where I lay in bed with flu, and where the kind Egyptian servants offered me their condolences as though I had suffered a personal loss. Indeed, like most Britons then, I felt I had.

Thorneycroft on the path up Spion Kop. He had conjured the
fearful beauties of Gallipoli. He had been Colonial Secretary,
Secretary of State for Ireland, First Lord of the Admiralty, Prime
Minister. He had soldiered for the Queen as a subaltern of cavalry,
for the King as a colonel of Fusiliers, and it was he whose edict
had sent Fisher's Grand Fleet to its war stations at Scapa Flow,
'like giants bowed in anxious thought'.

In his rhetoric, his humour and his rotund prose Churchill
had expressed the best and worst of imperial attitudes. After
Amritsar he had spoken nobly of 'the British way', which did not
include public murder as a deterrent, yet Gandhi he could describe
only as 'a miserable old man who has always been our enemy'. His
unerring opposition to Indian independence sent him into the
political wilderness between the wars: yet when Nehru called upon
him after independence Churchill saw him to the door in tears—
'we put that man in gaol for ten years, and he bears us no malice.
I could not have been so magnanimous.' His was the Empire that
was to last a thousand years, but the Statute of Westminster he
described as 'a repellent legalism'. For half a century he sent £2
every month to the Indian servant of his military youth, but he
displayed no jot of sympathy for the patriotism of the coloured
peoples, nor any fellow-feeling for subject leaders who sought to do
for their own countries just what he wished to do for England.

Churchill was only half English, his mother being American,
and devoted as he was to the British tradition, touchingly proud
of his forebear the Duke of Marlborough and his father Lord
Randolph Churchill, still intellectually he was more an inter-
nationalist than an imperialist. In this he illustrated one of the
recurring paradoxes of Empire—the wider it spread its frontiers, the
more parochial it became. Churchill was much more at home with
Americans than with Australians or Canadians. He was bored from
the start by the provincial hierarchy of Anglo-India ('a third-rate
watering-place', was his image for the garrison town of Bangalore,
'out of season and without the sea'). Imperial economics meant
nothing to him, colonial diplomacy was dull beside the grand sweep
of global activity which was his true metier, and he surprised King

George VI by the sang-froid with which, when the time came, he
adapted to the inevitable conclusion of the Raj.

Yet to another half of him the fact of Empire was the truth of
Britain's greatness. He loved the colour, the majesty, the idiom of
it—its 'valiant and benignant force', its 'fortress-islands', its
'scattered family of the Crown'. He was an aesthetic imperialist.
Holding no very strong moral views about it, believing as most of
his generation did that British rule was probably better than any
other, while he was no heady imperial idealist, he was no reformist
either. The detail of Empire bored and sometimes repelled him. It
was the idea of it that he found exciting, the spectacle of that immense estate enhancing the grandeur of England. Churchill was an
Anglo-American diplomatically, a European instinctively, an Englishman cerebrally: but emotionally he was an imperialist in the
classic High Victorian mould, loving Empire for its own sake, for
the swagger and the allegory.

2

Of all the charges of Empire, this simple dynamic had been the most
consistent. Economics, strategy, world politics had all contributed
to the British expansion, but the taste for glory had underpinned
them all, degenerating down the years into a hunger for prestige or
self-esteem, but still recognizable in 1965 as the same atavistic
tribal pride that animated the Diamond Jubilee at the apogee of
Empire.

Because it was fantasy, it was not unreal. To an astonishing
degree the world had been changed by its drive. The rise of the
Victorian Empire had acted as a gigantic prod or catalyst, stirring
dormant energies across the continents. It had been the principal
agent of an immense historical evolution, the distribution almost
everywhere of industrial civilization—which, having had its
beginnings in western Europe, was implanted in Africa and Asia
principally by this Empire. If it had not been done by the British,
it would have been done by somebody else: but still a combination
of chance, energy and geographical fact really had given to the

British people, as they liked to imagine during their evangelical years, a providential duty to perform. They really had been Chosen. That they fulfilled the mission with a mixture of motives and methods was irrelevant: they were simply the instruments of history or perhaps of biology, like the birds and beasts which, attracted by the gaudy appurtenances of sex, unwittingly perpetuate their species: there was never a mating dance like the dance of Empire, or plumage so seductive to its participants.

Some of the imperial achievements were indestructible, like the roads, railways and telegraph systems which provided the basis of the new industrial society, and which, though they would presently be superseded by more modern techniques, would leave their heritage of usefulness for ever. Others, often dearer to the British themselves, were less permanent. The Empire had not been a missionary venture, in the sense that it seldom tried to impose a religion upon its subject peoples, leaving them by and large to their own theological preferences. It did, however, diligently propagate a faith—faith in parliamentary democracy of the Westminster pattern, steady, evolutionary, rooted in the Rule of Law and the importance of the individual. Nothing gave the more earnest imperialists greater satisfaction than to observe their subjects honouring this vision in their turn. What nobler, they thought, than to see the ancient traditions of English life, hammered out so painfully over the centuries, translated to the distant dependencies of the Crown? Lionel Curtis saw the process as the grandest fulfilment of Empire, and his magnum opus *Civitas Dei*, published in 1938, envisaged it as the foundation of a true world order, a successor to *all* Empires and the predecessor of Paradise.

There seemed a time indeed when the emergent British Commonwealth might really become a community of cultists, dedicated to this creed. India was frequently boasted of, after independence, as the World's Largest Democracy, and Nigeria was said to be a very model of democratic rectitude. Kenya treated its white minority with perfect fairness, Ceylonese tea planters reported that nothing had much changed in the happy uplands of Nuwara Eliya. The faith soon shrivelled, though, and in a few years the ideology of the British Empire, such as it was, collapsed. The rule of law proved transitory when the imperial policemen were withdrawn, and tyrants more fierce than any colonial governor swept away the baubles of democracy. Nations gently nurtured into statehood fractured themselves in civil war, or were curdled in corruption. The fragile democratic flower soon wilted in climates like the Sudanese or the Zanzibari, and the principles of Common Law were tossed aside, as the subject peoples reverted to older standards, or devised new systems altogether.

Soon the old colonial empire had little more in common with Britain than with America or Soviet Russia—in great matters at least. In smaller ways the association proved more resilient. They still played cricket on the dust patches of Pakistan, among the Caribbean frangipani, or behind the yam-stores of New Guinea.[1] The Fiji Rugby XV still made its regular tour of Wales, and at the Commonwealth Games, so old imperialists liked to think, a recognizably higher standard of sportsmanship prevailed than at the Olympics. The huntsmen of Ootacamund, as of Montreal, still wore pink.[2] The regimental messes of the Indian Army still cherished their regimental silver, tarnished a little with the years perhaps, but still commemorating ancient triumphs of the imperial arms. In many a store and office the merchants of Empire survived, immensely rich in Hong Kong or Singapore, proud but seedy in Calcutta, astutely adapting to the times in Lagos, Mombasa or the Cayman Islands.[3]

A manner of thought survived here and there, and revived gracefully upon the arrival of a visiting Briton, when afternoon tea was poured in memory of the old days, a game of tennis was proposed, and the English slang of another generation was resuscitated

[1] Where the rules had been adapted to local conditions—they played fifty-nine a side.

[2] And the sherry barons of Jerez still went pig-sticking in the Cota Doñana, having been introduced to the sport by Anglo-Indians from Gibraltar.

[3] Especially the Cayman Islands. In 1953 they possessed one bank, and their population of 7,500 people produced an annual Government revenue of less than £200,000. Twenty years later their population had doubled, their annual revenue was £3½ million, they were host to 5,000 companies and 138 banks, they had their own airline and one Telex machine to every 200 inhabitants. Adaptability!

—'What a jolly nice surprise to see you! You're a brick to come all this way!' Among the Pathan soldiers of the Pakistan Army the command to stand at ease was still rendered 'Sundlies!', as the British had obligingly simplified it long before: among the soldiers of the Southern Yemen People's Republic the word 'dismiss' still meant 'screwdriver'—something that turns to the right, like a parade dismissed. One often felt, rather than actually identified, the traces of Britishness, when the flag had long been lowered and a generation had grown up who never knew the Empire: in the stance of a building, in a style of printing, in the posture of a sportsman, or in the echo of military music, half strange, half familiar among the teak trees or the deodars.

Victoria's Empire had created three nations more or less in the image of England, brought into being almost from scratch by the genius of the British, and faithful still to their doctrines. It left behind two universal achievements: the end of slavery, the freedom of the seas—the rules of the sea were imperially conceived, the slave trade was imperially abolished. It had given to the world its own language, English, one of the mightiest of all instruments of human intercourse. It had kindled the latent energies of many a people temporarily stagnant: 'he that wrestles with us', as Burke said, 'strengthens our will, and sharpens our wits—our antagonist is our helper.' It had done something, as Matthew Arnold said, to 'humanize men in society'—to curb the worst cruelties of primitives, and introduce people trapped in superstition and tradition to the idea that a man was a man for a' that. For a century it had sustained, as Carlyle said, 'a mighty Conquest over Chaos'. Here and there among the millions it had left, as Curzon hoped, 'a spring of patriotism, a dawn of intellectual enlightenment, or a stirring of duty where it did not exist before'. Through many generations, in many countries, it had been the peace-keeper and the law-maker, generally fair by its own standards, generally humane, and except in its own interests, which were always paramount, as impartial as a judge ever is. Now that it was gone, people often unexpectedly missed it.

3

Churchill died, and it died with him. It had lasted too long anyway: the subject peoples had outgrown its tutelage, and would progress much faster without it. For the British it was more of a wrench than they knew, for though by now the imperial idea seemed as antiquated as steam trains or antimacassars, the existence of Empire had impregnated all their lives. England without an Empire, Joseph Chamberlain had once said, would not be the England its people knew: and sure enough, when the imperial dimension was removed from the English national structure, nothing was ever quite the same again.

The British Empire had been haphazardly acquired, it had been hastily and sometimes sloppily discarded, but it had given the British people, seventy-odd years before, one brief moment of dazzle. The climax of Empire at the time of the Diamond Jubilee had come to the great mass of the citizenry as a revelation. It had seemed to them the greatest thing that ever happened to any nation, the duty and privilege of ruling a third of the world. Never mind the true motives and methods of imperialism—in the days of their imperial supremacy the British genuinely believed themselves to be performing a divine purpose, innocently, nobly, in the name of God and the Queen. The imperial dimension gave them a sense of scale and potential, and made them feel grand.

As Younghusband was permanently changed by his moment of ecstasy outside Lhasa, so the British were to be stamped with the stigmata of that passing conviction. It soured some of them, as they gradually realized it to be illusory. It matured many more, as they paid for it in blood and unhappiness. It confused them in the end, as rival visions overlaid their own. It made some of them proud in retrospect, some of them ashamed. It left its long trail of racial arrogance, so that long after Great Britain had retreated into the second ranks of the nations, the good-natured British people still sneered at blacks and laughed at foreigners. It gave them a cynical distaste for worldly power and influence, so transitory and treacherous. It embittered them a little, when they found there was nothing left to fire their hearts or their imaginations, that they were only another European people after all. But it made them, for better or for worse, a special people, one of those few peoples which, in the centuries of the nation States, were able to alter the face of the world—one of those peoples whose dust is left like a cloud in the air, as the Spaniard Miguel de Unamuno put it, when it goes galloping down the highroad of history.

In 1837, when Victoria came to the throne, the fact of Empire lay lightly upon Britain. Nobody had been greatly interested in the notion since the loss of the American colonies, and imperial symptoms were hard to find in Britain—only the nabobs and the sugar-kings built their country palaces, and the slave-ship ports flourished on the Triangular Trade. By 1965, when the last of the great Victorians left the stage, five generations of the imperial experience had changed all that, and the islands were thick with the accretions of imperialism, flavoured everywhere with its memories, tempered by its assumptions. By then every British town had its memorials of the enterprise, even those far from the sea, or on the periphery of political affairs, and every traveller to the islands was aware of the imperial past, embodied in relic, attitude or asset all around him.

Everywhere, ingrained, was Empire! The neo-classical bulks of the High Commissions, towering over Trafalgar Square—the gilded dome of Sezincote in the Cotswolds, that most enchanted of the naboberies—curry at Veeraswami's, founded to cater for Anglo-Indians on leave—the temple-memorial to Lord Durham, patron earl of Canada, that stood blackened and tremendous above Penshaw Moor. Livingstone of Africa lay in Westminster Abbey, the lady from the next-door cottage devotedly attended the grave of Younghusband at Lytchett Minster, every Australia Day the High Commissioners drove down from London to pay tribute to Admiral Arthur Phillip, first Governor of New South Wales, in his grave at Bath.

Stark but creeper-softened at the end of the Mall stood the Citadel, from whose bunkers the Lords of Admiralty directed the fleets of Empire in the Second World War, and down the road across the parade ground Clive stood in cocky effigy beside the old India Office. High in the mists of Knock Fyrish in Easter Ross loomed the great gates Sir Hector Munro of Foulis had erected to commemorate the capture of Negapatam; beneath the dome which Sir Herbert Baker had built for them in Oxford, surmounted by the mythical Zimbabwe bird, the Trustees of the Rhodes Foundation assembled to distribute the continuing largesse of Kimberley and the Rand.

United Africa Company—Imperial Chemical Industries—Anglo-Transvaal Trustees—Ionian Bank—Bank of the Middle East—across the City of London the brass plates announced the still generous legacies of Empire. In the House of Lords the names of Allenby, Dufferin and Ava, Kitchener of Khartoum, Mountbatten of Burma, Beatty, Strathcona, Napier of Magdala, Methuen, Lawrence, Fisher, Baden-Powell, Cromer, honoured, if only by the chance of hereditary succession, its champions. In Lancashire they still called the grandstand at a football ground 'The Kop'—two Lancashire regiments had fought at Spion Kop—and the jargon of the British Army was still rich in imperial derivatives, *bint, shuftee, char* or *bukshee*. Twenty thousand Asian seamen sailed in British merchant ships: a million black and brown immigrants, from Africa, the West Indies and the Indian sub-continent, had settled in the Mother-Country.

And in any street of any city, one could find more intimate relics of Greater Britain. In hundreds of thousands of homes there lay, wrapped in polishing cloth and occasionally threatened with auction sales, the medals of the imperial campaigners—120 were awarded altogether, and if the General Service Medal, 1939–45, cluttered up rather too many drawers, the Waziristan Campaign Medal was, by 1965, becoming rather rare. Curios of bronze or ivory abounded still in drawing-rooms and parlours, Kashmir rugs lay strewn on polished floors, on many a staircase framed groups of imperial sportsmen or soldiers gazed sternly across their moustaches. Sadly, sadly the ageing bourgeoisie of England watched the extinction of their patrimony, as the *Telegraph* chronicled, year after year, the retreat of Empire—'Taranji Minister of Posts and Telegraphs! He couldn't stick a stamp on an envelope. As for Jebajwi as Vice-Chancellor, well, words fail me. . . .' Sometimes they wrote despairing or indignant letters to the editor, deploring the outbreak of violence in some abandoned province, 'a part of the world I happen to know well, having served there as Assistant District Commissioner in 1933–6', and they were not always comforted when, looking out of their windows on summer afternoons, they saw the jet-black children of the immigrants playing exuberant cricket in the park.

4

But most of them, as in most periods of their history, disregarded it. They did not often think about the Empire. Families whose fortunes had been founded in Madras or Jamaica forgot the reasons for their continuing prosperity, and the great-grandsons of Boer War soldiers, the great-great-grandsons of Mutiny veterans, did not know where Lucknow or Ladysmith were to be found on the map. As the post-imperial generation advanced into middle age, the horizons of the nation contracted. No longer would one go to London to find the greatest living authority on the customs of the Shans, or irrigation problems in central Sudan, and one by one the men who knew all about Jordan, or Burma, or Basutoland, or the Andaman Islands, aged, died and were buried in village church-yards in the rain. The origin of the scimitar above the study door became uncertain—'Something to do with the Mad Mullah?'

The British preferred to forget it. 'We believe', Lord Caradon told the United Nations in 1965, 'that no nation and no race should be dominated by another. We believe that every nation should be free to shape its own destiny. We believe that colonialism should be ended as rapidly as possible.' They muffled the subject at schools. They put out of their minds the thought, cogently expressed by Ernest Bevin not so long before, that their own prosperity might be dispersed with the Empire. Their national fortunes slowly declined, their industrial pre-eminence was lost, their status in the world was diminished, but no politician dared blame it upon the loss of Empire.

Had it all been a colossal mistake? The profits had certainly been great. Control of raw materials had enabled the British to influence prices favourably to themselves; the flow of specie had kept the pound strong; the habit of Britishness, the familiarity of the structure, had given British exporters immense customary markets within the Empire. The possession of ports and coaling stations everywhere, the British dominance of things maritime and communicative, had made the islanders the world's greatest carriers, insurers and agents. The expertise of all sorts that came with Empire powerfully boosted the 'invisible exports' of the City of London, which for so many decades kept the national balance of payments in equilibrium.

But the cost of it all had been stupendous too. Supremacy was a speculative investment—one year might pay, the next show a loss—'divide the victories by the taxation'. In its prime it had been an economic stimulant, but in its decline it was debilitating, for it made things seem too easy. The British had not really been economically supreme for generations—perhaps since the 1870s, when it first became apparent that Germany and the United States had greater muscle-power, and when, partly in compensation, Disraeli first gave glamour to the imperial conception. The New Imperialism was an attempt to keep a medium-sized island State among the super-Powers, but it had failed: the possession of Empire eased the symptoms, but did not cure the cause—which was not a sickness at all, but merely the reality—the truth that the British were 50 million people more or less like their neighbours, less well endowed than most, and impelled into that century of greatness not by divine favour, but by circumstance and energy.

And by forfeit. Across the continents stood those tombs, from the triumphant mausolea of Anglo-Indian conquerors to the pathetic mementos of defeat, carved in prison workshops in the graveyards of Hong Kong and Singapore. Many expressed pride in death, many more contentment, but often the epitaphists were concerned less with the dead than with the living—

> *O! ye in the far distant place,*
> *O'er the infinite seas;*
> *When ye think of the sons of our race,*
> *Think deep upon these!*

or:

When you go home, tell them of us and say
For your tomorrows we gave our today.[1]

This is because they believed that there were lessons to be learnt
from the example of the imperial dead, that the Empire and the
world would be a better place because of the manner of their lives,
and the penalty of their deaths. Nobody who wandered among the
imperial gravestones, though, pondering the sadness of their
separate tragedies, could fail to wonder at the waste of it all, the
young lives thrown away, the useless courage, the unnecessary
partings; and the fading image of Empire, its ever dimmer panoply
of flags and battlements, seemed then to be hazed in a mist of
tears, like a grand old march shot through with melancholy, in a
bandstand by the sea.

5

The end of it was not surprising. Once the almost orgiastic splendour of its climax had been achieved, once the zest went out of it,
it became rather a sad phenomenon. Its beauty had lain in its
certainty and momentum, its arrogance perhaps. In its declining
years it lost the dignity of command, and became rather an exhibition of ineffectual good intentions. Its memory was terrific; it had
done much good in its time; it had behaved with courtesy as with
brutality, rapaciously and generously, rightly and in error; good
and bad had been allied in this, one of the most truly astonishing of
human enterprises. Now its contribution was over, the world had
moved on, and it died.

They performed its obsequies, with Sir Winston's, on a grey
London day in January, and for the last time the world watched a
British imperial spectacle. Melancholy though the occasion was,
intuitively though the British felt its deeper significance, they did it,
as Churchill wished, in the high old style. Big Ben was silenced for

---

[1] The first comes from a memorial at Wagon Hill, outside Ladysmith; the
second is from the famous memorial to the 14th Army at Kohima, and is an
English version of Leonidas' message from Thermopylae in 480 BC—*Go, tell
the Spartans, thou who passest by,/That here obedient to their laws we lie.*

the day. Mourning guns were fired in Hyde Park. The great drum-horse of the Household Cavalry, drums swathed in black crêpe, led the funeral procession solemnly through London to St Paul's, while band after band across the capital played the Dead March from *Saul,* and the soldiers along the way bared their heads and reversed their arms.

Five Field Marshals, four Prime Ministers, an Admiral of the Fleet and a Marshal of the Royal Air Force were among the pall-bearers when the coffin, draped with the flags of the Cinque Ports and of the Spencer-Churchill family, was carried up the great steps into the cathedral. A hundred nations were represented there, and twenty of them had once been ruled from this very capital. A bugle played the *Last Post* in the Whispering Gallery, another answered with *Reveille* from the west door, and after the funeral service they took the coffin down to the River Thames. There, as the pipers played 'The Flowers of the Forest', six tall guardsmen, their cold sad faces straining with the weight, carried it on board a river launch: and away up the London river it sailed towards West-minster, escorted by black police boats. 'Rule Britannia' sounded from the shore, fighter aircraft flew overhead, farewell guns fired from the Tower of London, and as the little flotilla disappeared upstream, watched by the great mourning crowd below the cathedral, all the cranes on the riverside wharves were dipped in salute. Everyone knew what was happening, even the enemies of Empire. 'The true old times were dead, when every morning brought a noble chance, and every chance a noble knight.'

In the afternoon they put the old statesman's body reverently on a train, for he was to be buried in the family churchyard in Oxfordshire: and so as dusk fell, with white steam flying from the engine's funnel, and a hiss of its pistons through the meadows, it carried him sadly home again, to the green country heart of England.

Envoi

IS that the truth? Is that how it was? It is *my* truth. It is how
Queen Victoria's Empire seemed in retrospect, to one British
citizen in the decades after its dissolution. Its emotions are coloured
by mine, its scenes are heightened or diminished by my vision, its
characters, inevitably, are partly my creation. If it is not invariably
true in the fact, it is certainly true in the imagination.

It has taken me ten years to write the trilogy, and I add this
epilogue now in the same beloved corner of Wales where I started
the work in 1966. During that time everyone's view of the imperial
idea has shifted. Some people have come to think, as Goethe did,
that injustice is preferable to disorder after all, while others have
recognized for the first time that there was cruelty to the conception even in its kindest forms. For myself, when I began to
write the book I thought I was describing something definitive in
human history, but I have ended it seeing the imperial story in
gentler but nobler terms, as a flicker of the divine progress.

In Canada one day, looking through old newspapers for relevant
material, I came across a report of a lecture given in Chantaqua,
Alberta, on a June evening in 1928. The lecturer was a Mr Walter J.
Millard, and he was talking to a women's club on 'The Relation
of Energy to Human Progress'. Preceded in the evening's entertainment by baritone solos and child-impersonations from Miss Jensen,
he had chosen to discuss the British Empire as an archetype of
historical energy, and at first sight his talk seemed to me an
imperialist address in the old convention of duty, privilege and far-
flung responsibility. When I read on, though, I found that Mr
Millard saw the Empire primarily not as an agency of development,
law and constructive order, but as an instrument of personal redemp-

tion. It represented above all, he said, 'the privilege of every man to find his scrap of truth and apply it to the advantage not of himself, but humanity'. *There*, he thought, lay the truest power of the enterprise.

At once, as I read these words, I found my own views clarified, for I too had lately come to view the Empire less in historical than in redemptory terms. I had been groping around Teilhard de Chardin's notion of 'in-furling'—that infinitely slow and spasmodic movement towards the unity of mankind. Teilhard saw love and knowledge as the twin impulses of that progress: Mr Millard, posthumously addressing me from the *Calgary Herald*, finally persuaded me that the British Empire too was a ripple in some cosmic urge to reconciliation. That 'scrap of truth' was all! The arrogance of the Empire, its greed and its brutality was energy gone to waste: but the good in the adventure, the courage, the idealism, the diligence had contributed their quota of truth towards the universal fulfilment.

'Are we to complain', wrote Nehru of the Empire, 'of the cyclone that uproots us and hurls us about, or the cold wind that makes us shiver? The British . . . represented mighty forces which they themselves hardly realized.' The wind dies, and is forgotten, but some of the seeds it blows about will be fertile in the end. Whenever I go to evensong in a cathedral of the old Empire, Lahore or Singapore, Auckland or Kingston, it always seems to end with the same last hymn. I am not a Christian really, but it never fails to move me. Often they sing it as the choir and clergy leave their stalls, and an old black verger, perhaps, precedes the Dean down the chancel, while the front-row ladies prepare their gloves for departure, or flowered and white-socked children of the Caribbean sing all the more lustily because supper is near. Night is falling through the slatted windows, jasmine, or magnolia, or tobacco flower hangs upon the air, clumsy black insects flounder about the dust-dimmed lights, or evade the whirring fan-blades.

The hymn was written in 1870 by John Ellerton, and was set to music by Clement C. Scholefield: and just as it ends those distant services upon a mingled note of gratitude and resignation, so I will quote its familiar words now to end my book, remembering equally all those whose lives may seem to have been wasted in the imperial cause, those who died to create the Empire and those who sacrificed themselves to end it:

> *So be it, LORD; Thy Throne shall never,*
> *Like earth's proud empires, pass away;*
> *Thy Kingdom stands, and grows for ever,*
> *Till all Thy creatures own Thy sway.*

Trefan Morys, 1978

BOOKS ABOUT THE EMPIRE

For those who would like to spend more time with the Victorian Empire, here are 100 books to read, most of them published both in London and in New York. I have selected them from an almost infinite choice, my criteria being the pleasure, the accuracy and where possible the up-to-dateness of them, and I have limited myself to one volume per topic, so that they do not include multi-volumed works like *The Cambridge History of the British Empire*, or series of volumes like the Corona Library. For a less hedonistic register I suggest Robin Winks' *A Historiography of the British Empire and Commonwealth*, 1968, or the *Cambridge History* itself: in the meantime, if you have enjoyed any particular chapter of my own books, you will find its matter more fully explored in one or another of these:

Five on Imperialism in general:
1 D. K. Fieldhouse, *The Colonial Empires*, 1966
2 Correlli Barnett, *The Collapse of British Power*, 1972
3 John Bowle, *The Imperial Achievement*, 1974
4 George Woodcock, *Who Killed the British Empire?*, 1974
5 Bernard Porter, *The Lion's Share*, 1975

One for its pictures of Empire:
6 *The Time-Life History of the British Empire*, 1970

Eight imperial classics:
7 Emily Eden, *Up the Country*, 1866
8 Charles Dilke, *Greater Britain*, 1869
9 John Seeley, *The Expansion of England*, 1883
10 Henry Yule & A. C. Burnell, *Hobson-Jobson*, 1886 (on Anglo-Indian speech forms)
11 Rudyard Kipling, *Kim*, 1902
12 T. E. Lawrence, *Seven Pillars of Wisdom*, 1926

13 E. M. Forster, *A Passage to India*, 1924

14 George Orwell, *Burmese Days*, 1935

Seventeen assorted biographies:

15 Elspeth Huxley, *White Man's Country*, 1935 (Lord Delamere)

16 Ronald Storrs, *Orientations*, 1937 (Autobiography)

17 A. L. Kennedy, *Salisbury*, 1953

18 Philip Magnus, *Gladstone*, 1954

19 Charles Carrington, *Rudyard Kipling*, 1955

20 Margery Perham, *Lugard*, 1956

21 Philip Magnus, *Kitchener*, 1958

22 J. G. Lockhart & C. M. Woodhouse, *Rhodes*, 1963

23 Joseph Lehmann, *All Sir Garnet*, 1964 (Garnet Wolseley)

24 Robert Blake, *Disraeli*, 1966

25 Anthony Nutting, *Gordon*, 1966

26 Geoffrey Ashe, *Gandhi*, 1968

27 David Dilks, *Curzon in India*, 1969

28 Jasper Ridley, *Lord Palmerston*, 1970

29 Ruddock F. Mackay, *Fisher of Kilverstone*, 1973

30 John E. Mack, *A Prince of Our Disorder*, 1976 (T. E. Lawrence)

31 John Marlowe, *Milner*, 1976

Fifteen about the imperial wars:

32 Rayne Kruger, *Good-bye Dolly Gray*, 1959 (The Boer War)

33 Robert Rhodes James, *Gallipoli*, 1965

34 Julian Symons, *England's Pride*, 1965 (the Gordon relief expedition)

35 Patrick Macrory, *Signal Catastrophe*, 1966 (the first Afghan war)

36 Donald Morris, *The Washing of the Spears*, 1966 (the Zulu war)

37 A. J. Barker, *The Neglected War*, 1967 (the Mesopotamian campaign)

38 Noel Barber, *Sinister Twilight*, 1968 (the fall of Singapore)

39 George Ludgate Bruce, *Six Battles for India*, 1969 (the Sikh wars)

40 Correlli Barnett, *Britain and Her Army*, 1970

41 Basil H. Liddell Hart, *A History of the First World War*, 1970

42 Stewart Perowne, *The Siege within the Walls*, 1970 (the siege of Malta)

43 Joseph Lehmann, *The First Boer War*, 1972

44 S. N. Sen, *Eighteen Fifty-Seven*, 1957 (the Indian Mutiny)

45 Byron Farwell, *Queen Victoria's Little Wars*, 1973
46 Mark Arnold-Forster, *The World at War*, 1974 (World War II)

Eight about British India:
47 Dennis Kincaid, *British Social Life in India*, 1938
48 Philip Woodruff, *The Men Who Ruled India*, 1953 (the ICS)
49 Michael Edwardes, *British India*, 1967
50 George Bruce, *The Stranglers*, 1968 (Thugee)
51 Erik H. Erikson, *Gandhi's Truth*, 1969
52 Charles Allen, *Plain Tales from the Raj*, 1975 (twentieth-century reminiscence)
53 Larry Collins & Dominique Lapierre, *Freedom at Midnight*, 1975
54 William Golant, *The Long Afternoon*, 1975

Sixteen on the white colonies:
55 Daisy Bates, *The Passing of the Aborigines*, 1938
56 Joseph Kinsey Howard, *Strange Empire*, 1952 (Louis Riel)
57 Leo Marquard, *The Story of South Africa*, 1955
58 Keith Sinclair, *A History of New Zealand*, 1959
59 Elizabeth Pakenham, *Jameson's Raid*, 1960
60 Douglas Pike, *Australia*, 1962
61 Pierre Berton, *Klondike*, 1963
62 Alan Moorehead, *Cooper's Creek*, 1963
63 L. L. Robson, *The Convict Settlers of Australia*, 1965
64 Douglas MacKay, *The Honourable Company*, 1966 (Hudson Bay)
65 Douglas Hill, *The Opening of the Canadian West*, 1967
66 Oliver Ransford, *The Rulers of Rhodesia*, 1968
67 Keith Dunstan (ed.), *Knockers*, 1972 (critical views of Australia)
68 Peter Firkens, *The Australians in Nine Wars*, 1972
69 Oliver Ransford, *The Great Trek*, 1972
70 David Davies, *The Last of the Tasmanians*, 1973

Seven on the British in Black Africa:
71 Reginald Coupland, *The Exploitation of East Africa*, 1939
72 Elspeth Huxley, *The Flame Trees of Thika*, 1959 (Kenya)
73 Alan Moorehead, *The White Nile*, 1960
74 R. Robinson, J. Gallagher & A. Denny, *Africa and the Victorians*, 1961

75 Michael Crowder, *The Story of Nigeria,* 1962
76 James Pope-Hennessy, *The Sins of the Fathers,* 1967 (slave trade)
77 Roy Lewis & Yvonne Foy, *The British in Africa,* 1971

Four on the British in the Middle East:
78 John Marlowe, *The Seat of Pilate,* 1959 (Palestine)
79 Elizabeth Monroe, *Britain's Moment in the Middle East,* 1963
80 Sarah Searight, *The British in the Middle East,* 1969
81 Peter Mansfield, *The British in Egypt,* 1971

Five about Ireland:
82 Cecil Woodham-Smith, *The Great Hunger,* 1962
83 Jules Abels, *The Parnell Tragedy,* 1966
84 A. T. Q. Stewart, *The Ulster Crisis,* 1967
85 Thomas M. Coffey, *Agony at Easter,* 1970 (the Easter Rising)
86 J. C. Beckett, *The Anglo-Irish Tradition,* 1976

Eight about other imperial territories:
87 Lawrence Durrell, *Bitter Lemons,* 1957 (Cyprus)
88 Arthur Grimble, *A Pattern of Islands,* 1962 (South Pacific)
89 James Pope-Hennessy, *Verandah,* 1964 (Hong Kong, Mauritius et al.)
90 Geoffrey Dutton, *The Hero as Murderer,* 1967 (Jamaica)
91 Arthur Foss, *The Ionian Islands,* 1969
92 Quentin Hughes, *Fortress,* 1969 (Malta)
93 George Woodcock, *The British in the Far East,* 1969
94 Cyril Hamshere, *The British in the Caribbean,* 1972

Five on imperial constructions:
95 James Leasor, *The Millionth Chance,* 1957 (the airship R101)
96 Alan Gowans, *Building in Canada,* 1967 (Canadian architecture)
97 J. M. Freeland, *Architecture in Australia,* 1968
98 Sten Nilsson, *European Architecture in India,* 1968
99 Mark Bence-Jones, *Palaces of the Raj,* 1973

And one for the curious fun of it:
100 John Colombo, *Colombo's Dictionary of Canadian Quotations,* 1973

ACKNOWLEDGEMENTS

The faults of this book are all mine, but a number of indulgent friends and colleagues vetted particular chapters and passages for me. Occasionally I was ungrateful enough to disregard their advice, but still I owe my heartfelt thanks to Colonel A. J. Barker, Professor C. J. Beckett, Mr David Dilks, General Anthony Farrar-Hockley, Sir Laurence Grafftey-Smith, Mr Kenneth Griffiths, Mrs Elspeth Huxley, Professor Robert Rhodes James, Mr Alistair Lamb, Mr Philip Mason, Miss Elizabeth Monroe and Sir Ronald Wingate. Finally Mr Donald Simpson, librarian of the Royal Commonwealth Society, drew upon his unrivalled knowledge of the Empire as a whole and most generously read the whole book in proof.

Mr Denys Baker drew all the maps; my friend and agent Julian Bach arranged nearly all the necessary journeys; the editors of *The Times* and *Encounter* (London), *Horizon* and *Rolling Stone* (New York) helped me with relevant commissions; for their loving and amused encouragement throughout this long endeavour I am indebted, now as always, to my family—Elizabeth, Mark, Henry, Twm and Susan Morris.

·

Index

After Oxford, and a Harkness Fellowship in the United States, James Morris spent ten years as a foreign correspondent for the London *Times* and *The Guardian*, an experience that led to Morris's historic coverage of the 1953 Everest expedition, and to numerous subsequent ventures to cover wars, revolutions, trials, elections, and political crises on every continent. In 1961, Morris left newspaper work to write books, and at one time or another lived in America, Venice, the French Alps, Egypt, Spain, and Wales, where Jan Morris is now settled.

Morris is the recipient of the George Polk Memorial Award for Journalism in America for 1961 and the Heinemann Award of the Royal Society of Literature in England for the same year.

In 1972, James Morris became Jan Morris; *Farewell the Trumpets* was the last of her works to appear under her original name.

JAMES MORRIS'S ENTHRALLING TRILOGY ON
THE RISE AND FALL OF THE BRITISH EMPIRE

Heaven's Command: An Imperial Progress

The British Empire from Queen Victoria's accession in 1837 to
her Diamond Jubilee in 1897, re-created in a series of character
studies, dramatic episodes, and astute political evaluations.
560 pp.; 32 pp. of illustrations.

Pax Britannica: The Climax of an Empire

An account of the British Empire at the height of its power in
1897: what it was, how it worked, what it looked like, and
how the British themselves saw it. 544 pp.; 32 pp. of illustrations.

Farewell the Trumpets: The Decline of an Empire

The waning of the greatest of empires, traced in the years
from Queen Victoria's Diamond Jubilee in 1897 to the death of
Winston Churchill in 1965. 576 pp.; 16 pp. of illustrations.